Chemistry Education in the ICT Age

Minu Gupta-Bhowon • Sabina Jhaumeer-Laulloo
Henri Li Kam Wah • Ponnadurai Ramasami
Editors

Chemistry Education in the ICT Age

Springer

Editors
Minu Gupta-Bhowon
University of Mauritius
Faculty of Science
Department of Chemistry
Réduit, Mauritius

Sabina Jhaumeer-Laulloo
University of Mauritius
Faculty of Science
Department of Chemistry
Réduit, Mauritius

Henri Li Kam Wah
University of Mauritius
Faculty of Science
Department of Chemistry
Réduit, Mauritius

Ponnadurai Ramasami
University of Mauritius
Faculty of Science
Department of Chemistry
Réduit, Mauritius

ISBN: 978-1-4020-9731-7 e-ISBN: 978-1-4020-9732-4
DOI: 10.1007/978-1-4020-9732-4

Library of Congress Control Number: 2009920158

Printed on acid-free paper

Springer is part of Springer Science+Business Media (www.springer.com)

Preface

The 20th International Conference on Chemical Education (20th ICCE), which had "Chemistry in the ICT Age" as the theme, was held from 3rd to 8th August 2008 at Le Méridien Hotel, Pointe aux Piments, in Mauritius. With more than 200 participants from 40 countries, the conference featured 140 oral and 50 poster presentations.

Participants of the 20th ICCE were invited to submit full papers and the latter were subjected to peer review. The selected accepted papers are collected in this book of proceedings.

This book of proceedings encloses 39 presentations covering topics ranging from fundamental to applied chemistry, such as Arts and Chemistry Education, Biochemistry and Biotechnology, Chemical Education for Development, Chemistry at Secondary Level, Chemistry at Tertiary Level, Chemistry Teacher Education, Chemistry and Society, Chemistry Olympiad, Context Oriented Chemistry, ICT and Chemistry Education, Green Chemistry, Micro Scale Chemistry, Modern Technologies in Chemistry Education, Network for Chemistry and Chemical Engineering Education, Public Understanding of Chemistry, Research in Chemistry Education and Science Education at Elementary Level.

We would like to thank those who submitted the full papers and the reviewers for their timely help in assessing the papers for publication.

We would also like to pay a special tribute to all the sponsors of the 20th ICCE and, in particular, the Tertiary Education Commission (http://tec.intnet.mu/) and the Organisation for the Prohibition of Chemical Weapons (http://www.opcw.org/) for kindly agreeing to fund the publication of these proceedings.

We hope that this collection of papers will serve as a useful reference set for teachers/researchers involved in chemical education.

February 2009

Gupta-Bhowon, M.
Jhaumeer-Laulloo, S.
Li Kam Wah, H.
Ramasami, P.
Department of Chemistry,
University of Mauritius

Contents

Chemistry and Chemical Education as a Bridge to Peace

Z. M. Lerman

Abstract Chemistry and chemical education can be important tools to advance the peace process, especially in the Middle East. The Middle East is a region in conflict for many years. This part of the world is of particular importance because it has a source of energy that is a strategic resource: fossil fuel. This nonrenewable source of energy not only fuels economic and political conflicts, but its worldwide use also places at risk the sustainability of life on Planet Earth, by polluting the environment and contributing to climate change. The Middle East also has major problems of air and water quality, which will require regional cooperation to solve. Geopolitical borders are only lines on a map; air and water do not recognize these lines. Therefore, any work concerning the environment - especially air and water quality - must be done in collaboration between nations. Chemistry is an international language. A chemist from Bethlehem, Pennsylvania in the USA, and a chemist from Bethlehem, Palestine, use the same chemical notations, and can communicate scientifically to one another without understanding each other's spoken language. Building on the international language of chemistry, three major international conferences called the "Malta Conferences" and formally titled "Frontiers of Chemical Science: Research and Education in the Middle East" were held in 2003 [1, 2, 3], in 2005 [4, 5, 6], and in 2007 [7, 8, 9]. In these conferences, chemists from 14 Middle East nations gathered to discuss solutions to the problems of air and water quality, energy resources, and chemical education in the Middle East. These collaborations between the chemists have yielded results that are a cornerstone for a bridge to peace.

1 Challenges Facing the Scientific Community

There are now over 6×10^9 people on Planet Earth consuming over 6×10^9 tons of fossil fuels each year; most of which is consumed by developed countries. With the expectation of 10×10^9 people by the year 2050 and with the increasing consumption

---

Z.M. Lerman (✉)
Columbia College Chicago, Institute for Science Education and Science Communication
600 S. Michigan Avenue, 60605, Chicago, Illinois, USA
e-mail: zafral@aol.com

M. Gupta-Bhowon et al. (eds.), *Chemistry Education in the ICT Age*,
DOI 10.1007/978-1-4020-9732-4_1, © Springer Science + Business Media B.V. 2009

of fossil fuels by developing countries, we will run out of these energy resources, cause irreversible environmental damage, adversely affect the food supply that depends on energy, and endanger the sustainability of life on Planet Earth.

We must develop new and clean energy resources. It is a simple fact that the amount of solar energy received by the surface of the Earth in 1 h is approximately equal to the total energy consumption of the entire planet in 1 year at present, and this is just **one** of the clean energy sources available to us.

The scientific community must promote the use of renewable sources of energy. In this way dependence on the energy resources in an unstable Middle East will be reduced. This will both decrease their contribution to detrimental climate change, as well as their value in the economic and political conflicts between countries.

Science is an international language. Borders are only lines on a map; nature and science do not recognize these lines. Therefore, any work concerning nature and the environment must be done in collaboration between nations. Science is probably the only field that contributes to the quality and longevity of life, but also has the ability to cut life short. Therefore, scientists are in a very special position compared to other professions. They have the responsibility to use their science and their status in society for the betterment of humankind and for all life on Earth. Specifically, I believe it is the responsibility of the scientific community:

- To promote the development and use of clean energy resources.
- To add the subject of sustainability to the curriculum to prepare future scientists with the background needed to preserve life on Planet Earth.
- To guarantee cross-border scientific collaboration and cooperation, even between countries whose governments are hostile toward each other.

Security is on the mind of every citizen of Planet Earth, and these concerns are closely tied to events in the Middle East. The political and economic climate currently shared by Middle Eastern nations is grave, and casts a shadow over the safety of everyone in the world. Events in the region burst into violence daily, consuming lives and resources while threatening a far wider conflagration.

Despite these unfortunate and tense circumstances, there is some light at the end of this dark tunnel [10].

We have learned from the past that scientists can continue to communicate with each other even when their respective governments are at odds. One example is the Pugwash conferences during the Cold War, which brought together scientists from the Soviet Union and from the USA.

I believe that scientists can again succeed where politicians have failed. In 2001 I decided that we chemists have an obligation to use our chemistry to solve the problems of the environment, water, energy, and education in the Middle East. Working together on problems will contribute to the peace process. I brought my idea of the conference "Frontiers of Chemical Sciences: Research and Education in the Middle East" to the board of the American Chemical Society (ACS) and received a go-ahead to organize a conference. An organizing committee was formed with representatives from ACS and from the following co-sponsoring organizations: the Royal Society of Chemistry (RSC), the International Union of

Pure and Applied Chemistry (IUPAC), and the German Chemical Society (GDCh). UNESCO became a co-sponsor in 2007. I traveled to the Middle East to recruit the participating chemists and personally persuaded the Nobel Laureates to join. In addition, various officials of the countries in the region had to be consulted; all of the negotiations and communications had to be carried out with sensitivity for and understanding of the cultures, religions, and the security issues involved [11].

2 Frontiers of Chemical Science: Research and Education in the Middle East

Three major international conferences titled "Frontiers of Chemical Science: Research and Education in the Middle East" have been held, with the fourth scheduled for 2009. The participating scientists came from 14 Middle East nations (Bahrain, Egypt, Iran, Iraq, Israel, Jordan, Kuwait, Lebanon, Libya, Palestine, Qatar, Saudi Arabia, Turkey, and United Arab Emirates) to discuss issues of scientific cooperation and collaboration.

Intensive workshops covered the following topics: Environment: Air and Water Quality, Science Education and Green Chemistry, Medicinal and Natural Products, Nanotechnology and Materials Science and Alternative Energy Sources. Each of the workshops came up with Future Action Plans.

The **Workshop on Air and Water Quality** discussed in detail the deterioration of water quality in Gaza and the fact that there is no clean drinking water in Gaza (see Fig. 1). The primary issues addressed were salinity, nitrate content, and health.

Fig. 1 Pollution by leaching waste water (Gaza Valley)

Salinity and Health:

- The quality of Gaza's municipal water supply is not acceptable.
- The chloride content in most wells fluctuates from 400 to 1,000 mg/L, which is double the recommended value by the World Health Organization (WHO), which is 250 mg/L (see Fig. 2).

Nitrate Content:

- Nitrate content in well-water is used as indicator, especially when salinity is low. Nitrate level in most of the wells is around 100–150 mg/L. This value is three times the recommended WHO value, which is 50 mg/L.
- This was reflected in the diseases that Palestinians in Gaza suffer, such as blue babies and renal failure.

The Workshop on Air and Water Quality reached the following conclusions:

- The Middle East has severe, generally transboundary, air and water quality problems.
- Without regional collaboration and cooperation, Middle Eastern air and water quality will continue to degrade.
- Regional collaboration generally requires an international "umbrella" to enable quick communication, travel, and action.
- The "Malta Conferences" have catalyzed regional collaboration on environmental issues through the establishment of a Middle East Air and Water Quality Forum, with IUPAC support.

The Air and Water Quality workshop participants committed themselves to working together to solve the problem of drinking water in Gaza and to develop methods to

Fig. 2 Pollution by leaching solid waste disposal

Fig. 3 Cairo/Tebbin South Area: Black plumes are from Mazout burning brick kilns; white plumes are from lead smelters

guarantee water quality in the region. In this workshop, the quality of air was also discussed (see Fig. 3) and cross-border collaborations were formed.

The **Workshop on Alternative Energy Sources** concentrated on alternative energy sources for the Middle East, citing solar energy as an ideal resource of energy for the area. The use of renewable alternative energy sources is growing rapidly and offers the best chance to replace fossil fuels and to develop a sustainable energy policy. The workshop developed specific short-term and long-term recommendations.

2.1 Short Term - up to 2020

- Energy Conservation (must be taken into account during construction); solar heating (both for water and space); and the use of waste heat from power plants.
- Recycling.
- Optimize the use of existing renewable technologies: wind; waves; hydroelectric; geothermal; solar conversion solutions; and solar-assisted, sustainable, clean fuel cycles (including advanced biofuels and partial artificial photosynthesis).
- Development and use of: "Clean Coal" technology, advanced geothermal technology, clean(er) nuclear breeders, and more efficient power stations.

2.2 Long Term

- Development and use of technology that is already theoretically available: 50 TW wind power; 12 TW geothermal; nuclear fusion; and widespread, efficient, cheap solar cells, and artificial photosynthesis.

Significant policy issues are related to the future of alternative renewable energy. These include financing research and development, access to energy sources, **education**, and community involvement. Alternative Energy Sources workshop participants recommended:

- Vigorous pursuit of Solar Energy options.
- Annual workshops involved in research and development of Alternative Energy, for senior researchers from the region and for students and young scientists, to reflect the challenge of developing alternative energy resources to achieve sustainability.

2.3 Energy, A Global Challenge

- Meeting this challenge requires coordinated action of the regional and world scientific communities, governments, and industries. Finding ways for efficient, cost-effective production of clean fuels can be the most important breakthrough in science in the twenty-first century - but to help it happen, we need to start working NOW.

The **Workshop on Science Education and Green Chemistry** (see Fig. 4) discussed the need for workshops and other educational programs to foster: development of alternative energy; sustainability and Green Chemistry; disposal of chemical waste; discussion of scientific method and ethics of science in the Middle East; strategic plans to attract students to scientific careers; and to educate scientists from multiple disciplines in the fields of pharmacology, toxicology, pharmacy, and clinical chemistry.

Science Education and Green Chemistry workshop participants made the following recommendations:

- Centers of excellence should be developed for chemical analysis and structure determination of natural products. Programs should be instituted to enable short-term exchange visits by faculty and students.
- A Middle East Virtual Campus should be established to facilitate exchanges of ideas among Middle East scientists. Web-based resources are needed, including a directory of laboratory equipment and expertise plus weblinks connecting to freely available databases and software.
- Newly developed theories in chemical education should be integrated into Middle East curricula. Green Chemistry, energy, nanotechnology, medicinal chemistry should be combined with the Systematic Approach in Teaching and Learning Chemistry for assessing secondary and tertiary students' skills.

Fig. 4 Workshop on Chemical Education led by Nobel Laureate Roald Hoffmann

Fig. 5 A poster on Water Quality in Gaza

In addition to the workshops, the scientists presented their results in poster sessions that stimulated interaction between scientists from different countries (see Figs. 5 and 6).

Fig. 6 Scientists from 14 Middle Eastern countries participated in the poster sessions

3 Specific Outcomes from the "Malta Conferences"

As a result of the "Malta Conferences," many concrete examples of cooperation among the various nations in the region have already taken place to address the recommendations of the different workshops [12]

- A research grant was received for participants from Bethlehem University (Palestine), from Bar Ilan University (Israel), and from the Weizmann Institute of Science (Israel) to work together on a joint water purification research project. This project is ongoing.
- Collaboration is underway between professors in Gaza and Israel for heavy metals analysis (ICP analysis) of water samples brought from Gaza to be analyzed at the Technion-Israel Institute of Technology.
- Samples of TiO_2 in different dimensions prepared in Bar Ilan University (Israel) are being used in Bethlehem University (Palestine) for the joint research between these universities.
- As a result of a question raised by a conference working group concerning training of scientists in the use of the synchrotron being built in Jordan with UNESCO support, Nobel Laureate Yuan T. Lee (Taiwan) offered six full 1-year scholarships for Middle Eastern scientists to train on the Taiwanese synchrotron facility; all six have completed the training and are now back in the Middle East.
- The Malta-III scientists from Iraq who work on water requested to visit the Water Center in the Technion - Israel Institute of Technology in order to work

with the Israelis and learn different techniques to solve the severe problems with water in the region.

- Improved communications between Palestinian and Israeli universities has led to an agreement between Palestinian institutions and the Weizmann Institute of Science (Israel). Palestinian students are now accepted to study for the M.Sc. and Ph.D. degrees in the Weizmann Institute's graduate school.
- The president of the Technion - Israel Institute of Technology offered to provide three full Technion scholarships for students from an Arab country.
- Several collaborations began as a result of the "Malta Conferences" between two groups of scientists in Israel: Arabs who work in Arabic colleges, and Jewish scientists in Israeli universities. Scientists from the Weizmann Institute of Science are collaborating on environmental issues with scientists from the Arab Academic College for Education.
- As a result of the "Malta Conferences," the National Science Foundation made a grant of $134,000 to Nobel Laureate Professor Roald Hoffmann to hold workshops for Middle Eastern graduate students and young chemists under 35. Two workshops were held in Jordan, and the third was held in Egypt.
- Numerous ongoing collaborations are occurring between scientists in Palestine, Israel, Kuwait, Iran, Jordan, and Egypt, especially in producing a database for water purification.
- The participants of Malta-III unanimously adopted a resolution calling on world governments to address the critical shortage of clean water in the Gaza strip (see box for full text of the Resolution).

Resolution on water in the Gaza Strip from the Conference on "Frontiers of Chemical Sciences III: Research and Education in the Middle East"

As scientists from throughout the Middle East, with some of their colleagues from other parts of the world, we wish to draw immediate and urgent attention to one such issue. Water is of central importance to human life; water in the Gaza Strip is of particular concern in terms of quantity and quality, threatening the health of every inhabitant regardless of their political inclination.

We urge governments to look beyond the present conflicts and disagreements that afflict the region. As with some other treaties, where difficult conflicts are set aside for future consideration, we urge that the interested governments and agencies ignore their current disagreements and, by drawing on scientific expertise, urgently address the issue of water in the Gaza Strip, taking into account the whole cycle from collection to reuse [13].

Scientists and governments alike have recognized the uniqueness and effectiveness of this conference. The National Science Foundation recognized the "Malta Conferences" as one of the ten best examples of "thinking outside the box." United States Senator

Richard Durbin made a speech on the "Malta Conferences" from the floor of the US Senate, in which he labeled the conferences as "truly historic" [10].

Nobel Laureate Walter Kohn wrote: "Our great Middle East chemistry meeting in Istanbul ... was for me - an elderly 'freshman' - a thrilling experience. The unfailingly friendly and cooperative tone; the excellent presentations, from basic science to urgent local and regional problems; the enjoyable banquets with opportunities for informally meeting colleagues with very different backgrounds and perspectives - yet all of us united by our love of science and commitment to its use for the benefit of mankind."

The opinions of the participants, however, matter most. The general consensus among attendees in each of the conferences has been that the achievements of the conference far exceeded all expectations, prompting Malta-III participants to vote unanimously to hold a Malta-IV conference in 2009 in Jordan. For the first time since the inception of the "Malta Conferences," ten Middle Eastern chemists have agreed to serve on the conference organizing committee, in addition to the existing international organizing committee. The convictions of the conference attendees and the concrete results of these conferences have borne out my initial vision: collaborations between scientists of the region will be the bridge to peace in the Middle East.

References

1. Freemantle, M. (2004) "Rendezvous in the Mediterranean" *Chemical and Engineering News*, 82, 2, pp. 36–39.
2. Langer, S. (2004) "Middle East talks hailed as 'Pugwash' for chemical sciences at Malta convention" *Chemistry World*, 1, 1 January, p. 6.
3. Malin, J.M. (2004) "Frontiers of Chemical Sciences – Research and Education in the Middle East" *Chemistry International*, 26, 3, pp. 7–9.
4. Lerman, Z.M. and Wade, J.S. (2006) "Frontiers of Chemical Sciences II: Research and Education in the Middle East" *Chemistry in Israel: Bulletin of the Israel Chemical Society*, Issue 21, pp. 21–23.
5. Malin, J.M. (2006) "Frontier Science in the Middle East" *Chemistry International*, 28, 2, pp. 9–11.
6. Ritter, S.K. (2005) "Science for Peace in the Middle East" *Chemical and Engineering News*, 83, 46, p. 15.
7. Everts, S. (2008) "Middle East Connections" *Chemical and Engineering News*, 86, 4, pp. 59–61.
8. Everts, S. (2007) "Conference uses science to build bridges in the Middle East" *Chemical and Engineering News*, 85, 51, p. 10.
9. Begitt, K., HopfRingsdorf, H. and Ringsdorf Ringsdorf, H. (2008) "Schritte über Grenzen" *Nachrichten aus der Chemie* 56, February, pp. 201–202.
10. Durbin, R. (2004) "Chemists Working Cooperatively" *United States Congressional Record*, 150, 66, May 12.
11. Velasquez, R. (2006) "Middle Eastern Scientists Promote Peace at Malta Conference" *Washington Report on Middle East Affairs*, 25, 1, p. 71.
12. Malin, J. (2008) "Middle East Chemists Meet in Istanbul" *Chemistry International*, 30, 3, pp. 31–34.
13. (2008) "Water in the Gaza Strip" *Chemistry International*, 30, 2, p. 18.

"Jeopardy" in the Inorganic Classroom – Teaching Descriptive Chemistry Using a Television Game Show Format

J. Van Houten

Abstract Descriptive chemistry is difficult to teach in the traditional manner. However, it is an important topic in any chemist's training, as it is at the heart of many important processes and applications such as: Industrial Processes, Analytical Applications, the Environment, Health, Consumer Products, Atmospheric Cycles, Geochemical Cycles, Biological Cycles, etc. Descriptive chemistry forms the historical basis of the periodic table. Nevertheless, it is often omitted from many undergraduate courses. The approach described herein allows the instructor to teach descriptive chemistry without "teaching" in the traditional manner. It has been adapted effectively using a series of weekly quizzes. This talk describes an alternative approach based on the format of the TV game show "Jeopardy." This not only enlivens the classroom, but it also provides opportunities for instant feedback and elaboration on topics covered. The rules of the game are modified so that points are awarded, rather than cash, and all students have an opportunity to earn points for all questions.

1 Introduction - Why Teach Descriptive Chemistry?

Teaching descriptive chemistry has always been a challenge. The subject does not lend itself well to the lecture/discussion format commonly employed in classrooms today; and the subject matter is, regrettably, not considered to be "appropriate" for modern courses at the undergraduate level. Nevertheless, descriptive chemistry lies at the heart of much (if not all!) of what we, as chemists, do. Descriptive chemistry includes an understanding of the nomenclature (including popular, historical, and "trivial" names) and properties of the elements and their common compounds, the major sources and uses of common substances, and typical reactions - in particular those reactions that are important to the production and use of common substances in everyday life and in the laboratory.

The question is: "Why has descriptive chemistry fallen into disfavor?" Perhaps it is because descriptive chemistry is difficult to "teach" in the traditional manner.

J. Van Houten (✉)
Saint Michael's College, Colchester, Vermont, 05439, USA
e-mail: jvanhouten@smcvt.edu

M. Gupta-Bhowon et al. (eds.), *Chemistry Education in the ICT Age*,
DOI 10.1007/978-1-4020-9732-4_2, © Springer Science + Business Media B.V. 2009

This paper describes an approach to teaching descriptive chemistry in a nontraditional manner to third- and fourth-year college undergraduates as part of our required inorganic chemistry course.

The history of chemistry as a modern science is generally traced to the late eighteenth and early nineteenth century when chemistry emerged from the shadow of alchemy as Dalton, Lavosier (to name just the two best-known examples) and others began to describe and catalog the results of their systematic and quantitative observations of the properties of matter. The work of that time was largely descriptive as chemists were beginning to discover the combining properties of the elements and the relationships between chemical structures and their properties. Figure 1 shows Dalton's depiction of atoms and molecules. That work remains at the heart

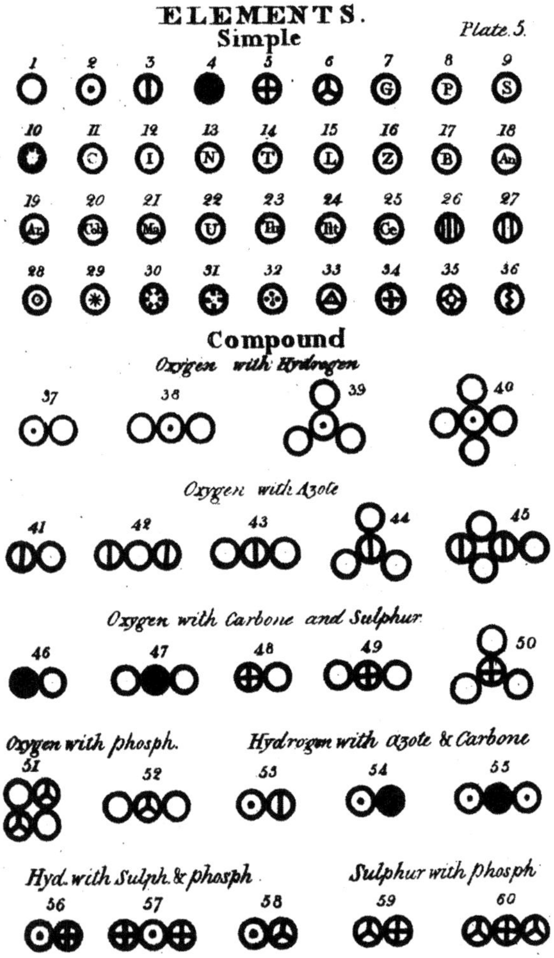

Fig. 1 Plate depicting atoms and molecules in John Dalton's *A new System of Chemical Philosophy* (Manchester 1808–1810) (Reproduced from [1], p. 171. With permission)

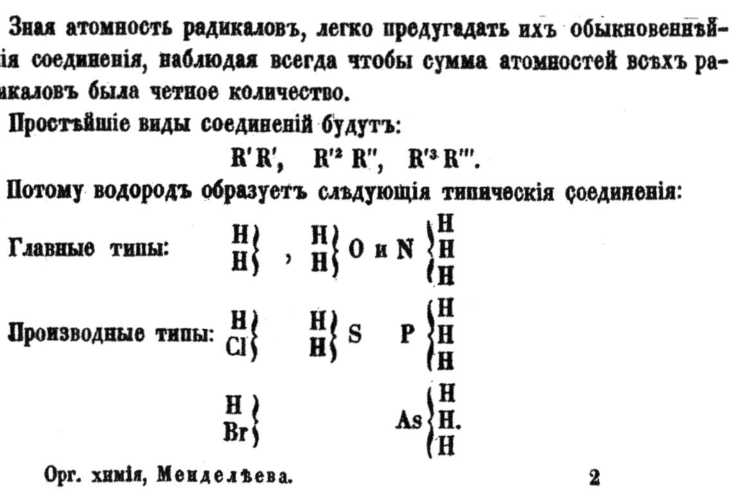
Зная атомность радикаловъ, легко предугадать ихъ обыкновеннѣйшія соединенія, наблюдая всегда чтобы сумма атомностей всѣхъ радикаловъ была четное количество.

Простѣйшіе виды соединеній будутъ:

R'R', R'² R'', R'³ R'''.

Потому водородъ образуетъ слѣдующія типическія соединенія:

Главные типы: H H , H H O и N H H H

Производные типы: H Cl H H S P H H H

H Br As H H. H

Орг. химія, Менделѣева. 2

Fig. 2 Dimitri Mendeleev's depiction of simple hydrogen compounds presented in the second Russian edition of his text on organic chemistry, *Organicheskaja Khimia*, Saint Petersburg, 1863. (Reproduced from [1], p. 215. With permission.)

of what chemists and scientists in related fields do in the twenty-first century. Therefore, it is critical that our students understand and appreciate the beauty and elegance as well as the importance of descriptive chemistry. A worker in a modern nanotechnology lab would not be very successful if he or she believed silver chloride was a pale green gas (to quote an oft-cited misconception).

An early example of how descriptive chemistry formed the basis for the teaching of chemistry can be found in Fig. 2 taken from Dimitri Mendeleev's text on organic chemistry (*Organicheskaja Khimia*, Saint Petersburg, Russia, 1863). On page two of his textbook Mendeleev shows a table containing formulas for H_2, HCl, HBr, H_2O, H_2S, NH_3, PH_3, and AsH_3 arranged in columns according to the number of hydrogen atoms in each compound. Furthermore, the formulas are written in a manner indicating that Mendeleev understood and appreciated the acid-base and electronegativity characteristics of the elements; although those concepts were not formally described until many decades later by Arrhenius, Brønsted, Lowry, Lewis, and Pauling (among others). Of course it is well-known to most students of introductory chemistry that Mendeleev's understanding of the descriptive chemistry of the elements led him to arrange them in what has become our modern periodic table, which he first published in 1869. A later version of Mendeleev's table (see Fig. 3), published in his book *Grundlagen der Chemie* (Saint Petersburg, 1891), has the groups labeled with the typical hydride for each group (RH_4, RH_3, etc.) and with the highest salt-forming oxide (R_2O, RO, R_2O_3, etc.), again emphasizing the importance of descriptive chemistry at this important point in the development of chemistry as a science and in the teaching of the subject.

In addition to being the original basis of the periodic chart, descriptive chemistry remains at the heart of all that we, as chemists, do today. It is central to the various

GRUPPE:	I.	II.	III.	IV.	V.	VI.	VII.	VIII.
Reihe: 1	• H			RH^4 •	RH^5	RH^2	RH	Wasserstoffverbindugen.
» 2	Li	Be •	B	C •	N •	O •	F •	
» 3	• Na	Mg	• Al	• Si	• P	• S	Cl	
» 4	K •	Ca •	Sc •	Ti •	V •	Cr •	Mn •	Fe. Co. Ni. Cu.
» 5	• (Cu)	• (Zn)	• Ga	• Ge	• As	• Se	• Br	
» 6	Rb. •	Sr •	Y •	Zr •	Nb •	Mo •	– •	Ru. Rh. Pd. Ag.
» 7	• (Ag)	• Cd	• In	• Sn	• Sb	• Te	• J	
» 8	Cs •	Ba •	La •	Ce •	Di? •	– •	– •	– – – –
» 9	• –	• –	• –	• –	• –	• –	• –	
» 10	– •	– •	Yb •	– •	Ta •	W •	– •	Os. Ir. Pt. Au.
» 11	• (Au)	• Hg	• Tl	• Pb	• Bi	• –	• –	
» 12	– •	– •	– •	Th •	– •	U •	– •	
	R^2O	R^2O^2 RO	R^2O^3	R^2O^4 RO^2	R^2O^5	R^2O^6 RO^2O^8	R^2O^7	Höchste salzbildende Oxyde RO^4

Fig. 3 Periodic table from Dimitri Mendeleev's book *Grundlagen der Chemie* (Saint Petersburg, 1891) (Reproduced from [1], p. 215. With permission)

atmospheric, geological, and biological cycles described in our textbooks. It is incorporated (somewhat) into our organic courses when we talk about the general properties of the various functional groups such as alcohols, ketones, etc. Most general chemistry textbooks include a section on descriptive chemistry, but many instructors tend to gloss over those sections or to skip them completely in favor of more quantitative concepts, thereby turning the general chemistry course into "baby P-chem." Descriptive chemistry, and the concepts arising from it, underlies all of the techniques of "wet" chemical analysis; but, again, those techniques have fallen into disfavor as colleges and universities tend to emphasize "modern" instrumental methods. Nevertheless the "old-fashioned" wet techniques remain widely used in many industrial, environmental, and forensic applications. Finally, descriptive chemistry appears in the development and use of consumer and industrial products, biomedical applications, and all the other aspects that make chemistry such an important part of our daily lives and our economic vitality.

2 Discussion - The "Jeopardy" Approach

For all the reasons discussed above, descriptive chemistry remains an important part of the chemical curriculum, and we are doing our students a great disservice if we fail to instill in them an appreciation for the properties of the elements at the most fundamental level, for it is only through that that they will truly appreciate and understand what they will learn in their advanced courses and what they will encounter in their later careers once they leave school. The question, then, is: "How best to teach the subject matter of descriptive chemistry?" The answer I have found is not to attempt to teach it in the traditional manner but to have the students learn the material on their own in preparation of a series of weekly exercises.

When I was a student studying inorganic chemistry with John Burmeister at the University of Delaware, those weekly exercises took the form of a written quiz, with one quiz per week covering a different group of the periodic table. When I began my own teaching career almost 3 decades ago, I followed Burmeister's example; however in more recent years I have abandoned the written quiz in favor of a format based on the popular television game show, "Jeopardy." On the TV show, contestants compete for cash prizes by answering questions based on knowledge of current events, popular culture, geography, history, etc. In the inorganic classroom students receive grade points rather than cash. The inorganic version's questions stress understanding of basic facts and periodic trends. Categories are typically:

- Properties
- Sources and Uses
- Definitions and Nomenclature
- Common Reactions

Students are required to study written material in preparation for the weekly exercise. The material consists of a short (5- to 10-page) chapter typically covering one group of the periodic table, so the entire table can be covered in one academic term. Many General Chemistry texts contain this material at the appropriate level, discussing the topics shown in the categories above. The weekly exercise - written quiz or Jeopardy game - takes about 10–15 min at the beginning of a class period.

3 Rules of the "Inorganic Jeopardy" Game

As mentioned above the rules of the television show must be modified somewhat to adapt it to the classroom environment. On the television show, individual contestants vie for the opportunity to attempt to answer a question by pressing a buzzer. Only one person can answer and earn cash for each question. In order to allow all students in a class an opportunity to earn points for each question, this procedure has been modified as follows:

- Rotate around class, students choosing a category and value in turn.
- As each question is revealed, all students write their answers on an answer sheet set up with a line for each category and question. The correct answer is then revealed by the instructor.
- If "chooser" gets the correct answer he/she gets full credit and others with the correct answer get half-credit. The half-credit is facilitated by having all questions have an even-numbered value. Higher value questions can have multiple parts - e.g., a $600 (i.e., 6 point) question could have three parts with each part being worth $200 (2 points).
- If "chooser" gives an incorrect answer he/she gets zero and all other students with the correct answer get full credit.
- Grade is total points for each student.

- Depending on the size of the class, the instructor may keep a running score for each student in the class.
- Students turn in their answer sheets at the end of the exercise.

When I first began using the Jeopardy approach, the questions were printed on overhead transparencies, covered with paper, and revealed by removing the paper. In recent years (i.e., the ICT Age) the questions are typed into cells of an Excel spreadsheet. After being typed, the font color is changed to match the cell background color, thereby rendering the question invisible. It is revealed at the appropriate time by changing either the font color or the background color. The same can be done with the answers. Table 1 shows a set of questions for the Properties category of periodic table Group 3.

Table 1 Sample quiz on Group 3 Elements

Properties	
$200	What is the most abundant metal in the earth's surface?
$400	Which element melts in your hand?
$600	Describe the acid-base properties of boron compounds.
$800	A. ($400) What has the widest liquid temperature range known for any main-group element?
	B. ($200 each) What are the melting and boiling points of that element?
Sources and Uses	
$200	What is bauxite?
$400	What is a use for an indium compound?
	Identify the compound.
$600	A. What is alumina?
	B. State a use for alumina.
$800	Describe the Hall process, the source of a major industrial material.
Definitions & Nomenclature	
$200	Give the structure of diborane.
$400	What is a "three-center bond orbital"? Give an example.
$600	A. What is the chemical composition of Drano?
	B. How does it work?
$800	A. What is borax?
	B. How is it converted to boric acid?
Reactions	
$200	Give an example of a thermite reaction
$400	A. LiH(sol'n) + $AlCl_3$(sol'n) → ___ + ___
	B. What solvent?
$600	A. Boric acid + water → ___ + ___.
	B. This shows that boric acid is a ___ acid.
$800	A. Give a reaction that is part of the Bayer process.
	B. Give the formula for the ultimate product, cryolyte.

4 Advantages (and Disadvantages) of the Jeopardy Approach

The game show format has a number of advantages over the written quiz. Most obviously it enlivens the classroom. In addition it allows for instant feedback. It also allows the instructor to correct misconceptions and to expand on the answer if desired. One possible disadvantage is that some students may not like the loss of privacy that comes with answers and individual scores being revealed and discussed in the open classroom.

5 Conclusion

The Jeopardy approach provides a novel method of teaching descriptive chemistry. It avoids what some view as the "drudgery" of lecturing on this routine material. Instead it provides a lively classroom environment in which learning can take many forms. My inorganic chemistry course has become legendary among our students for this aspect of it.

Reference

1. Greenburg A., *A Chemical History Tour*, Wiley, New York, 2000.

Teaching Thermodynamic Relations Using a Story and Two-Dimensional Cartesian Coordinate System

Y. H. Chung

Abstract Thermodynamic energy functions are related to six variables such as volume, pressure, temperature, entropy, chemical potential and amount of substance. They are rather cumbersome and perplexing to undergraduates who start to learn their relations. With a story and the two-dimensional Cartesian coordinate system, most of thermodynamic relations could be obtained in addition to the Maxwell relations for a reversible change in a closed system only in the presence of pressure-volume work and heat.

1 Introduction

Thermodynamic energy functions related to internal energy, U or E [1], enthalpy, H, Gibbs energy, G, and Helmholtz energy, A are affected by the six variables such as volume, V, pressure, P, temperature, T, entropy, S, amount of substance, n and chemical potential, μ. Their relations are in general cumbersome and perplexing to undergraduates who start to learn them.

Various mnemonics have been reported to help students to be familiar with thermodynamic relations [2–5]. Most of them are rather direct notation and demand their memorization. Teaching the pertinent thermodynamic relations to them could be consummated with a simple story displayed in the two-dimensional Cartesian coordinate system for a reversible change in a closed system without composition change in the absence of any other work except pressure-volume work.

Y.H. Chung (✉)
Department of Chemistry, Hallym University, Korea
e-mail: yhchung@hallym.ac.kr

M. Gupta-Bhowon et al. (eds.), *Chemistry Education in the ICT Age*,
DOI 10.1007/978-1-4020-9732-4_3,

2 Methodology

In teaching the thermodynamic energy functions and their relations to their pertinent variables, a simple story displayed in the two-dimensional Cartesian coordinate system could be used for a reversible change in a closed system without composition change in the absence of any other work except pressure-volume work. In the beginning stage of teaching, students are asked to imagine their honeymoon on a cruise ship. When they are on board, they are treated as VIPs and led to a table at the first class dining room in the center of the ship. Various dishes are swiftly served without rest and all drinks are delivered right on their request. Their service is immaculate during the cruise. When they leave the ship, they receive the bills with value added tax. Handsome tips are given for the fantastic service. This story can be summarized as follows:

> I, as a ViP SiTting down at the center of the SHiP, was NiMbly SErVed without rest; a bill with VAT was received and a TiP Given.

The six variables in the story are paired in terms of energy: volume, V and pressure, P from 'ViP', entropy, S and temperature, T from 'SiTting', and amount of substance, n and chemical potential, μ from 'NiMbly'. The notation of the chemical potential is substituted by 'M' in the story since μ is converted to m in English. The product of each pair is expressed in energy and is easy to be recalled since they are familiar words.

'i' is located at the centre of the two-dimensional Cartesian coordinate system and shown in Fig. 1. 'ViP' and 'SiT' imply that V and P lie in the x-coordinate and S and T in the y-coordinate in the two-dimensional Cartesian coordinate system. 'SHiP' indicates that the enthalpy, H in the first quadrant between S and P is a function of S and P, expressed in $H(S,P)$. 'SErVed without rest' shows that the internal energy, E in the second quadrant between S and V is a function of S and V, expressed in $E(S,V)$. 'VAT' implies that the Helmholtz energy, A in the third quadrant between T and V is a function of T and V, expressed in $A(T,V)$. 'TIP Given' refers that the Gibbs energy, G in the fourth quadrant between T and P is a function of T and P, expressed in $G(T,P)$.

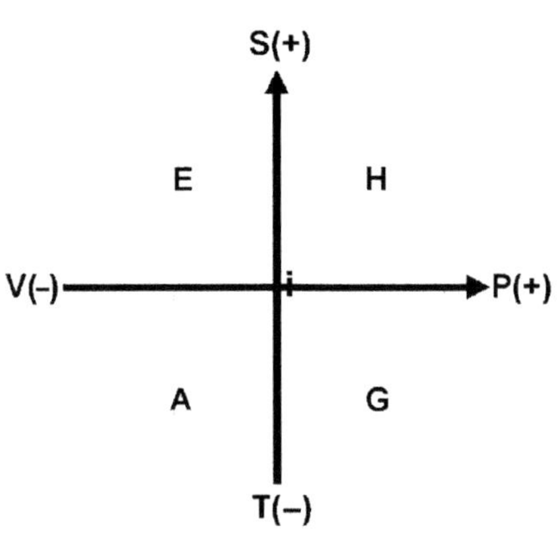

Fig. 1 Thermodynamic energy functions with the four variables shown in the two-dimensional Cartesian coordinate system

The four energy functions related to internal energy, enthalpy, Gibbs energy and Helmholtz energy are state functions and have their corresponding exact differentials. Their exact differentials dE, dH, dG, and dA could be expressed by summing two pairs d(ST) and d(PV) for a reversible change in a closed system without composition change in the absence of any other work except pressure-volume work. From $E(S,V)$, dE can be expressed by adding two energy pairs, TdS and PdV. Since E is in the second quadrant in Fig. 1, dV is replaced by –dV. Consequently, dE = TdS – PdV. Similarly, $H(S,P)$ in the first quadrant gives dH = TdS + VdP. From $G(T,P)$, dG can be expressed by adding two pairs, SdT and VdP. Since G is in the fourth quadrant in Fig. 1, dT is replaced by –dT. Consequently, dG = –SdT + VdP. From $A(T,V)$, dA can be shown by adding two pairs, SdT and PdV. Since A is in the third quadrant in Fig. 1, dT and dV are replaced by –dT and –dV, respectively. Consequently, dA = –SdT – PdV.

From the equations dE = TdS – PdV and dE = $(\partial E/\partial S)_VdS$ + $(\partial E/\partial V)_SdV$, one can have

$$(\partial E/\partial S)_V = T;\ (\partial E/\partial V)_S = -P \tag{1}$$

Since E is a path-independent state function, it satisfies its pertinent Euler's relation

$$\partial^2 E/\partial V \partial S = \partial^2 E/\partial S \partial V \tag{2}$$

From equations (1) and (2), the following relation can be deduced as

$$(\partial T/\partial V)_S = -(\partial P/\partial S)_V \tag{3}$$

The equation (3) is one of the four Maxwell relations [6, 7]. This relation can be obtained directly from Fig. 2. The partial derivative, $(\partial T/\partial V)_S$ can be obtained by clockwise choosing three sequentially adjacent variables, T, V and S. Since the first

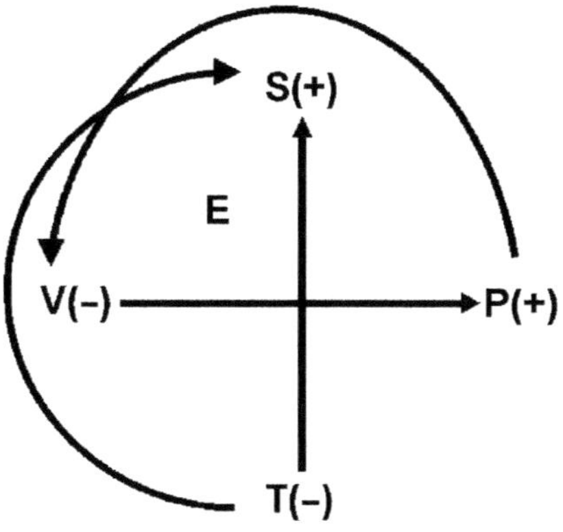

Fig. 2 Partial derivatives chosen for $-(\partial T/\partial V)_S = (\partial P/\partial S)_V$ from the internal energy relations

variable is negative in the y-coordinate, T is replaced by $-T$ and the resultant partial derivative becomes $-(\partial T/\partial V)_S$. Its counterpart of the Maxwell relations, $(\partial P/\partial S)_V$ can be attained by counterclockwise choosing three variables, P, S and V. By equating these partial derivatives, one can get the equation (3).

Similarly, from the equations $dH = TdS + VdP$ and $dH = (\partial H/\partial S)_P dS + (\partial H/\partial P)_S dP$, one can have

$$(\partial H/\partial S)_P = T; \; (\partial H/\partial P)_S = V \tag{4}$$

Since H is a state function, it satisfies its pertinent Euler's relation

$$\partial^2 H/\partial P\partial S = \partial^2 H/\partial S\partial P \tag{5}$$

From equations (4) and (5), the following relation can be deduced as

$$(\partial T/\partial P)_S = (\partial V/\partial S)_P \tag{6}$$

The equation (6) is also one of the four Maxwell relations, which can be obtained from Fig. 3. The partial derivative, $(\partial V/\partial S)_P$ can be obtained by clockwise choosing three sequentially adjacent variables, V, S and P. Since the first variable is negative in the x-coordinate, V is replaced by $-V$ and the resultant partial derivative becomes $-(\partial V/\partial S)_P$. Its counterpart of the Maxwell relations, $(\partial T/\partial P)_S$ can be attained by counterclockwise choosing three variables, T, P and S. Since the first variable is negative in the y-coordinate, T is replaced by $-T$ and the resultant partial derivative becomes $-(\partial T/\partial P)_S$. By equating these partial derivatives, one can get the equation (6).

From the equations $dA = -SdT - PdV$ and $dA = (\partial A/\partial T)_V dT + (\partial A/\partial V)_T dV$, one can have

$$(\partial A/\partial T)_V = -S; \; (\partial A/\partial V)_T = -P \tag{7}$$

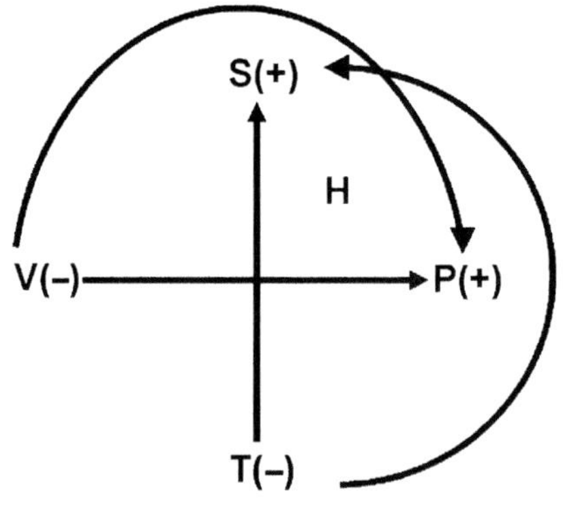

Fig. 3 Partial derivatives chosen for $-(\partial V/\partial S)_P = -(\partial T/\partial P)_S$ from the enthalpy relations

Since A is a state function, it satisfies its pertinent Euler's relation

$$\partial^2 A/\partial V\partial T=\partial^2 A/\partial T\partial V \tag{8}$$

From equations (7) and (8), the following relation can be deduced as

$$(\partial S/\partial V)_T=(\partial P/\partial T)_V \tag{9}$$

The equation (9) is one of the four Maxwell relations, which can be obtained from Fig. 4. The partial derivative, $(\partial P/\partial T)_V$ can be obtained by clockwise choosing three sequentially adjacent variables, P, T and V. Its counterpart of the Maxwell relations,$(\partial S/\partial V)_T$ can be attained by counterclockwise choosing three variables, S, V and T. By equating these partial derivatives, one can get the equation (9).

From the equations $dG=-SdT+VdP$ and $dG=(\partial G/\partial T)_P dT+(\partial G/\partial T)_T dP$, one can have

$$(\partial G/\partial T)_p=-S;\ (\partial G/\partial P)_T=V \tag{10}$$

Since G is a state function, it satisfies its pertinent Euler's relation

$$\partial^2 G/\partial P\partial T=\partial^2 G/\partial P\partial T \tag{11}$$

From equations (10) and (11), the following relation can be deduced as

$$-(\partial S/\partial P)_T=(\partial V/\partial T)_P \tag{12}$$

The equation (12) is one of the four Maxwell relations, which can be obtained from Fig. 5. The partial derivative, $(\partial S/\partial P)_T$ can be obtained by clockwise choosing three sequentially adjacent variables, S, P and T. Its counterpart of the Maxwell relations, $(\partial V/\partial T)_P$ can be attained by counterclockwise choosing three variables,

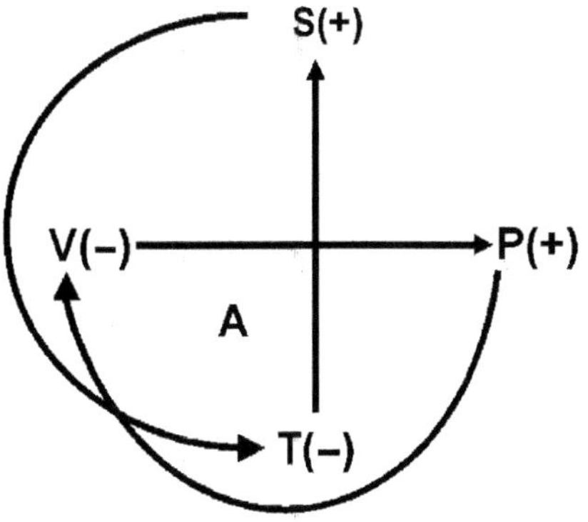

Fig. 4 Partial derivatives chosen for $(\partial P/\partial T)_V = (\partial S/\partial V)_T$ from the Helmholtz energy relations

Fig. 5 Partial derivatives chosen for $(\partial S/\partial P)_T = -(\partial V/\partial T)_P$ from the Gibbs energy relations

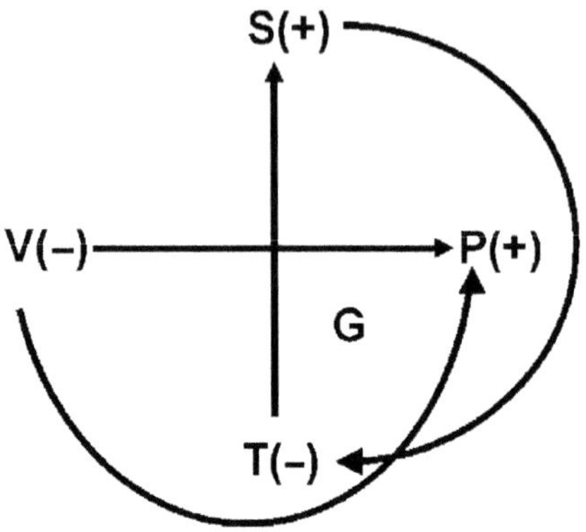

V, T and P. Since the first variable is negative in the x-coordinate, V is replaced by $-V$ and the resultant partial derivative becomes $-(\partial V/\partial T)_P$. By equating these partial derivatives, one can get the equation (12).

The teaching process will be concluded by asking each of the students to narrate the story, deduce the pertinent thermodynamic relations from it and write them down on the blackboard. This will show how much the students become familiar with them.

3 Results and Discussion

The simple story displayed in the two-dimensional Cartesian coordinate system has been used in teaching chemistry students thermodynamic energy functions and their pertinent relations involving variables for a reversible change in a closed system without composition change in the absence of any other work except pressure-volume work. They were asked to narrate it, deduce the thermodynamic relations from it and write them down on the blackboard. In this way their familiarity with thermodynamic relations could be checked. Most students were able to get the exact differentials of internal energy, enthalpy, Gibbs energy and Helmholtz energy without difficulty in addition to all the Maxwell relations. Furthermore, they immediately named three energy pairs, PV, ST and μn from ViP, SiT and NiMbly, respectively, even without projecting the story into the two-dimensional Cartesian coordinate system.

It would in general take considerable amount of time and pains for students to be familiar with any new concept. When the students knew the thermodynamic relations, it seemed much more effective to teach them more complex concepts. This simple scheme would facilitate teaching the thermodynamic energy functions and their pertinent relations to chemistry undergraduates or graduates.

4 Conclusion

For a reversible change in a closed system without composition change and any other work except pressure-volume work, the story displayed in the two-dimensional Cartesian coordinate system could be used to facilitate teaching the thermodynamic functions and their relations to the variables to chemistry undergraduates. For a system involving composition change, an additional energy pair could be added simply to the known relation of its internal energy change and shown as

$$\mathrm{d}E = -P\mathrm{d}V + T\mathrm{d}S + \mu dn$$

Similarly, in the case of a system under electrical work, an energy pair could be added using electromotive force, E and charge, Q as

$$\mathrm{d}E = -P\mathrm{d}V + T\mathrm{d}S + \mathrm{E}\mathrm{d}Q$$

In the magnetic case, the corresponding energy pair will be used with magnetic field strength, H and magnetization, M as

$$\mathrm{d}E = -P\mathrm{d}V + T\mathrm{d}S + \mathrm{H}\,\mathrm{d}M$$

Since students have to be familiar with fundamental chemical relations, more diverse schemes could be employed in elaborately teaching them rather perplexing and cumbersome concepts that are new to them.

References

1. Brown TL, Lemay HE Jr, Bursten HE (1994) Chemistry. 6th edn. Prentice Hall, Englewood Cliffs
2. Koenig FJ (1935) Families of thermodynamic equations. I The method of transformations by the characteristic group. *J. Chem Phys* 3:29–35
3. Gilvarry JJ (1951) Maxwell's thermodynamic relations. *Am J Phys* 19:131–132
4. Gallen HB (1985) Thermodynamics and an introduction to thermostatistics. Wiley, New York
5. Pate SF (1999) The thermodynamic cube: A mnemonic and learning device for students of classical thermodynamics. *Am J Phys* 67:1111–1113
6. Moore WJ (1972) Physical chemistry. 4th edn. Prentice-Hall, Englewood Cliffs
7. Atkins PW (1998) Physical chemistry. 6th edn. Oxford, Oxford

Heralding Calamity of Global Warming and Chemistry Role Through a Chorus

Y. H. Chung

Abstract Global warming needs imminent attention in searching for its optimal resolution. Chemistry could pave the way to prevent its foreseeable calamity. In order to publicize chemistry's role to chemistry undergraduates and the public, CHEM2future formed at the Chemistry Department, Hallym University, has performed a chorus and a CHEMopera for 2 years. Three female undergraduates performed a chorus related to the global warming. The performance accompanied a narration of chemistry's role in a seminar and two music tunes representing the current status of the global warming. It would promote chemistry undergraduates to be dedicated to correcting devastating environmental disparities in the future.

1 Introduction

Global warming has derailed the mechanism of environmental restoration and brought unwanted calamity. Since it needs imminent attentions of people in every walk of life, its gravity could be publicized to chemistry students to be interested in searching for its resolution. For this purpose, a club named CHEM2future was organized at the Chemistry Department, Hallym University 2 years ago. It has performed a sing-along and a CHEMopera for 2 years [1, 2].

In this work, three female students of the CHEM2future performed a chorus in a program titled "Chorus and Seminar on Calamity of Global Warming and Chemistry Role." The message of the imminent crisis of the global warming was delivered to the students through the chorus of a song composed for this purpose. Several chemical reactions were introduced in its lyrics to shed light on chemistry's role. The seminar following the chorus was focused on researches of carbon dioxide

Y.H. Chung (✉)
Department of Chemistry, Hallym University, Korea
e-mail: yhchung@hallym.ac.kr

M. Gupta-Bhowon et al. (eds.), *Chemistry Education in the ICT Age*,
DOI 10.1007/978-1-4020-9732-4_4,

reduction and fuel cell to motivate the participants to be interested in pursuing diverse solutions for global warming in the future.

2 Methodology

The program "Chorus and Seminar on Calamity of Global Warming and Chemistry Role" employed in this work consisted of four segments: two viola solo tunes representing the current status of global warming, chorus, seminar on global warming and chemistry's role, and sing-along.

Two music tunes, Beethoven's "Virus" and Offenbach's "Jacqueline's Tears" were chosen to be played by a violist to depict the rate and gloom of the global warming. After the play, there were pop quizzes to name their titles in order to encourage the audience's participation.

The CHEM2future performed the chorus of the song "We have a dream [3]." It was composed for this purpose 2 years ago and its lyrics are as follows:

This world has all sorts of problems.
CO_2 and methane cause global warming.
Glaciers in the South and North Pole are melting daily,
And sea level's rising up!
It causes flood and weather change.
El Niño is a frequent visitor.
Hurricanes like Katrina visit us frequently, leading to Pandemonium.
We couldn't curb 'em by the present science and tech.
Dream of making a better future by chemistry.
We have a dream, where chemistry would solve all sorts of problems.
Haber caught N_2 in air and with H_2 made ammonia.
Ammonium sulfate, fertilizer becomes a solution of food shortage.
With metal like zinc, HCl makes H_2.
It meets with NaOH to become water and salt.
In a world without water and salt we couldn't live.
Our forefathers used to wear white cotton clothes.
They're worn out easily.
It's natural that they be stitched.
Adipic acid and diaminohexane turn into nylon, lasting fabric of stockings,
Revealing beauty of lady's legs.
Ethylene glycol and terephthalic acid become polyester,
Turning to PET bottles or clothes according to its usage.
Dream of making a better future by chemistry.

A seminar on global warming and chemistry's role was given. In the seminar chemistry's role was focused to various solutions for carbon dioxide reduction and energy shortage.

All the participants learned to sing "We have a dream" in sing-along with the CHEM2future members along its viola tune. They read its lyrics to notice its message.

3 Results and Discussion

The rate of the global warming was depicted in terms of Beethoven's "Virus," implying that it speeds up. Its gloomy aftermath was described in Offenbach's "Jacqueline's Tears," indicating that melting glaciers were compared to the tears shed in the earth's suffering. After the two viola tunes, several participants were eager to name their titles. Most of the participants seemed to grasp a sensuous image of the global warming. This observance manifested that the ongoing environmental crisis could be expressed not only in words but also with music more effectively. One could elaborately develop this type of programs to convey educational messages to students.

As shown in Fig. 1, the chorus performed by three female members of the CHEM2future followed the viola solo tunes. The lyrics of the song "We have a dream" delivered to the audience the messages on the global warming and chemistry's role depicted by various chemical reactions. Its lyrics and score were displayed on a screen during the chorus. The spectators could visually recite it along the chorus.

The seminar on global warming and chemistry's role focused on the current status of energy shortage and reutilization of carbon dioxide. It drove many students to ask the speaker diverse questions related to alleviation of the global warming and energy alternative.

In the sing-along, the members of the CHEM2future sang together with the audience "We have a dream" along the tune played by the violist. During the session, all the participants read the lyrics of the song and recognized the essential chemical reactions.

Fig. 1 Three female members of the CHEM2future performing a chorus

4 Conclusion

In the program "Chorus and Seminar on Calamity of Global Warming and Chemistry Role," the classical musical tunes were employed in conveying to the public chemical messages related to the global warming. In the chorus the lyrics and score displayed on the screen captured the audience's attention and helped them to be absorbed into it. The seminar given with the broad chemical information facilitated the audience's understanding of the gravity of the global warming. All the participants became harmonized in unison along the sing-along.

More diverse programs amalgamated with chorus, seminar, and play could be developed to draw the audience's attention and effectively deliver the chemical messages to them in the future.

References

1. Chung, Y. H. 2006. *Performance of Chemical Education in Sing-along and Play*, 19th International Conference on Chemical Education, Seoul, Korea, August 12–17.
2. Chung, Y. H. 2006. *CHEMopera Combined with Chemistry Magic Show*, 232nd National Meeting of the American Chemical Society, San Francisco, USA, September 10–14.
3. Hwang, B.-R., Kim, Y.-W. and Chung, Y. H. 2006. We have a dream. HYC2future, Chuncheon.

Using the Arts and Computer Animation to Make Chemistry Accessible to All in the Twenty-First Century

Z. M. Lerman and D. Morton

Abstract The twenty-first century is truly the century of science and technology. If we will not make science and technology accessible to all, we will form a two-class society, divided not by royalty or economic status, but by knowledge of science and technology. It is our tenet in life that science education is a human right that belongs to all. Therefore, it is essential that we employ every method possible to make chemistry accessible to all in the twenty-first century [1, 2]. A Chinese proverb states: "I hear and I forget; I see and I remember; I do and I understand." At the Science Institute, Columbia College Chicago, we believe strongly in this proverb, and we incorporate visualization and art in the teaching of chemistry. Students remember and understand abstract chemistry concepts best by creating their own artistic projects. Through this process, students take an active role in the learning process, instead of only being passive observers. The students can produce visualization projects using the media of their choice, from computer animation (High Tech) to dance and drama (No Tech). These projects are used as an alternative assessment method where the evaluation is done in a constructive way by the whole class and not just by the instructor. This method has been proven successful with undergraduate students, with science teachers, and with middle and high school students. Many institutions in the USA and around the world have adopted this method.

1 Narrative

A key element of the innovative science curriculum used at Columbia College in teaching, learning, and assessing science to its non-science major student body requires all students to prepare projects to express their knowledge of science in an original and creative way [3], making use of the skills and talents in each individual's

Z.M. Lerman (✉) and D. Morton
Institute for Science Education and Science Communication, Columbia College Chicago
600 S. Michigan Avenue, Chicago, Illinois, 60605, USA
e-mail: zafral@aol.com

M. Gupta-Bhowon et al. (eds.), *Chemistry Education in the ICT Age*,
DOI 10.1007/978-1-4020-9732-4_5, © Springer Science+Business Media B.V. 2009

particular major field of study, personal interest, or cultural background. These projects have taken the form of videotapes, paintings [4], sculptures, audiotapes, dances, "rap" songs, and scripts for theater. Students present their research and investigation results using the medium in which they feel most competent, comfortable, and talented [5]. It has been observed that students better understand and retain longer the scientific information they present in their projects. The entire class benefits from the scientific content of the project presentations: each presentation inspires in the other students new ideas and new ways of looking at scientific material. This form of assessment allows the teacher, the entire class, and the student presenter to engage in a cooperative assessment. Furthermore, the projects often result in a permanent record, similar to portfolio assessment, which is both useful for its scientific and educational content, and valuable from a dissemination standpoint. This approach works outstandingly well with undergraduate non-science majors, with teachers and students of grades K-12, and with people who have developed hostility or anxiety toward science and standardized tests. This method, if properly utilized, can yield enormous benefits to many generations of students. Not only can it help students become citizens who are well-informed about science, it can also encourage more of them to choose science and science-related fields as possible career choices.

2 Examples of Student Projects

Abstract scientific concepts have been made concrete through paintings, sculptures, dances, and plays. A group of high school students from a very low-income neighborhood who participated in our program were invited to perform in a Gordon Research Conference on science visualization [6]. The Gordon Research Conferences are very prestigious scientific conferences; the students showed the conference participants how they visualize scientific concepts through dance, and how they are able to grasp the abstract concepts after visualizing them though dance. They performed dances on the Periodic Table, the Ionic Bond, and the Effect of Water on the Ionic Bond as part of the visualization process (Figs. 1 and 2). A group of middle school children performed a dance on the depletion of the Ozone Layer by CFCs. This dance was aired globally by CNN, and in the USA by NBC, ABC and WGN television stations (Fig. 3).

A group of college students in the course "From Ozone to Oil Spills: Chemistry, the Environment and You" produced a video on the formation of NH_3 (Fig. 4), in which they named nitrogen as "Agent 007," and the three hydrogens as "The Hydrogen Triplets." The video describes the covalent bond and the ionic bond of table salt in a satiric, but scientifically accurate, way. An art student [7] presented his understanding of the fission reaction and a nuclear explosion through an art book (Fig. 5), while another art student expressed in a painting her concern for the environment and the adverse effects of air pollution on our cities and on the human body (Fig. 6). An animation student created a DVD showing glacial melting as a

Fig. 1 Periodic Table dance performance in a Gordon Research Conference on science visualization

Fig. 2 Periodic Table dance performance in a Gordon Research Conference on science visualization

result of climate change (Fig. 7), and the destruction of the Ozone Layer by CFCs was presented using computer animation (Fig. 8). The "Science Classroom of the Future" was visualized by a group of students (Fig. 9) who presented abstract scientific concepts through computer animation. A movie depicting the discovery of Ra and

Fig. 3 Depletion of the Ozone Layer dance performance by Middle School children

Fig. 4 Student-produced video on the formation of NH_3

Po by Madame Curie, as well as her presentations in front of the French Science Academy, was produced by a team of students (Figs. 10 and 11).

An art student truly represented the danger we face by forming a two-class society. He presented a drawing of two groups of students resembling the two-class society divided by knowledge of science and technology. One group is busy working on their computers, while the other group is sad and deprived (Fig. 12).

Fig. 5 Images from an art book produced by a Columbia College Chicago student on the fission reaction and a nuclear explosion

Fig. 6 An art student's painting on the adverse effects of air pollution on our cities and on the human body

Fig. 7 Image from an animation student's DVD showing glacial melting as a result of global warming

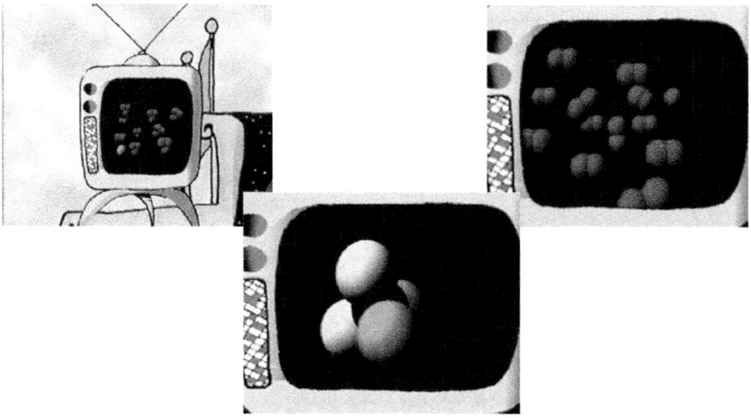

Fig. 8 Computer animation on the destruction of the Ozone Layer by CFCs, produced by a Columbia College Chicago student and presented on a television screen

3 Summary

This method of teaching chemistry at all levels has been proven successful in many institutions in the USA and around the world. Much of the funding for this curriculum was received from the National Science Foundation (NSF) and required an independent evaluator. The results of the evaluation showed that in

Fig. 9 A scene from the project "The Science Classroom of the Future" as visualized by a group of students

Fig. 10 A scene from a student-produced movie, showing Madame Curie in her lab

grades 5 through 10, students who studied using these methods achieved on standardized tests an average of 20% higher scores than their peers. An NSF site visitor to Columbia College Chicago, after attending one of our classes that was funded by NSF and developed in collaboration with Princeton University, wrote:

Fig. 11 A scene from a student-produced movie, showing Madame Curie with her equations

Fig. 12 A two-class society for children

"The class meeting I sat in on was scintillating. I have rarely been in a classroom where students had such energy and enthusiasm. It is quite important that the results of this project be shared with faculty at as many institutions as possible; this will certainly help faculty at other institutions to adopt/adapt this successful approach."

Acknowledgement Acknowledgement is given to the National Science Foundation for financial support of this work.

References

1. Lerman, Z. M. 1986. Chemistry for Art and Communication Students: *J. Chem. Ed.*, *63*, 142.
2. Lerman, Z. M. 1986. Energy for Art and Communication Students: *J. Chem. Ed.*, *63*, 520.
3. Gardner, H. 1993. *Multiple Intelligences: The Theory in Practice*. HarperCollins, New York.
4. Hoffmann, R. and Torrence, V. 1993. *Chemistry Imagined: Reflections on Science*. Smithsonian Institution Press,Washington, DC.
5. Lerman, Z. M. 2000. Chemistry for the People who will Shape our Future. *Chemical Education Journal*, *4*, (1), Special Issue on the 8ACC Symposium on Chemical Education; http://chem.sci.utsunomiya-u.ac.jp/v4n1/indexE.html/
6. Lerman, Z. M. 2001. Visualizing the Chemical Bond. *Chemical Education Journal*, *5*, (1), Special Issue on Pacifichem 2000; http://ce.t.soka.ac.jp/cei/v2n1/ZLerman/index.html
7. Lerman, Z. M. 2003. Using the Arts to Make Chemistry Accessible to Everybody. *J. Chem. Ed.*, *80* (11), 1234.

Chemical Potential from the Beginning

R. Rüffler and G. Job

Abstract The chemical potential μ, commonly defined as a partial derivative of a quantity which involves energy and entropy, is often regarded as difficult to grasp – not only by students. As a fast and easy way, without frightening mathematics, it is suggested to introduce this quantity by a phenomenological definition and direct measuring procedure, a method usual for various basic quantities such as length, time and mass. The proposed approach is elementary, does not require any special previous knowledge and immediately leads to results which can be utilised practically. For example, it is possible to predict by means of the chemical potential alone whether a considered reaction is possible or not. Its temperature and pressure dependence allows calculation of phase transition and decomposition temperatures, phase diagrams and so on. Furthermore, the chemical potential takes a key position in dealing with physicochemical problems and starting from this central quantity, it is possible to explore many other fields up to quantum statistics.

1 Introduction

The chemical potential μ, commonly defined as a partial derivative of a quantity which involves energy and entropy, is often regarded as difficult to grasp – not only by students. On the other hand it is unquestionably very useful for the accurate description of the chemical and physical behaviour of substances. Therefore, as a fast and easy way, without frightening mathematics, we propose to introduce this quantity to first-year students (and even pupils) by a phenomenological definition and – if desired – direct measuring procedure. That means, the chemical potential is first characterised by its typical and easily observable properties comparable to a kind of "wanted poster" of μ that allows it to be identified as a measurable physical quantity. The phenomenological definition may be followed by a direct measuring

R. Rüffler (✉) and G. Job
Job-Foundation, University of Hamburg, Institute of Physical Chemistry, Grindelallee 117
20146, Hamburg, Germany
e-mail: regina.rueffler@job-stiftung.de

M. Gupta-Bhowon et al. (eds.), *Chemistry Education in the ICT Age*,
DOI 10.1007/978-1-4020-9732-4_6,

procedure, a method usual for various basic quantities such as length, time and mass. The proposed approach is elementary, does not require any special previous knowledge and immediately leads to results which can be utilised practically.

A thermodynamic course in which the chemical potential was introduced in the manner described was first proposed in 1972 by G. Job [1–4]. Since then, the approach has been successfully applied in introductory lectures in thermodynamics at the Universities of Hamburg and Karlsruhe, Germany. It was also adopted in H.U. Fuchs' textbook "The Dynamics of Heat" [5]. Because of the elementary intuitive interpretation of the quantity the concept can be easily adapted to all levels of education. It is already a part of textbooks for schools in Germany [6] and Switzerland [7]. It also plays an important role in the textbook "Physical Chemistry – An Introduction with New Concept and Numerous Experiments" [8] for undergraduates now in preparation. For strengthening of the understanding theory is complemented by more than a hundred illustrative, simple and safe demonstration experiments.

2 Main Characteristics of the Chemical Potential

Every substance shows a more or less pronounced tendency to change. This was already stated by the Greek philosopher HERACLITUS with the words "Everything flows – Nothing stands still". For example, wet dishes, moist laundry, fresh bread, etc. dry in the air of a room, iron rusts when it is exposed to air and water, sugar dissolves in tea and so on. The perishing of foodstuffs in unopened cans and even pure chemicals in sealed bottles indicates that the cause for the ubiquitous change of substances is not an interaction between different reaction partners (as chemists believed in former times), but is an intrinsic property of each substance itself.

Let us now have a closer look at the most important characteristics of the chemical potential, compiled into a short outline, a kind of "wanted poster":

- The tendency of a specific substance – let us call it A – to change, i.e.,
  - to *react* with other substances
  - to *transform* into another state
  - to *redistribute* in space

  can be expressed by one and the same quantity – its chemical potential μ_A.
- The magnitude of this tendency, meaning the numerical value of μ_A:
  - is determined by the *nature* of the substance (composition, state of aggregation, crystal structure, etc.),
  - as well as by its *milieu* (temperature, pressure, in the case of a solute concentration and kind of solvent, etc.)

  but *not* by the nature of the reaction partners or the resulting products.
- A *reaction*, *transformation*, *redistribution*, etc. can only proceed voluntarily if the tendency for the process is more pronounced in the initial state than in the final state.

When a substance disappears during the chemical process, one or even several substances are produced from it, or the substance reappears in another location. The produced substances, however, also show a tendency to change just like the reactants, so the direction in which a certain process will run depends upon which side has the stronger tendency. Therefore, chemical processes resemble a *competition* between the substances on either side of the reaction equation.

An image commonly used for this competition is the relationship between things on the right and left pan of a balance scale (Fig. 1). The direction in which the scale tips depends solely upon the sum of the "weights" G on each side of it. Even negative weights are allowed if objects floating upwards (maybe balloons) are attached to the scale.

This behaviour can also be expressed formally: The left side wins, i.e., the objects A′, A″,... on the left side of a balance scale or seesaw are successful against the objects B′, B″,... on the right side in their attempt to sink downward if

$$G(\mathrm{A}') + G(\mathrm{A}'') + \ldots > G(\mathrm{B}') + G(\mathrm{B}'') + \ldots, \tag{1}$$

equilibrium rules when the sum of the weights on the left and right side of the scale are just equal,

$$G(\mathrm{A}') + G(\mathrm{A}'') + \ldots = G(\mathrm{B}') + \mathrm{G}(\mathrm{B}'') + \ldots\,. \tag{2}$$

The statements made here for weights correspond completely to the role of chemical potentials in substance conversion. It makes no difference whether it is a reaction between several substances or a transformation of a substance into another state, or just a change of location. The direction in which such a process progresses, for example the reaction

$$\mathrm{A}' + \mathrm{A}'' + \ldots \rightarrow \mathrm{B}' + \mathrm{B}'' + \ldots,$$

depends solely upon the sums of the chemical potentials μ of all the substances on either side. The substances on the left side prevail in their attempt to react if

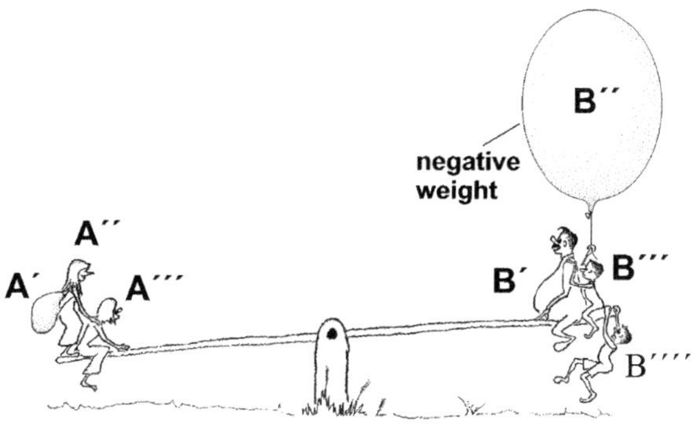

Fig. 1 Weight as model. The competition between the potentials μ of the reactants A′, A″, ... and of the products B′, B″, ..., participating in a chemical process can be compared to the effects of objects A′, A″, ... and B′, B″, ... on the two sides of a balance scale or a seesaw.

$$\mu(\mathrm{A}') + \mu(\mathrm{A}'') + \ldots > \mu(\mathrm{B}') + \mu(\mathrm{B}'') + \ldots, \tag{3}$$

(see Fig. 2), equilibrium rules if the sum of the conversion tendencies of the substances on both sides is the same:

$$\mu(\mathrm{A}') + \mu(\mathrm{A}'') + \ldots = \mu(\mathrm{B}') + \mu(\mathrm{B}'') + \ldots . \tag{4}$$

Therefore, each realisable reaction is comparable to a kind of scale which allows the comparison of chemical potentials or their sums, respectively. But the measurement is often impossible due to any *inhibitions*, i.e., the scale "is jammed." If there is a decline in potential from the left to the right side, that only means that the process *can* proceed in this direction in principle; however, it does not mean that the process will actually run. Therefore, a potential drop is a necessary but not sufficient condition for the reaction considered. The problem of inhibitions can be overcome if appropriate catalysts are available or indirect methods including chemical (using the mass action law), calorimetric, electrochemical and others can be used. Because we are interested in a first knowledge of the chemical potential, we assume for the moment that all these difficulties have been overcome and consider the values as given, just as we would consult a table when we are interested in the mass density or the electric conductivity of a substance.[1]

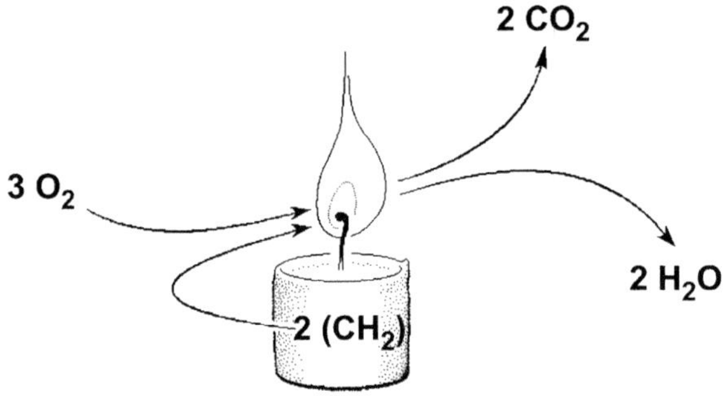

Fig. 2 Burning candle. The process takes place because the initial substances combined (in this case atmospheric oxygen and paraffin wax, formula approximately (CH_2) have a higher chemical potential than the products (in this case carbon dioxide and water vapour) $3\mu(O_2) + 2\,\mu((CH_2)) > 2\,\mu(CO_2) + 2\,\mu(H_2O)$

[1]The chemical potential of a substance corresponds at 298 K and 100 kPa (1 bar) to its molar Gibbs energy of formation $\Delta_f G$ if the zero points of the potential scales are appropriately chosen. Therefore, when looking for the chemical potential, tables in which this quantity is listed can be used.

3 Reference Point and Values of Chemical Potentials

Chemical tables never contain absolute values of chemical potentials, although the potential can be assigned an absolute zero value, just as temperature can. The reason is that the corresponding values would be enormous. In order to work with the tiny differences in potentials common in chemical and biological reactions, at least 11 digits would be necessary (the ratio between the potential differences and the absolute values is around one to one billion!). This alone would lead to numbers that are much too unwieldy not to mention that the absolute values are not known accurately enough for this to be feasible.

However, the heights of mountains are not referred to the geocentre but to the sea level (Fig. 3). Everyday temperatures are not referred to absolute zero, but are given as Celsius temperatures based upon the freezing point of water.

It is similarly practical to choose a convenient level of reference for the values of the chemical potential because differences of μ can be determined much more precisely than absolute values. Moreover, because we only need to compare potential values or their sums, it does not matter what the unit is at first. The μ values could be expressed in different scales similarly to how temperature can be expressed (Celsius, Fahrenheit, Kelvin, etc.). We will use "*Gibbs*", abbreviated to G, following a proposition of Wiberg [9] to honour Josiah Willard Gibbs (1839–1903) who first introduced the concept of chemical potential. For avoiding complications in further considerations we choose the unit to be SI-consistent, which means 1 G corresponds to 1 $J \cdot mol^{-1}$.

Next we enter into the question what reference states are suitable for the measurement of the potential differences. It is useful to refer to the conventional basic materials in chemistry, the elements. Because it is not possible to transform one element into another by chemical means (i.e., nuclear reactions are excluded), the values of the various elements are not related to each other. Therefore, the reference level for any

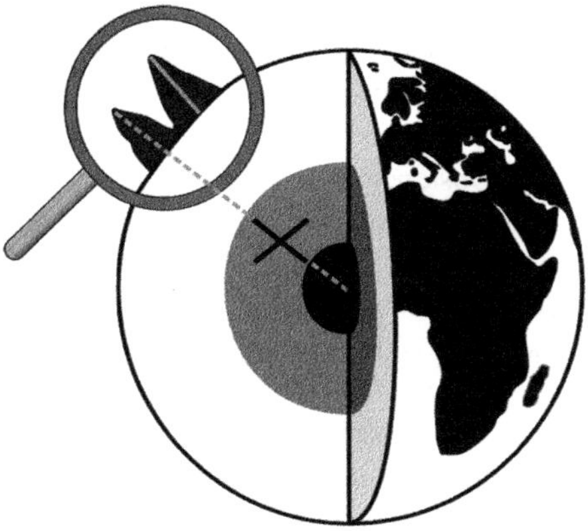

Fig. 3 Geographical elevation. Levelling in geography is an example for the selection of an appropriate reference point.

element could in principle be chosen arbitrarily without this having an effect upon the potential differences measured. For the sake of simplicity the value 0 is allocated to all elements in their most stable modification and under standard conditions (meaning 298 K and 100 kPa and labelled by$^{\ominus}$) (An exception is phosphorus where the more accessible white (or red) modification is preferred to the more stable, but difficult to produce, black modification).

The chemical potential μ of an arbitrary substance itself depends upon temperature and pressure. Therefore, it is evident – as usual otherwise in chemistry – to tabulate the potentials of substances (referred to the elements that form them) in the form of *standard values* $\mu^{\ominus}$·, i.e., for 298 K and 100 kPa. For *dissolved substances* the concentration c, for which the tabular values will be valid, must be specified in addition to p and T. Water is normally assumed to be the solvent unless otherwise stated. The usual reference value is 1 $kmol \cdot m^{-3}$ (= 1 $mol \cdot L^{-1}$). Ions can be assigned a chemical potential as well. They can be treated as "compounds" of atoms or molecules with a positive or negative number of electrons. Therefore, the amount n of electrons has to be conserved in chemical processes like that of the elements, i.e., the electrons can be treated as a kind of additional element. However, electrons in a free state play no role in chemistry. Therefore, a value for $\mu^{\ominus}(e^-)$ has been arbitrarily chosen so that the most commonly appearing type of ion H^+ (in an aqueous solution and at standard conditions) receives the $\mu^{\ominus}$ value of zero. In Table 1 we find standard values for several common substances.

The chemical potential of a pure substance depends on its state of aggregation, its crystal structure, etc. For example, liquid water and water vapour or diamond and graphite have different potentials at the same temperature and pressure. In order that the μ values are unambiguously given the aggregation state of the substance concerned is added to the formula by a vertical line and the abbrevations *s* for solid, *l* for liquid and *g* for gaseous; modifications can be characterised, for example, by their names. In the case of a solute, the solvent used can be labelled in the same manner; for water the abbreviation *w* is chosen.

Several further insights about the chemical potential can be gained just by considering such a table. The chemical potential of elements like iron is 0 G per definition but this does not mean that iron has no "tendency to change" in a chemical process but only that we have used its potential as the zero level to base the values of other iron-containing substances upon. Moreover, we state that the chemical potential of most existing substances is negative. This is not surprising: A substance with negative chemical potential can be produced spontaneously from the elements because it has a weaker tendency to change than the elements it is produced from. However, this also means that this substance shows no tendency to decompose, it is *stable*. A substance with a positive chemical potential on the other hand tends to decompose into its elements, it is *unstable* and therefore difficult to prepare and conserve or *metastable* at best. Often it reacts violently, especially when the value of μ is very large. However, a positive μ does not mandatorily mean that the substance is explosive. For example, benzene remains rather stable in spite of its μ value of +125 kG, because the decomposition process is inhibited, i.e., the substance is metastable.

Table 1 Chemical potentials under standard conditions (298 K, 100 kPa, dissolved substances at 1 kmol·m^{-3} (idealized))

Substance	Formula	$\mu^{\ominus}$/kG
Pure substances		
Iron	Fe|s	0
Graphite	C|graphite	0
Diamond	C|diamond	+3
Oxygen	O_2|g	0
Ozone	O_3|g	+163
Water	H_2O|l	–237
Water vapour	H_2O|g	–229
Table salt	NaCl|s	–384
Limestone	$CaCO_3$|s	–1128
Cane sugar	$C_{12}H_{22}O_{11}$|s	–1544
Benzene	C_6H_6|l	+125
Carbon dioxide	CO_2|g	–394
Ammonia	NH_3|g	–16
Ethyne (acetylene)	C_2H_2|g	+210
In water		
Ammonia	NH_3|w	–27
Carbon dioxide	CO_2|w	–386
Cane sugar	$C_{12}H_{22}O_{11}$|w	–1552
Hydrogen(I)	H^+|w	0
Iron(II)	Fe^{2+}|w	–79
Calcium(II)	Ca^{2+}|w	–553
Chloride	Cl^-|w	–131

4 Application in Chemistry

The most important application for the chemical potential μ is that it enables us to predict whether a change of substances *can* happen spontaneously or not (the role of possible inhibitions is left aside).

As first example we consider the dissolution of limestone or marble in hydrochloric acid (Fig. 4).

The demonstration experiment suggests that a decline in potential between the reactants and the products has to exist. Indeed, this can be confirmed by using the tabulated potential values (assuming an acid concentration of 1 kmol·m^{-3}).

$$CaCO_3|s + 2\,H^+|w \rightarrow Ca^{2+}|w + CO_2|g + H_2O|l$$

$$\mu^{\ominus}/\text{kG:} \quad \underbrace{-1128 + 2\cdot 0}_{-1128} \; > \; \underbrace{-553 + (-394) + (-237)}_{-1184}$$

Also, the solution behaviour of substances in a solvent can be described with the help of the concept of potentials. Whether a substance dissolves easily or not in water, alcohol, benzene, etc. is a result of the difference of its chemical potential in

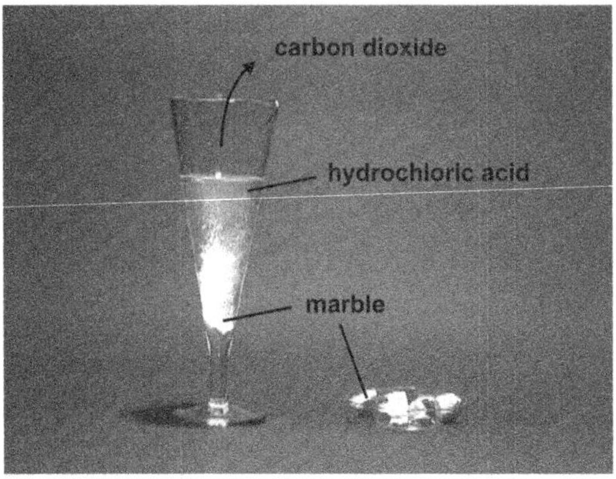

Fig. 4 Dissolution of marble in hydrochloric acid. If hydrochloric acid, an aqueous solution of hydrogen chloride, HCl, is poured over marble, foam develops that contains carbon dioxide. HCl is a strong acid and is entirely dissociated into hydrogen and chloride ions, H^+ and Cl^-. The H^+ ions are responsible for the reaction while the Cl^- ions remain more or less inactive.

the pure and dissolved state. Even the solubility of gases like that of ammonia in water can be easily described in this way.

$$NH_3|g \rightarrow NH_3|w$$

$$\mu\cdot/\text{kG}: -16\}\} > -27$$

Consequently, ammonia is very easily dissolved in water. An impressive way of showing this excellent solubility is with a so-called fountain experiment (Fig. 5).

As discussed a reaction always runs in the direction of a potential drop. This might give students or pupils the impression that substances with a positive potential cannot ever be produced by normal reactions of stable substances (with negative μ). The production of ethyne (acetylene) with a high positive chemical potential from calcium carbide and water (Fig. 6) shows that this is not the case.

Both water and carbide have negative chemical potentials which might give the impression of the reaction running "uphill," against the potential gradient. But the very low chemical potential of calcium hydroxide on the product side makes sure that the necessary potential drop exists, even though μ(ethyne) is > 0 is positive.

$$CaC_2|g + 2\ H_2O|l \rightarrow Ca(OH)_2|s + C_2H_2|g$$

$$\mu^\ominus/\text{kG}: \quad \underbrace{-68 + 2\cdot(-237)}_{-542} > \underbrace{(-898) + (+209)}_{-689}$$

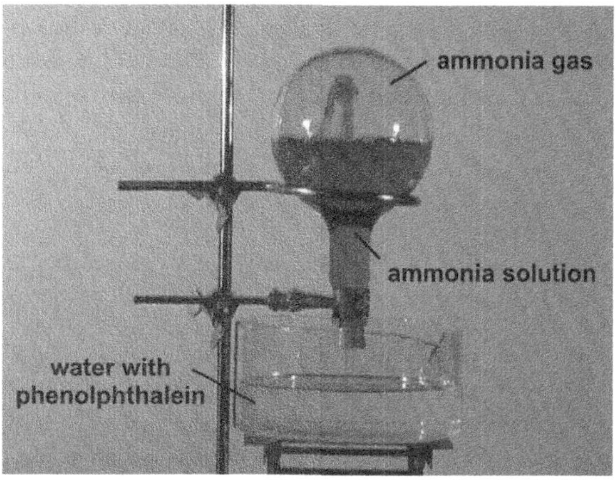

Fig. 5 Ammonia fountain. NH_3 gas dissolves so readily in water that just a few drops are enough to decrease the pressure in the flask so drastically that water is drawn upward into it in a strong jet.

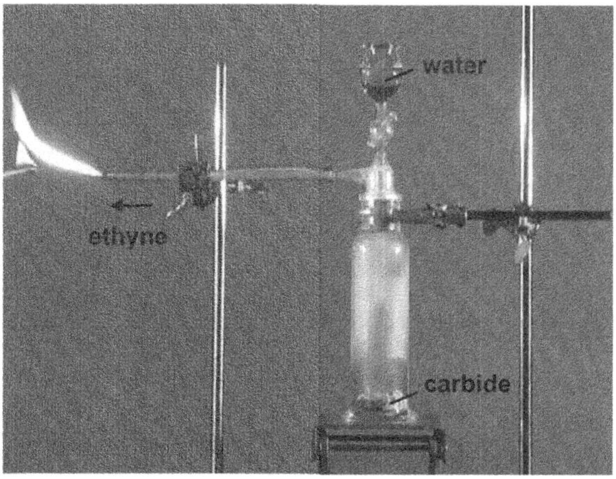

Fig. 6 Carbide lamp. Ethyne (acetylene) was used in power miners' lamps and bicycle lights in earlier times because of its bright flame. It is still used today for welding because of its high combustion temperature. The inflammable gas is produced when water is poured onto the carbide.

5 Temperature and Pressure Dependence

Until now, the tabular values we have used were the so-called standard values based upon room temperature and standard pressure (298 K and 100 kPa). For dissolved substances, the standard concentration is 1 kmol·m^{-3}. So far, our statements about the possibility of a chemical change have been valid for these conditions only.

However, temperature and pressure often have a decisive influence upon the chemical potential and thereby, upon the course of chemical processes. Water freezes in the cold and evaporates in the heat, ice melts under the blades of ice-skates, butane gas (the fuel of a cigarette lighter) becomes liquid when compressed and so on. That means the chemical potential μ is not a material constant but depends upon temperature, pressure, etc., summarily upon all the influences of its milieu.

For describing the temperature and pressure dependence of μ, linear approximations are in many cases sufficient[2]:

$$\mu = \mu_0 + \alpha(T - T_0), \tag{5}$$

$$\mu = \mu_0 + \beta(p - p_0). \tag{6}$$

Here, μ_0 characterises the initial value of the chemical potential. This represents a value at an arbitrarily chosen temperature T_0, pressure p_0 and concentration c_0 (in contrast to the standard value $\mu^{\ominus}$·). But standard values often serve as the initial values of such calculations, so that in these cases, $\mu_0 = \mu^{\ominus}$. α represents the *temperature coefficient* and β the *pressure coefficient* of the chemical potential.

The values of α and β of a number of substances are listed in Table 2 together with the chemical potential.

A basic rule (for which there are only a few exceptions) states that the temperature coefficient α is always negative, and the pressure coefficient β is always positive. This means that the chemical potential falls with increasing temperature, but rises with increasing pressure.

Another rule becomes apparent when the α and β values are compared for phase changes from solid to liquid and finally to the gaseous state[3]:

Table 2 Chemical potentials μ as well as the corresponding temperature and pressure coefficients α and β for standard conditions (298 K, 100 kPa)

Substance	Formula	$\mu^{\ominus}$/kG	α/G·K^{-1}	β/μG·Pa^{-1}
Iron	Fe|s	0	−27.3	7.1
	Fe|l	11.1	−34.3	7.3
	Fe|g	370.7	−180.5	24.4·10^3
Graphite	C|graphite	0	−5.7	5.4
Diamond	C|diamond	2.9	−2.4	3.4
Water	H_2O|s	−236.6	−44.8	19.7
	H_2O|l	−237.2	−69.9	18.1
	H_2O|g	−228.6	−188.7	24.4·10^3
Ammonia	NH_3|l	−11.0	−123.1	28.2
	NH_3|g	−16.4	−192.1	24.4·10^3

[2]For gases, the linear approximation is applicable only in a small interval of pressure ($\Delta p/p < 10\%$). For greater intervals, a logarithmic approximation has to be used. But the so-called "mass action" will be the topic of a subsequent paper.

[3]The temperature coefficient α corresponds to the negative molar entropy and the pressure coefficient β to the molar volume according to MAXWELL relations. Anticipating these relationships can help with remembering the mentioned rules but for a basic knowledge and most applications they are not mandatory.

$$0 > \alpha(\mathrm{B|s}) > \alpha(\mathrm{B|l}) >> \alpha(\mathrm{B|g}),$$
$$0 < \beta(\mathrm{B|s}) < \beta(\mathrm{B|l}) <<<< \beta(\mathrm{B|g}).$$

One of the rare, but important exceptions of the above rule is water where $\beta(\mathrm{H_2O|s}) > \beta(\mathrm{H_2O|l})$. The jump in the values corresponding to the second transition is considerably greater than the one corresponding to the first one which is represented by multiples of the signs < or >.

Already these simple qualitative rules allow many useful conclusions. For example, we can expect that all solid substances will melt and finally vaporise at high enough temperature (because with increase in temperature $\mu(\mathrm{B|l})$ decreases faster than $\mu(\mathrm{B|s})$ and $\mu(\mathrm{B|g})$ decreases still faster).

With the aid of tabular values also quantitative considerations are possible and for example, phase transition temperatures can be calculated. If a substance like lead is solid at room temperature, this is because its chemical potential has its lowest value in this state. But the chemical potential decreases with warming and this happens more quickly in the liquid state than in the solid as specified by the mentioned rule ($0 > \alpha(\mathrm{Pb|s}) > \alpha(\mathrm{Pb|l})$) (Fig. 7). For this reason, the curves must intersect at some point, say at the temperature T_{sl}. This T_{sl} is the *melting point* of lead because below T_{sl}, the most stable state of lead is the solid state, above T_{sl}, however, it is the liquid state. In order to indicate the phase transformation in question, the symbols for the corresponding aggregation states are inserted.

We can calculate the temperature T_{sl} as follows: T_{sl} is the temperature at which the chemical potentials of solid and liquid phase are equal, $\mu_{\mathrm{s}} = \mu_{\mathrm{l}}$. At this temperature, the two phases are in equilibrium. The temperature dependence of μ can be expressed by the linear approach:

$$\mu_{0,\mathrm{s}} + \alpha_{\mathrm{s}}(T_{\mathrm{sl}} - T_0) = \mu_{0,\mathrm{l}} + \alpha_{\mathrm{l}}(T_{\mathrm{sl}} - T_0) \qquad (7)$$

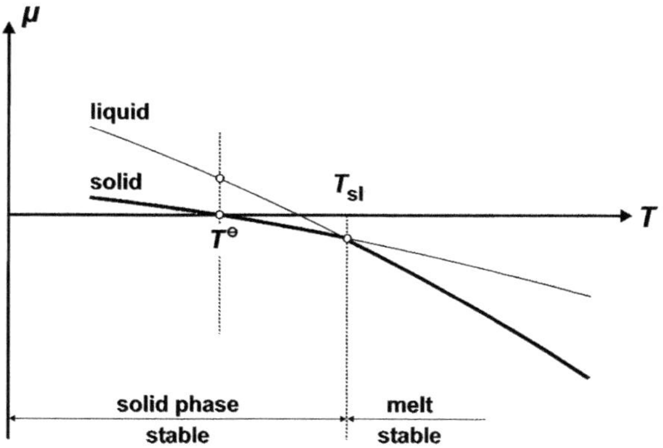

Fig. 7 Temperature dependence of the chemical potentials of the solid and liquid phase of a substance (the lowest chemical potential for each is highlighted).

By transforming this we obtain:

$$T_{\mathrm{sl}} = T_0 - \frac{\mu_{0,\mathrm{s}} - \mu_{0,\mathrm{l}}}{\alpha_{\mathrm{s}} - \alpha_{\mathrm{l}}}. \tag{8}$$

Of course, strictly speaking, our result cannot be accurate because our formula for temperature dependence is only an approximation. The smaller $(T_{\mathrm{sl}} - T_0)$ is, the more exact the calculated value will be. The experimentally measured melting point of lead for example is 601 K. Based on the tabular standard values, our calculation yields

$$T_{\mathrm{sl}} = 298\mathrm{K} - \frac{0 - 2220 \quad \mathrm{G}}{(-64.8) - (-71.7)\ \mathrm{G/K}} \approx 620\mathrm{K}.$$

This result is surprisingly good for the rough approximation used.

But also "real" chemical reactions can be dealt with in the same way. As an example we consider the thermal decomposition of silver oxide which can be described by:

	$2\ Ag_2O\|s \rightarrow$	$4\ Ag\|s\ +$	$O_2\|g$
μ/kG:	$2\cdot(-11.2) <$	$4\cdot0$	0
$\alpha/\mathrm{G.K^{-1}}$:	$2\cdot(-121)$	$4\cdot(-43)$	-205

The decomposition does not take place at room temperature because the chemical potential of the reactant is smaller than the combined potentials of the products. However, since a gas that means a substance with strongly negative temperature coefficient α should be formed, we expect that this process will start at a high enough temperature (Fig. 8). The minimum decomposition temperature T_{D} is obtained from the condition that the combined chemical potentials of the initial and final substances have to be equal. The calculation yields to a similar equation as equation (8). Inserting the μ and α values results in $T_{\mathrm{D}} \approx 465$ K.

Next let us consider the pressure dependence. Because of the mentioned qualitative rule an increase in pressure results in an increasing chemical potential, but the increase varies for the different states of aggregation with the smallest change in the solid state. Therefore, at high pressure the solid state is preferred compared to the others. Conversely, a pressure reduction results in a preference of the gaseous state. For example, lukewarm water can boil at low pressure (Fig. 9), although, at standard conditions, $\mu(H_2O|g) > \mu(H_2O|l)$, liquid water is the stable phase.

	$H_2O\|l \rightarrow$	$H_2O\|g$
μ/kG:	$-237.2 <$	-228.6
$\beta/10^{-5}\ \mathrm{G.Pa^{-1}}$:	1.81	2446

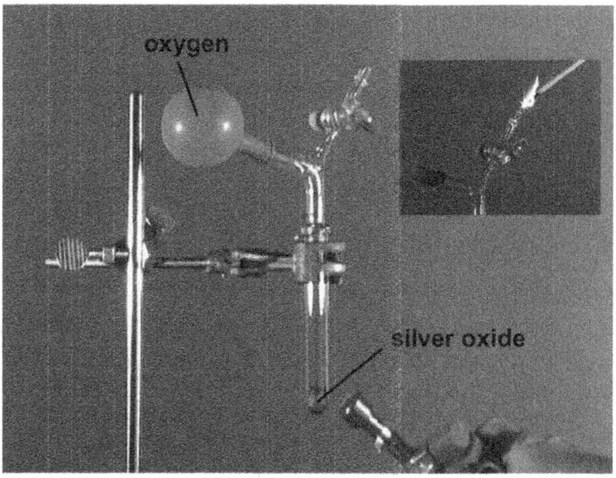

Fig. 8 Annealing of silver oxide. When the blackish brown silver oxide is moderately heated by a burner the formation of a gas is detectable by the slow blowing up of the balloon. The substance gradually changes its colour to whitish. Finally, the gas can be identified as oxygen with a glowing splint. White shiny silver metal remains in the test tube

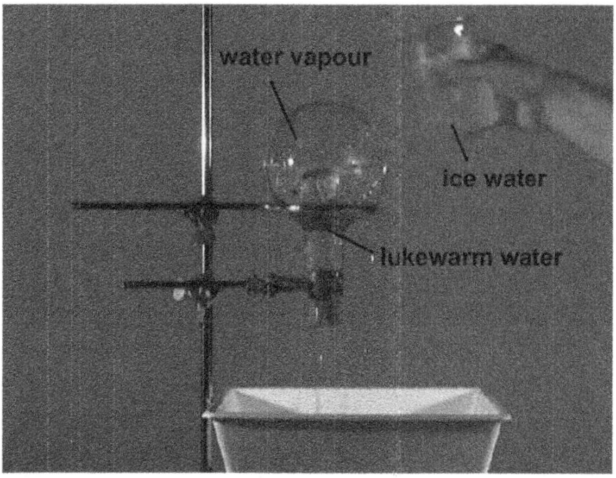

Fig. 9 Boiling by cooling. When ice water is poured over a round-bottomed flask filled with lukewarm water and water vapour, the water in the flask begins to boil heavily. The low pressure necessary for the process originates from condensation of a part of the vapour

A reduction of pressure results in an especially strong decrease of the chemical potential $\mu(H_2O|g)$ because the pressure coefficient β is very high for the gaseous state. At some point, the "boiling pressure" p_{lg}, $\mu(H_2O|g)$ will fall below $\mu(H_2O|l)$, and the liquid water begins to transform into water vapour by boiling.

Finally, a simultaneous temperature and pressure dependence can be described by the following combined equation,

$$\mu = \mu_0 + \alpha \cdot (T - T_0) + \beta \cdot (p - p_0). \quad (9)$$

By use of this equation the phase diagram of a substance can be constructed if the equilibrium condition is considered.

6 Conclusion and outlook

The chemical potential μ is a quantity that is in fact indispensable to chemistry but also to physics. It takes a key position in dealing with physicochemical problems (Fig. 10).

We have only discussed two of the sixteen fields given in the figure, the prediction of the direction in which a reaction can proceed spontaneously by means of the chemical potential and the temperature and pressure dependence of μ and its application. A next step would be to go over to "mass action," i.e., the concentration dependence of μ. This leads directly to the deduction of the mass action law, calculation of equilibrium constants, solubilities, and many other data. An expansion of the concept to colligative phenomena, diffusion processes, surface effects, electrochemical processes, etc., is easily possible. Furthermore, the same tools allow solving problems even at the atomic and molecular level that are usually treated by quantum statistical methods.

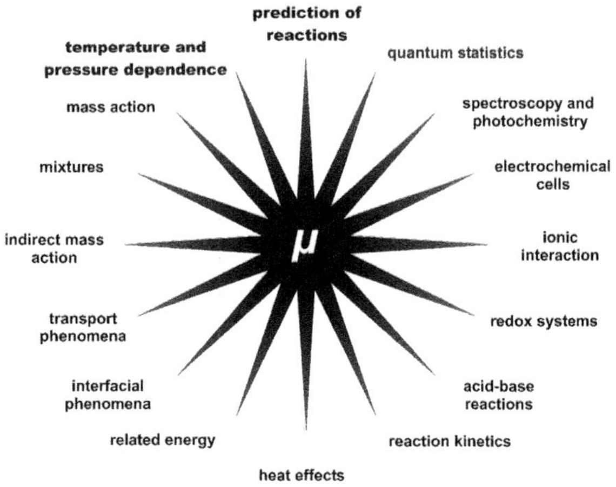

Fig. 10 Key position of the chemical potential μ. Figuratively speaking, the chemical potential forms the central pivot through which macroscopic and microscopic data coming in from and going out in every direction can be distributed and related

References

1. Job G (1972) Neudarstellung der Wärmelehre (A new concept of thermodynamics). Akademische Verlagsgesellschaft, Frankfurt (English translation available)
2. Job G (1978) Das chemische Potential im Physik- und Chemie-Elementarunterricht. In: Falk G, Herrmann H (ed) Konzepte eines zeitgemäßen Physikunterrichts, Heft 2. Schroedel, Hannover, 67–78
3. Job G (1981) Chemische Reaktionen physikalisch beschrieben. In: Falk G, Herrmann H (ed) Konzepte eines zeitgemäßen Physikunterrichts, Heft 4. Schroedel, Hannover, 14–31
4. Job G (1992) Teaching thermodynamics: Chemical potential from the beginning. *Proc. Taormina Conf. on Thermodynamics, Classe I di Scienze Fis. Mat. e Nat.* 70: 385–409
5. Fuchs HU (1996) The dynamics of heat. Springer, New York
6. Herrmann F (2003) Der Karlsruher Physikkurs (The Karlsruhe physics course). Aulis-Verlag, Köln (English translation available)
7. Borer T, Frommenwiler P, Fuchs HU (2005) Physik – Ein systemdynamischer Zugang für die Sekundarstufe II, 2nd edn. h.e.p, Bern.
8. Job G, Rüffler R. (2009) Physical chemistry – an introduction with new concept and numerous experiments, Vieweg+Teubner, Wiesbaden
9. Wiberg E (1972) Die chemische Affinität. de Gruyter, Berlin

The Chemistry of Carbenes and Their Metal Complexes: An Undergraduate Laboratory Experiment

J. P. Canal and T. Ramnial

Abstract Carbenes contain a formally divalent carbon center and are increasingly becoming an important class of compounds. The study of carbenes and their complexes introduces students to novel bonding modes of carbon atoms not normally encountered in traditional organic courses. We have developed a straightforward undergraduate experiment which isolates an air-stable silver carbene complex using the standard equipment found in most teaching laboratories. This experiment provides a convenient entry point for undergraduates to investigate carbenes and their transition metal complexes.

1 Introduction

The importance of the olefin-metathesis reaction was highlighted in 2005 with the awarding of the Nobel Prize in Chemistry to Chauvin, Grubbs, and Schrock [1–3]. The development of this process, which offers a "Greener" synthetic route to a host of organic products, paralleled the isolation of increasingly effective catalysts by Grubbs and Schrock [4, 5]. High catalytic activity was achieved through Grubbs' "Second Generation Catalyst" (Fig. 1), a ruthenium-based catalyst which contains a *N*-heterocyclic carbene (NHC) ligand (Fig. 1:1). Carbenes are an extremely reactive class of compounds which contain a formally divalent carbon atom (Fig. 1:2–5), and have been the focus of many review articles [6–10]. Interest in these compounds has been extensive partially due to the carbenes role in a wide variety of catalysts. Carbenes are used as catalysts in synthetically important reactions such as aryl aminations [7, 11, 12], hydrosilations [13], and Suzuki coupling [7, 11, 12].

J.P. Canal
Department of Chemistry, Simon Fraser University, Canada

T. Ramnial (✉)
Applied Chemistry, PowertechLabs, BC Hydro, Canada
e-mail: taramatee.ramnial@powertechlabs.com

M. Gupta-Bhowon et al. (eds.), *Chemistry Education in the ICT Age*,
DOI 10.1007/978-1-4020-9732-4_7,

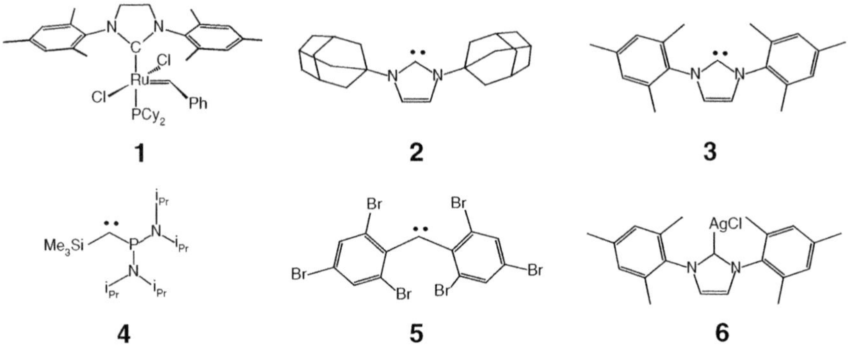

Fig. 1 Grubbs second-generation catalyst (1), examples of carbenes (2–5), and a carbene transfer agent (6). Compounds 2, 3 are examples of NHCs

In the late 1960s Öfele and Wanzlick independently isolated the first examples of transition metal complexes containing a NHC ligand [14–16]. In the following years many more examples of metal-stabilized carbenes were isolated, including the classic Fischer and Schrock type carbene complexes [17]. The first stable free carbene was not reported until 1988 when Bertand et al. published the isolation of the singlet carbene, *bis*(diisopropylamino)phosphino[trimethylsilylcarbene] (Fig. 1:4) [18]. In 1991, Arduengo et al. revisited the pioneering work of Öfele and Wanzlick and isolated 1,3-diadamantylimidazol-2-ylidene, the first imidazol-2-ylidene (Fig. 1:2) [19]. The first stable triplet carbene was reported by Tomioka et al. in 1995 (Fig. 1:5) [20].

Due to the extreme sensitivity of carbenes to both moisture and oxygen, technically demanding procedures and specialized equipment are often required for their isolation. Typical methods to produce free NHC include (i) the deprotonation of imidazolium ions via NaH, KH [21], or K [22] and (ii) the desulfurization of imidazole thiones using potassium metal [23]. In recent years, research has occurred to circumvent these reaction conditions through the development of air-stable carbene transfer agents [9, 24]. Although carbenes are important to many reactions, their air and moisture reactivity has limited their inclusion in the undergraduate laboratory curriculum. Presented here is the synthetic steps for the isolation of a carbene transfer agent, imidazol-2-ylidene-silver(I) chloride (Fig. 1:6), an experiment appropriate for the undergraduate curriculum [25–27].

This laboratory experiment was developed to introduce undergraduate students to the synthesis of metal complexes containing NHC ligands (Fig. 2). It was designed to be included into a second-year (or higher) inorganic course. The experiment is completed over two 4-h laboratory periods, where the first session concludes midway through the second reaction. The final step of day 1 requires the product to be stirred and left unattended until the next laboratory period. High yields are obtained with a stirring time of 3–7 days. The second laboratory period requires the completion of steps 2 and 3, as well as the determination of the melting points and obtaining NMR spectra (Fig. 2) [25–27].

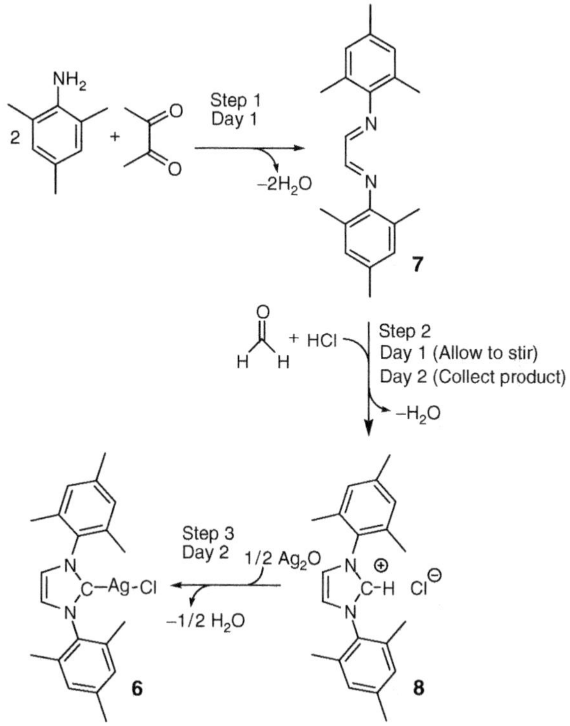

Fig. 2 Full synthetic sequence for the production of the carbene transfer agent, 6

2 Methodology

This experimental procedure involves a three-step synthesis of a carbene transfer agent, imidazo-2-yildene-silver(I) chloride (6) (Fig. 2). In the first step glyoxal-*bis* (2,4,6-trimethylphenyl)imine (7) is produced, which is converted into 1,3-*bis*(2,4,6-trimethylphenyl)-imidazolium chloride (8) in step 2. This is the main reactant in step 3, which ultimately yields 6 [25–27].

Step 1: Preparation of glyoxal-*bis*(2,4,6-trimethylphenyl)imine (7). To a 100 mL round-bottom flask fitted with a stir bar was added 2,4,6-trimethylaniline (10 mL), 40% glyoxal in water (4 mL) and ethanol (45 mL) as well as a few drops of 88% formic acid to initiate precipitation. After the solution was stirred for an hour the previous amber solution contained yellow crystals, which were collected by vacuum filtration using a Büchner funnel. The crystals were washed with cold ethanol (2 × 7 mL) then cold diethyl ether (7 mL). Yield 6.76 g (65%). m.p. 157–160°C, ^{1}H NMR (400 MHz, $CDCl_3$): δ 2.16 (s, 12 H, *ortho*-CH_3), 2.29 (s, 6 H, *para*-CH_3), 6.91 (s, 4 H, *meta*-H), 8.10 (s, 2 H, CH) ppm.

Step 2: Preparation of 1,3-*bis*(2,4,6-trimethylphenyl)imidazolium chloride (8). To a 50 mL round-bottom flask fitted with a stir bar was added glyoxal-*bis*(2,4,6-trimethylphenyl)imine (1.50 g), paraformaldehyde (0.15 g), and toluene (30 mL).

A water-cooled condenser was added to the reaction flask and the solution heated to 100°C for 15 min and then allowed to cool to 40°C. Once cooled, 4 M HCl in dioxane (1.0 mL) was added drop-wise to the reaction solution, the round-bottom flask topped with a stopper and the solution was left to stir for a week. (A minimum of 3 days is required. The stirring of the solution is essential to obtain a product but short interruptions (1–2 h), which allow the stir plates to be used by other students, do not appear to have a negative effect.) The resulting precipitate was collected by vacuum filtration using a Büchner funnel and the solid washed with cool tetrahydrofuran (2 × 5 mL). Yield 1.11 g (65%). m.p. 349–352°C, ^{1}H NMR (400 MHz, $CDCl_3$): δ 2.16 (s, 12 H, ortho-CH_3), 2.32 (2, 6 H, para-CH_3), 7.01 (s, 4H, meta-H), 7.62 (d, 2 H, im-H), 9.72 (s, 1 H, im-H) ppm.

Step 3: Preparation of imidazol-2-ylidene-silver(I) chloride (6). To a 100 mL round bottom flask fitted with a stir bar was added compound 8 (0.50 g), dichloromethane (30 mL) then silver(I) oxide (0.20 g). A water cooled condenser was attached to the reaction flask and the solution was heated to reflux for 2 h. The excess silver(I) oxide was separated by gravity filtration and the solution concentrated on a rotary evaporator. Cooling the solution resulted in a crop of clear crystals of the carbene transfer agent, imidazo-2-yildene-silver(I) chloride. Yield 0.32 g (50%). m.p. 280–282°C, ^{1}H NMR (500 MHz, $CDCl_3$): δ 2.07 (s, 12 H, ortho-CH_3), 2.35 (s, 6 H, para-CH_3), 6.99 (s, 4H, meta-H), 7.13 (d, 2H, im-H) ppm, $^{13}C\{^{1}H\}$ NMR (100.61 MHz, $CDCl_3$) δ (ppm) 17.8 (s, o-CH_3), 21.3 (s, p-CH_3), 122.9 (s, NCC), 129.7 (s, Ar-C-3,5), 134.7 (s, Ar-C-2,6), 135.4 (s, Ar-C1), 139.8 (s, Ar-C-4), 183.6* (d,^{107}Ag-C), 186.1* (d, ^{109}Ag-C) ppm (*only observed in ^{13}C enriched sample).

3 Results and Discussion

The reactivity of free NHCs to both moisture and oxygen has driven the development of air-stable carbene transfer agents, such as 6. Air-stable carbene transfer agents can avoid the tedious preparation of a free carbene, yet can adopt the same role of the carbene. The lability of the C-Ag single bond of 6 in concert with the high lattice energy of AgCl are the driving force for the transfer of the NHC to a variety of transition metals, such as rhodium, iridium, palladium, and gold (Fig. 3) [24]. The lability of the C-Ag bond can also be utilized for catalytic reactions. Through the thermal decomposition of the NHC-AgCl complex, active carbene catalysts can be used for reactions such as the transesterification of small molecules [10, 28].

The lability of the C-Ag bond results from the nature of the metal-carbon interaction in the NHC-metal complexes, which differs from those found in both the Fischer and Schrock type carbene complexes. Both the Fischer and Schrock type carbenes have a double bond between the metal and the carbon (Fig. 4) comprising of both σ and π components. This orbital overlap differs in the NHC-metal complexes. The *p* orbital on the carbeneic carbon, which was used to π bond in the Fischer and Schrock carbenes, receives electron density from the *p* orbital of the neighboring N atoms, thus the carbeneic *p* orbital is unavailable to form a significant π bond with the orbitals

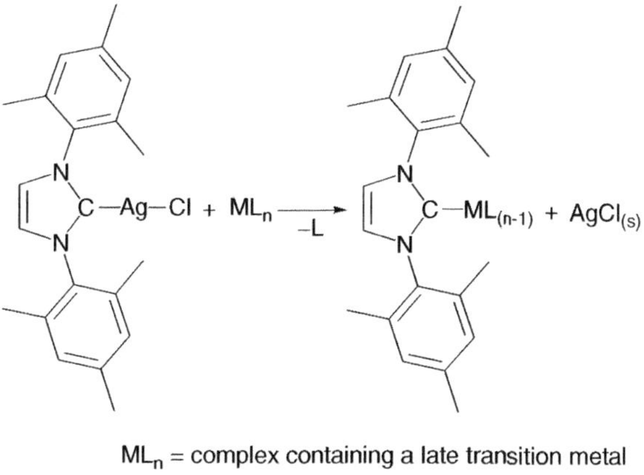

Fig. 3 Transfer of a NHC from Ag to another transition metal

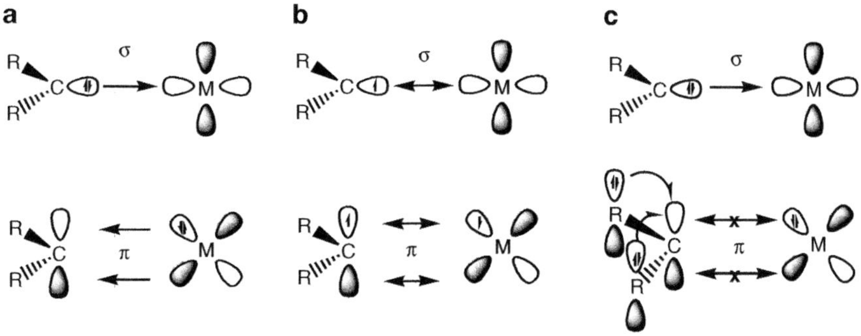

Fig. 4 The σ and π contribution to the metal-carbon bond of (**a**) a Fischer carbene complex, (**b**) a Schrock carbene complex and (**c**) a NHC-metal complex

from the metal [24, 29]. Therefore metal-carbon bonds lengths more consistent with a single metal-carbon bond are observed. This delocalization of the electrons from the N atoms to the carbeneic carbon stabilizes the NHC [9].

The chemistry presented by this experiment was well received by the students. They preferred learning about an active and a growing field of research rather than stagnant area with reactions perfected many years ago. The students also liked the idea of incorporating components of both organic and inorganic chemistry into one experiment. This experiment was also beneficial to the students as it further developed the skills they would have already acquired, such as interpretation of multinuclear NMR spectra and the laboratory techniques required for a multistep experiment, such as heating to reflux, and vacuum and gravity filtration. Melting point determination was used to provide a convenient means to make an initial determination of the purity of the product, which was confirmed by ^{1}H and $^{13}C\{^{1}H\}$ NMR spectroscopy studies. This laboratory experiment introduces students to an important class of compounds through an experiment which can be adopted into any inorganic chemistry laboratory course.

4 Conclusion

A straightforward preparation for an air-stable silver carbene complex using a standard wet chemistry kit was presented. This experiment removed the tedious requirements of a carbene synthesis and allows for the inclusion of carbenes and carbene transfer agents in the undergraduate laboratory curriculum.

Acknowledgments We would like to thank Dr. J. A. C. Clyburne and Dr. C. D. Abernethy for their contribution to this work as well as the Natural Sciences and Engineering Research Council of Canada (NSERC) and Simon Fraser University for funding.

References

1. Chauvin Y (2006) Olefin metathesis: the early days (Nobel Lecture). Angew Chem Int Ed 45:3741–3747
2. Schrock RR (2006) Multiple metal-carbon bonds for catalytic metathesis reactions (Nobel Lecture). Angew Chem Int Ed 45:3748–3759
3. Grubbs RH (2006) Olefin-metathesis catalysts for the preparation of molecules and materials (Nobel Lecture). Angew Chem Int Ed 45:3760–3765
4. Rouhi AM (2002) Olefin metathesis: Big-deal reaction. Chem Eng News 80(41):29–33
5. Rouhi AM (2002) Olefin metathesis: The early years. Chem Eng News 80(41):34–38
6. Canal JP, Ramnial T, Dickie DA et al. (2006) From the reactivity of N-heterocyclic carbenes to new chemistry in ionic liquids. Chem Commun 1809–1818
7. Herrmann WA (2002) N-heterocyclic carbenes: a new concept in organometallic catalysis. Angew Chem Int Ed 41:1290–1309
8. Cowley AH (2001) Some aspects of the Lewis base and ligand behavior of N-heterocyclic carbenes. J Organomet Chem 617:105–109
9. Bourissou D, Guerret O, Gabbaï FP et al. (2000) Stable Carbenes. Chem Rev 100:39–92
10. Marion N, Diez-Gonzales S, Nolan SP (2007) N-heterocyclic carbenes as organocatalysts. Angew Chem Int Ed 46:2988–3000
11. Clavier H, Nolan SP (2007) N-heterocyclic carbenes: advances in transition metal-mediated transformations and organocatalysis. Ann Rep Prog Chem Sect B 103:193–222
12. Hillier AC, Grasa GA, Viciu MS et al. (2002) Catalytic cross-coupling reactions mediated by palladium/nucleophilic carbene systems. J Organomet Chem 653:69–82
13. Dragutan V, Dragutan I, Delaude L et al. (2007) NHC-Ru complexes: Friendly catalytic tools for manifold chemical transformations. Coord Chem Rev 251:765–794
14. Öfele K (1968) 1,3-Dimethyl-4-imidazolinyliden-(2)-pentacarbonylchrom ein neuer übergangsmetall-carben-komplex. J Organomet Chem 12:P42–P43
15. Wanzlick HW, Schönherr HJ (1968) Direct synthesis of a mercury salt-carbene complex. Angew Chem Int Ed Engl 7:141–142
16. Luger P, Ruban G (1971) Crystal structure of a mercury salt-carbene complex. Acta Crytallogr Sect B 27:2276–2279

17. Elsenbroich C, Salzer A (1992) Organometallics: A concise introduction, 2nd ed. VCH, New York
18. Igau A, Grutzmacher H, Baceiredo A et al. (1988) Analogous.α,α'-bis-carbenoid, triply bonded species: synthesis of a stable λ3-phosphino carbene-λ5-phosphaacetylene. J Am Chem Soc 110:6463–6466
19. Arduengo AJ III, Harlow RL, Kline M (1991) A stable crystalline carbene. J Am Chem Soc 113:361–363
20. Tomioka H, Watanabe T, Hirai K et al. (1995) 2,2′,4,4′,6,6′-Hexabromodiphenylcarbene. The first stable triplet carbene in fluid solution at low temperature and in the crystal state at room temperature. J Am Chem Soc 1995 117:6376–6377
21. Arduengo AJ III, Krafczyk R, Schmutzler R (1999) Imidazolylidenes, imidazolinylidenes and imidazolidines. Tetrahedron 55:14523–14534
22. Gorodetsky B, Ramnial T, Branda NR et al. (2004) Electrochemical reduction of an imidazolium cation: a convenient preparation of imidazol-2-ylidenes and their observation in an ionic liquid. Chem Commun 1972–1973
23. Kuhn N, Kratz T (1993) Synthesis of imidazol-2-ylidenes by reduction of imidazole-2(3H)-thiones. Synthesis 561–562
24. Garrison JC, Youngs WJ (2005) Ag(I) N-heterocyclic carbene complexes: synthesis, structure, and application. Chem Rev 105:3978–4008
25. Canal JP, Ramnial T, Langlois LD et al. (2008) A three-step laboratory sequence to prepare a carbene complex of silver(I) chloride. J Chem Ed 85:416–419
26. Ramnial T, Abernethy CD, Spicer MD et al. (2003) A monomeric imidazol-2-ylidene-Silver(I) chloride complex: synthesis, structure, and solid state ^{109}Ag and ^{13}C CP/MAS NMR characterization. Inorg Chem 42:1391–1393
27. Canal JP, Ramnial T, Clyburne JAC (2008) A carbene transfer agent. In: Roesky HW, Kennepohl D (eds) Experiments in sustainable chemistry, Wiley-VCH Verlag GmbH & Co. KGaA, Weinheim (in press).
28. Sentman AC, Csihony S, Waymouth RM et al. (2005) Silver(I)-carbene complexes/ionic liquids: novel N-heterocyclic carbene delivery agents for organocatalytic transformations. J Org Chem 70:2391–2393
29. Lin IJB, Vasam CS (2004) Silver(I) N-heterocyclic carbenes. Comments Inorg Chem 25:75–129

SATL, Learning Theory, and the Physiology of Learning

J. J. Lagowski

Abstract It is argued that the demonstrable success of the Systemic Approach to Teaching and Learning (SATL) arises from the relationship of the SATL techniques to constructivist learning theory and the most current interpretation of brain function which, from one point of view, also operates on constructivist principles.

1 Introduction

The Systemic Approach to Teaching and Learning (SATL) is a tool designed to help teachers teach and students learn, and which has been used successfully in a variety of disciplines over the last ten (10) years. Here our goal is to link SATL methods to the constructivist theory of learning and to the current views of brain function. We begin with learning theory because there is an obvious and natural connection to the current ideas on brain function, which is also of interest to us here.

2 Constructivist Learning Theory

Several theoretical frameworks for how learning occurs have been developed by cognitive and educational psychologists that are displacing behaviorist notions [1]; the latter are often the basis of many current instructional methods. One of the more useful of these newer theories of learning for our purposes here is *constructivism.* Historically, constructivist theory has multiple roots going back to Piaget [2] and Ausubel [3]; constructivist theory also has several manifestations [4]. We choose here to concentrate on that thread of constructivist theory that goes back to Ausubel [3] who describes the learning process of students as taking the new knowledge to

J.J. Lagowski (✉)
Department of Chemistry, University of Texas at Austin, Austin, Texas, USA
e-mail: jjl@mail.utexas.edu

M. Gupta-Bhowon et al. (eds.), *Chemistry Education in the ICT Age,*
DOI 10.1007/978-1-4020-9732-4_8,

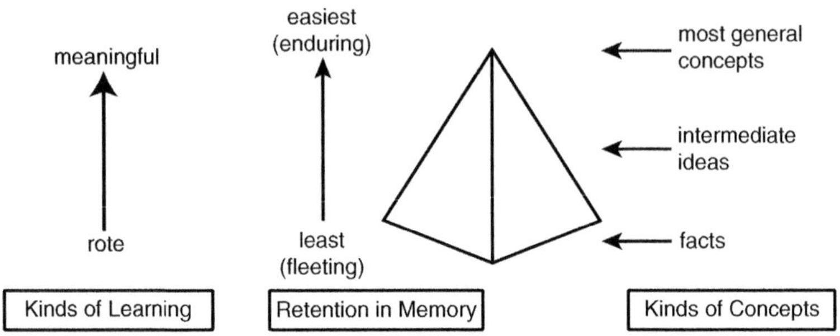

Fig. 1 A diagrammatic relationship among the kinds of concepts and learning, and the ability to retain these in memory

be learned and assimilating it – incorporating it – into what the learner already knows. In Ausubel's view, successful students take possession of knowledge *actively*, by seeking explicit conceptual linkages between the new concepts they learn and those they already possess. By "possess" Ausubel means "deeply known" which he distinguishes from "rote learning" or memorization. The process called "assimilation" creates personal meaningful knowledge by restructuring the already existing conceptual frameworks that the learner possesses to accommodate to the new concepts being learned. Ausubel's constructivist ideas are summarized in Fig. 1. "Facts" are the most numerous concepts. They are learned by rote and are the most fleeting, being least easily retained in memory. At the other extreme, the most general concepts are the most meaningful and are most easily retained in memory; they are the most enduring concepts available because they subsume all the facts.

3 Concept Maps

Novak [5], using Ausubel's ideas of how learners construct meaning, developed *concept maps* as a tool to represent the concept/propositional framework for domain-specific knowledge. In a concept map, labels representing concepts are arranged in hierarchical order and are connected by linking verbs forming propositions. See Fig. 2 for an example of a simple map involving the concepts of periodic table, atomic number, periodic groups, and periods. The resulting two-dimensional organization of concepts reveals the cognitive structure of the map's maker. This learner has taken the "Periodic Table" as the key concept of the group and, accordingly, it is placed highest in the hierarchy above the other concepts. The concepts of " # of protons" and "noble gases" elaborate in different ways on the "atomic number" and "groups" concepts, respectively. After these concepts are arranged as indicated, the connecting verbs (in the ovals) then establish the propositions. A different learner might produce a different concept map by choosing a different key concept; starting with a given collection of concepts it is possible that several "acceptable" concept maps can be

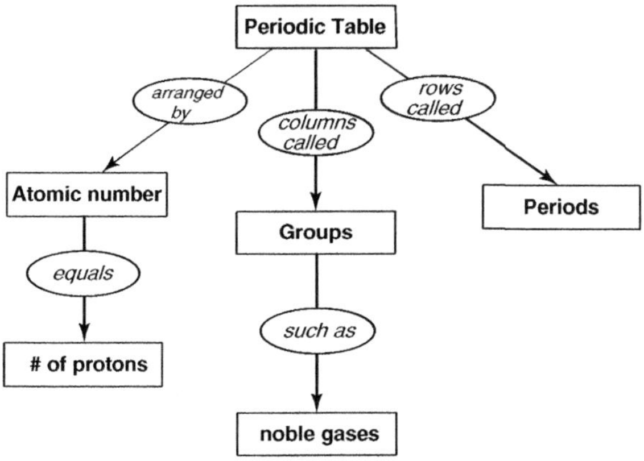

Fig. 2 An example of a concept map

constructed. Thus, concept maps become a vehicle by which teachers can describe the key relationships among concepts and, in doing so, reveal to students the way they think about those concepts. Concept maps produced by learners provide an insight to the teacher of the learners' understanding of a knowledge domain.

Novak [5a] defines a concept as a "perceived regularity (a pattern) in events or objects or records of events or objects, designated by a label." The basic unit of a concept map is a proposition, which consists of two concepts connected by a linking word, e.g.

- 

- 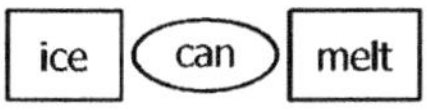

- 

- 

A concept map is a collection of concepts organized as propositions in a hierarchical manner (see Fig. 2 for an example). The most general concepts are at the highest rank; the lower ranks are filled with examples. The most efficient process of producing concept maps from, say 15–20 concepts, is to start with the most general concepts at the top followed by a group that is less general, forming propositions with appropriate linking words. The lower ranks of concepts are very often specific examples of the more general concepts. The final step in creating a concept map is to establish cross-links or relationships in, and between, different sections of the map. The original concept map can be elaborated as the learner experiences new concepts. Thus, the creator of a concept map incorporates within it a construction process that reflects one of the basic tenets of constructivist theory, i.e., that knowledge is attained by a learner by assimilating (or integrating) it into his/her understanding of a basic knowledge structure.

Meaningful learning, from Ausubel's point of view, is expressed in the most general concepts which are the most enduring in memory and which are constructed from (related to) all other units of knowledge in a given domain. The process of creating (constructing) a knowledge structure in a concept map by the learner produces meaningful learning which is the most enduring (Fig. 1) and which allows the learner to transfer knowledge to novel settings. Progressively greater skills development in solving novel problems comes from the process of the continual refinement of a learner's knowledge base through constructivist methods.

4 Systemic Approach to Teaching and Learning (SATL)

The relationship between SATL methods and Constructivist Theory which is the focus of this section goes through the idea of concept mapping. We do not intend to address the details of the SATL method here; this has been done by others [6] more effectively than I have time or space to do here. A quick review of the key SATL ideas is, however, appropriate for the purposes of establishing the relationship between SATL, Constructivist Theory, and concept maps. For the purposes of orientation, we observe that concept maps and SATL techniques share some common ideas.

Recall that, in concept maps, the concepts are arranged hierarchically and in two dimensions; concepts are connected with connecting phrases to produce propositions. As an example, Fig. 3 is one concept map involving the concepts compounds,

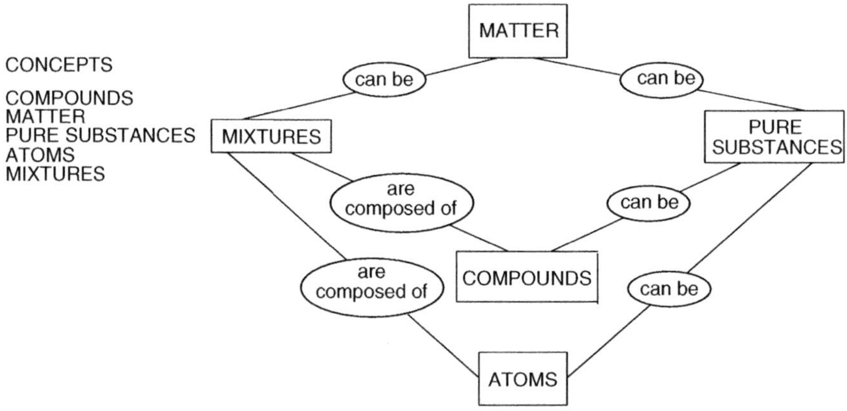

Fig. 3 A representation of a closed concept map involving the concepts listed at the left. Closed concept maps that share common concepts are the basis for producing overall systemic diagrams as illustrated in Fig. 6

matter, pure substances, atoms, and mixtures. In general, several different concept maps may exist for a given collection of concepts; one concept map is not necessarily "better" than another, except by personal preference. As the number of concepts increases, the complexity of the corresponding map increases. Note that the concept map shown in Fig. 3 could be expanded to include other concepts such as formulas, atomic mass, molecular weight, chemical symbols, and chemical equations. In other words, we could *start* with the concept map shown in Fig. 3 and make a larger, more encompassing map that includes these new concepts. This kind of cycle could repeat any number of times, each time the previous concept map "grows," to include the new concepts. The addition of new concepts to a previously established concept map is, in its essence, a manifestation of constructivist ideas.

The key structural element of the SATL method is the *systemic diagram* which has all of the attributes of a *closed* concept map. A closed concept map is limited by the number of relationships. Let's now relate these ideas to the basic unit of learning in the SATL technique. Figure 4 is a simple systemic diagram that covers a *part* of the chemistry of organic acid chlorides.

The SATL approach involves the creation of a series of interlocking closed concept maps that will, ultimately, be a part of the overall systemic diagram for a given domain of knowledge. The overall systemic diagram is a representation of the way a teacher views the concepts in question. Starting with a prerequisite closed concept map where all the relationships are known, the learner works through a series of associated closed concept maps containing unknown (to the learner) relationships to be learned until all the unknown relationships are known. By "associated closed concept maps," we mean closed maps that share at least one concept.

The overarching idea of systemic diagrams is that new knowledge – understanding – is constructed upon previous knowledge that the learner possesses (Fig. 5). Thus, the SATL technique starts with a systemic diagram – a closed concept map – that incorporates previously known relationships represented by SD_0 in Fig. 5. Then the

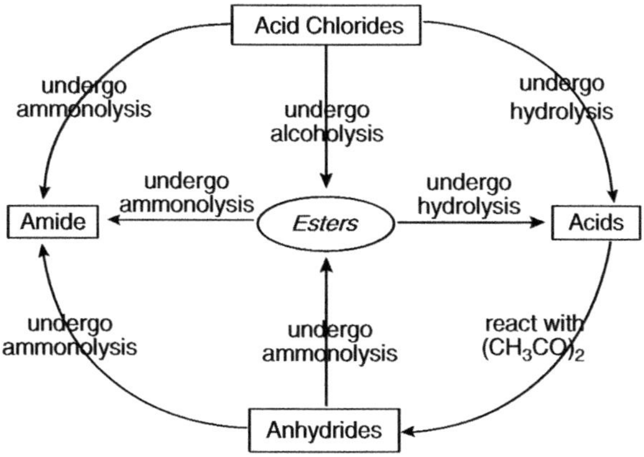

Fig. 4 A simple systemic diagram describing the chemistry of organic chlorides

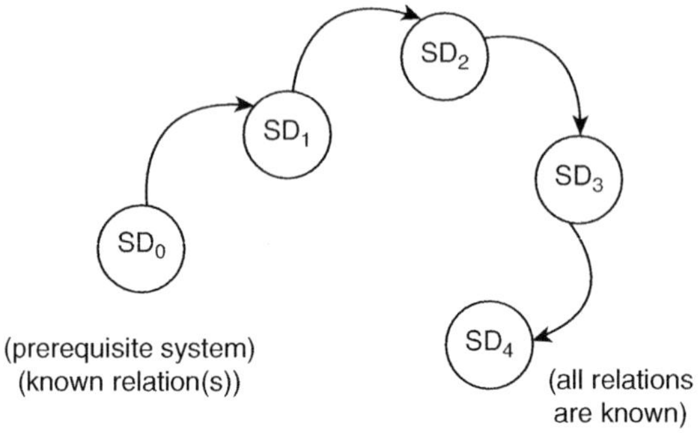

Fig. 5 Individual small systemic diagrams, each of which incorporates stored concepts, are the basis of producing an overall systemic diagram of a domain of knowledge

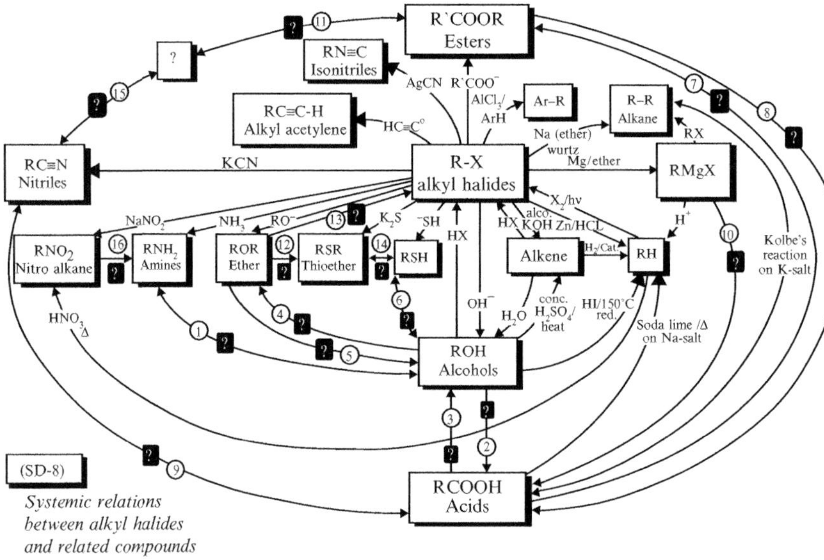

Fig. 6 An overall systemic diagram that represents the chemistry of alkyl halides. The numbers represent one sequence of systemic diagrams that can be used to reveal the overall systemic diagram. The question marks (?) represent the unknown relationships that are revealed during this sequence

relationships to be learned are systematically assimilated into the overall systemic diagram. Using other words, the learner constructs the final systemic diagram upon his/her previous knowledge using the process shown in Fig. 5.

Overall, systemic diagrams can become very complex as shown in Fig. 6, which covers the essential chemistry of alkyl halides. The SATL approach requires that the teacher produces the final systemic diagram (e.g., Fig. 6) and breaks it up into the smaller systemic diagrams (e.g., Fig. 5). In doing so, the teacher's view of the knowledge domain is revealed to him/herself as well as eventually to the students. The teacher determines the prerequisite systemic diagram (e.g., SD_0 of Fig. 5), which also includes the links to the next systemic diagram; this, then, becomes the new prerequisite diagram, which has a new link, etc., etc., until the entire systemic diagram is revealed (e.g., Fig. 6). By this general process, the teacher becomes a guide to the learner's learning process through potentially very complex relationships using constructivist principles.

SATL methods have been shown, empirically, to be successful in helping students learn in a variety of settings – pre-college, college, and graduate systems of formal education as well as adult education – in a variety of disciplines; the sciences, (chemistry, biology, physics); mathematics; engineering disciplines; medical-related disciplines; and linguistics. A number of statistical studies involving student achievement indicate that students involved with SATL methods taught by teachers trained in those methods achieve at a significantly higher level than those taught by standard linear methods of instruction. Finally, more than 60,000 teachers have been trained in SATL methods in Egypt, Libya, and Jordan, countries that were seeking to reform their educational systems. Thus, SATL methods clearly are universally useful, irrespective of the discipline or level of education.

5 Brain Function

Research on the way the brain functions has involved, in the recent past, the study of the electrochemical impulses that occur when the brain is actively engaged in various tasks [7]. Non-evasive probes that have been employed in establishing brain behavior include:

- Computed Tomography (CT)
- Computer Axial Tomography (CAT)
- Magnetic Resonance Imaging (MRI)
- Functional Magnetic Resonance Imaging (fMRI)
- Positron Emission Tomography (PET)
- Single-photon Emission Computer Tomography (SECT)
- Diffuse Optical Imaging (DOI)
- Event Related Optical Signal (EROS)
- Electroencephalograms

Using such techniques, the functions of the different areas of the brain have been identified. One current view of the human brain is that it has a modular organization consisting of identifiable component processes that participate in the generation of a cognitive state. The five senses – sight, smell, touch, hearing, and taste – are the gateways to the brain (Fig. 7). Our view of the world is *constructed* by our brain, as it interprets the signals from these five senses coming through the gateways. Although much is known about the details of how the chemical and electrical signals from the five senses are created and pass into the various areas of the brain, these details are not important for our purposes here. The totality of these methods and the results of other experiments produce a representation of the major parts of the brain as well as detailed information of how these are believed to interact with each other.

Our current knowledge produces the following model of how the brain works – how it does what it does. The information input in the brain is *not* stored in a single part of the brain. The brain does *not* store information like an encyclopedia – to be retrieved as a unit on demand. Rather, the data suggest that information is distributed in different networks of neurons, which are the basic elements of brain activity (Fig. 7). Thus, when someone perceives a *tiger*, all the sensual characteristics of the tiger – the snarl, the stripes, the stealthy movement, the cat-like odor, etc., are stored in different, but appropriate neuron networks (Fig. 8). Retrieving the concept of the tiger from memory corresponds to the interaction of all the specialized networks that contain the tiger-related characteristics, which are then reassembled by the brain into the memory of a tiger.

The human mind creates a number of categories for the kinds of information it stores. About 20 have been identified and there are probably a very large number more (Fig. 9). Notice how the categories listed have strong components associated with the senses, because these are the only signals that reach the brain. So, it appears that this kind of information storage in the brain is genetically encoded since humans have only five (5) senses with which to learn about the world in which they live. So, from one point of view, the human brain is automatically (genetically hard-wired) a knowledge-seeking entity. The knowledge is that associated with the world in which the brain will exist.

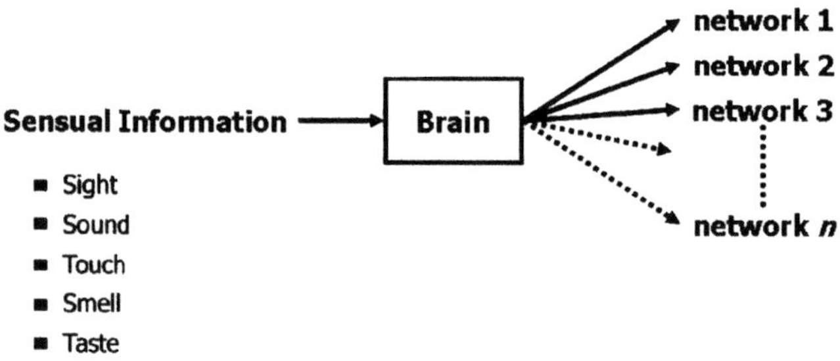

Fig. 7 A representation of how the brain takes input from sensual information and deposits its components in various neural networks

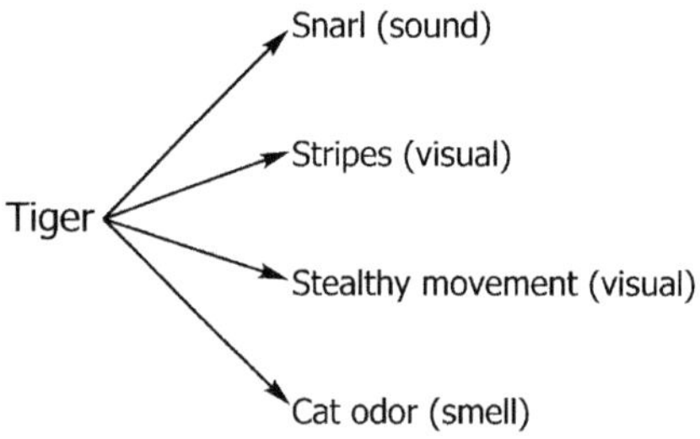

Fig. 8 Some of the sensual information that is associated with the concept "tiger." This information is obtained by the brain and parsed to be deposited in appropriate neural networks

KINDS OF CATEGORIES IDENTIFIED FOR KNOWLEDGE ORGANIZATION

- **Fruits and vegetables**
- **Plants**
- **Animals**
- **Body parts**
- **Colors**
- **Numbers**
- **Letters**
- **Nouns**
- **Verbs**
- **Proper names**
- **Faces**
- **Facial expressions**
- **Several different emotions**
- **Several different features of sound**

⋮

etc.

Fig. 9 Kinds of categories identified for knowledge organization

The distributed information is stored in appropriate networks of neurons that exist in many parts of the brain. The networks are probably interconnected so that the retrieval of the distributed information can start from many places. Many experiments indicate that information is stored in distributed forms, which is then reassembled or reconstructed upon retrieval. It must be noted that "reassembled" and "reconstructed" represent processes that are synonymous with the constructivist model of learning.

6 Conclusion

We believe we have shown here that the unprecedented success in using SATL techniques to help students learn a variety of disciplines stems from modern learning theory (constructivism) and the current ideas of brain function (Fig. 10). In effect, the arguments made here represent the theoretical basis for the effectiveness of SATL methods. From this point of view, it is, perhaps, not surprising that the SATL

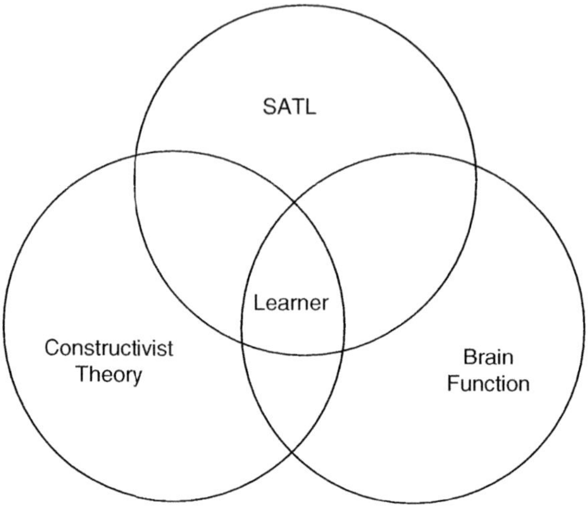

Fig. 10 A representation of the relationship of SATL methods, Constructivist Theory, and brain function with learner activities in learning new knowledge

methods have proven to be so successful irrespective of the discipline in which they have been expressed. Success in learning new knowledge, whatever it may be, comes from teachers teaching in a systemic way and learners trained in the use of systemics learning in a systemic way.

References

1. Skinner BF (1980) About behaviorism. Random House, New York.
2. (a) Herron JD (1975) Piaget for chemists. Explaining what good students cannot understand. J Chem Educ 52:146; (b) Herron JD (1978) Piaget in the classroom, J Chem Educ 55:165; (c) Goodstein MP, Howe AC (1978) Application of Piagetian theory to introductory chemistry instruction. J Chem Educ 55:171.
3. (a) Ausubel DP, Novak JD, Hariesian H (1968) Educational psychology: A cognitive view. Reinhart & Winston, New York; (b) Ausubel DP, Novak JD, Hanesian H (1978) Educational psychology: A cognitive view. Holt, Rinehart, & Winston, New York.
4. (a) Bodner G, Klobuchar M, Geelan D (2001) The many forms of constructivism. J Chem Educ 78:1107; (b) Bodner G J (1986) Constructivism: A theory of knowledge. J Chem Educ 63:873.
5. (a) Cardellini L (2004) Conceiving of concept maps to foster meaningful learning: An interview with Joseph D. Novak. J Chem Educ 81: 1303; (b) Novak JD, Gowin DB (1984) Learning how to learn. Cambridge University Press, Cambridge.
6. Fahmy AFM, Lagowski JJ (1999) The use of systemic approach in teaching and learning for 21st century. J Pure Appl Chem [15th ICCE, Cairo, August 1998].
7. Koslow SH, Huerta MF (1997) Neuroinformatics: An overview of the human brain project. Laurence Erlbaum Hillsdale, NJ; (b) Toga AW, Mazziotta JC (2002) Brain mapping: The methods. Academic Press, New York; (c) Arbib MA (ed) (2002) The handbook of brain theory and neural networks. MIT Press, Cambridge, MA; (d) Pedlura, CM, Martin JB (1991) Mapping the brain and its functions. National Academy Press, Washington, DC

A Rubric to Guide Curriculum Development of Undergraduate Chemistry Laboratory: Focus on Inquiry

L. B. Bruck, S. L. Bretz, and M. H. Towns

Abstract The term "inquiry" is used ubiquitously in the chemistry and science education literature. However, there are inconsistencies in how the modifiers of inquiry such as "guided," "structured," or "open" are used. We developed a rubric to classify levels of inquiry based upon laboratory characteristics and student independence. We validated the rubric and used it to analyze 229 undergraduate chemistry laboratories in 13 manuals. Our findings indicated that nearly 90% were highly structured laboratory activities with little opportunity for student independence. The rubric can be used to analyze a laboratory course or departmental curricula thus playing a role in departments that are attempting to modify the level of inquiry in their curriculum. To demonstrate how the level of inquiry can be changed, we offer an example of a laboratory written to support varying degrees of student independence. By connecting terms that describe inquiry to levels of student independence this rubric can be used by researchers to place their definition of inquiry on a firm foundation and by practitioners to analyze and modify practices.

1 Introduction

Schwab [1] and Herron [2] first developed and presented a rubric for identifying the amount of inquiry in high school laboratory activities. Their "Levels of Openness" rubric used three characteristics to determine the type of inquiry facilitated by a given activity. Each laboratory activity was analyzed via the "problem," the question that would be investigated in the laboratory, the "ways and means," the procedure used in the activity, and "answers," the findings or conclusions. To determine the level of inquiry, each characteristic was coded as "given" or "open" [1]. Table 1 displays the levels of inquiry yielded by this rubric.

L.B. Bruck and M.H. Towns (✉)
Department of Chemistry, Purdue University, West Lafayette, Indiana, USA
e-mail: mtowns@purdue.edu

S.L. Bretz
Department of Chemistry and Biochemistry, Miami University, Oxford, OH, USA

M. Gupta-Bhowon et al. (eds.), *Chemistry Education in the ICT Age*,
DOI 10.1007/978-1-4020-9732-4_9, © Springer Science + Business Media B.V. 2009

Table 1 Levels of openness in the teaching of inquiry [1,2]

	Problem	Ways and means	Answers
Level 0	Given	Given	Given
Level 1	Given	Given	Open
Level 2	Given	Open	Open
Level 3	Open	Open	Open

Since the publication of Schwab [1] and Herron [2], other rubrics for analyzing and describing inquiry in laboratory activities have been added to the literature. Such examples include analysis of the Biological Science Curriculum Study [3,4] and high school-level activities [5]. In addition, a rubric was developed to assess how similar a laboratory activity is to scientific practice [6]. However, none of these rubrics were developed for application to undergraduate laboratories. In addition, the use of the word "inquiry" in the researcher and practitioner literature is ambiguous with authors defining terms such as "guided-inquiry" or "open-inquiry" inconsistently.

Thus, we set about using the literature as a foundation for the development of a rubric that could be used to characterize the level of inquiry facilitated in undergraduate chemistry laboratories. We also sought to tie levels of student independence to specific terms describing inquiry such as "structured-inquiry" or "guided-inquiry" in order to standardize the use of these terms amongst researchers and practitioners.

2 Methodology

Using the literature as a foundation, we began developing a rubric that could be used to determine the level of inquiry in undergraduate chemistry laboratories. We proceeded by collecting undergraduate chemistry laboratory manuals published in the USA for evaluation. These manuals were chosen for their innovations (green chemistry), impact on teaching practices (cooperative learning), and/or to represent a broad range of chemistry content areas. Additional manuals were chosen based upon literature references that discussed inquiry.

Our sample included 13 laboratory manuals: six general chemistry manuals, two GOB (general, organic, biochemistry) manuals, four organic manuals, and one biochemistry manual. Among general chemistry laboratory manuals we either evaluated nearly all the experiments [7–9], laboratories that were in common across the manuals [10–12], laboratories currently in use at the institution (Purdue), or a smaller subset of an innovative laboratory manuals [13]. Laboratory manuals in other areas of chemistry were sampled similarly.

Laboratory activities were analyzed in the following manner. For each specific section of the laboratory – the problem or question, theory or background, procedures, analysis of results, communication of results, and conclusions – the researchers determined if that portion of the laboratory provided specific guidance to the students. The permutations of providing or not providing specific guidance produced seven levels of inquiry across six characteristics or specific laboratory sections, where the most highly structured lab provided students with guidance at every step, and the least structured was completely open to student decision-making. To generate the levels of inquiry, we began with Schwab [1] and Herron's [2] model of levels zero through three, and integrated Levels ½, 1 ½, and 2 ½ to correspond to the seven permutations of providing student guidance.

Carrying out these procedures across 229 laboratories produced evidence for five levels of inquiry that are detailed in the results section. There were laboratories that fit into Level ½, but no laboratories were found that fit into Level 1 ½ or 2 ½. Thus those levels were removed.

An inter-rater reliability (IRR) study was carried out with 36 of the 229 laboratories. The IRR value was 83% agreement across researchers, which is well above the 70% standard to establish reliability.

3 Results and Discussion

The sampling method employed allowed us to build upon the research literature and develop a more fine-grained rubric as shown in Table 2. It uses six characteristics that originated from the terminology used in the laboratory manuals to describe five levels of inquiry grounded in the data that we collected and analyzed.

Each level of inquiry is tied to a specific phrase from the literature describing inquiry as shown in Table 3, thus this rubric can be used to reduce the ambiguity present in the literature pertaining to inquiry.

One advantage of this rubric to researchers and practitioners is that it provides operational definitions of widely used terms or modifiers of the word "inquiry." Thus, it places on firm footing the characteristics of laboratories that are stated to be "guided-inquiry" or "structured-inquiry."

Table 2 A rubric to characterize inquiry in the undergraduate laboratory. In each cell P = provided to the student, and NP = not provided to the student

	Level 0	Level ½	Level 1	Level 2	Level 3
Characteristic	Confirmation	Structured inquiry	Guided inquiry	Open inquiry	Authentic inquiry
Problem or question	P	P	P	P	NP
Theory or background	P	P	P	P	NP
Procedures or design	P	P	P	NP	NP
Analysis of results	P	P	NP	NP	NP
Communication of results	P	NP	NP	NP	NP
Conclusions	P	NP	NP	NP	NP

Table 3 Descriptions of each level of inquiry

Level of inquiry	Terminology and Description
0	*Confirmation*: The students confirm a principle or outcome that is known in advance. A highly structured lab that provides the question, theory, procedures, analysis, communication, and conclusions to the student.
1/2	*Structured inquiry:* The students use a prescribed procedure to gather and analyze data answering an instructor provided question. The question, background, procedure, and methods of analysis are provided. The method of communicating the results is left open to the student and the conclusions or results are not known in advance.
1	*Guided inquiry:* The students use a prescribed procedure to collect data that guides them in addressing an instructor provided question. The question, background, and procedure are provided to the student, but the method of analysis, interpretation, and communication are left open to the student.
2	*Open inquiry:* The students develop their own procedure, analysis, communication, and conclusions to address an instructor provided question. The instructor provides the question and background, but the remaining characteristics are left open to the student.
3	*Authentic inquiry:* The students craft their own question for research and subsequently design procedures to collect, analyze, and interpret the data, and to communicate the results of their investigation. This laboratory activity provides the least structure and all characteristics are left open to the student.

3.1 Analysis of Laboratory Texts

The rubric we developed and validated was applied to 229 laboratories in 13 manuals to characterize the level of inquiry within a laboratory experiment or activity. The results of this analysis are displayed in Table 4.

Although many laboratory manuals incorporate novel concepts, modern instrumentation, and new techniques, these advances in curriculum development were not accompanied by a corresponding shift in pedagogy towards incorporating inquiry. The analysis of 229 individual laboratory activities revealed that nearly 90% of the experiments were highly structured Level 0 or Level ½ laboratories, as shown in Table 4. The vast majority of experiments, $n = 191$ out of 229 or 83%, were classified as Level ½ structured inquiry. The only Level 2 open inquiry chemistry experiments ($n = 5$) we found were contained in *Inquiries into Chemistry* [10]. While the K-12 science curriculum in the USA moves towards the incorporation of inquiry, we found that the college chemistry laboratory curriculum remains entrenched in confirmation and structured inquiry levels.

Table 4 Evaluation of levels of inquiry for chemistry laboratory texts

	Level of inquiry						
Laboratory texts	0	1/2	1	2	3	Experiments in manual	Experiments analyzed
LASER Experiments for Beginners [13]	8	–	–	–	–	29	8
Cooperative Chemistry Laboratory Manual [8]	2	4	9	–	–	15	15
Laboratory Inquiry in Chemistry [7]	2	9	12	–	–	29	23
Purdue University CHM 115 Lab Manual, Fall 2006 [14]	–	7	–	–	–	23	7
Working with Chemistry: A Laboratory Inquiry Program [9]	–	24	–	–	–	26	24
Inquiries into Chemistry [10]	–	5	–	5	–	63	10
Laboratory Manual for General, Organic, and Biological Chemistry [12]	–	12	–	–	–	42	12
Modern Projects and Experiments in Organic Chemistry: Miniscale & Standard Taper Microscale [15]	–	13	–	–	–	43	13
Green Organic Chemistry: Strategies, Tools, and Lab Experiments [16]	–	19	–	–	–	19	19
Exploring Chemistry: Laboratory Experiments in General, Organic, and Biological Chemistry [11]	–	19	–	–	–	33	19
Organic Chemistry Laboratory with Qualitative Analysis: Standard and Microscale Experiments [17]	–	29	–	–	–	45	29

(continued)

Table 4 (continued)

Laboratory texts	Level of inquiry 0	1/2	1	2	3	Experiments in manual	Experiments analyzed
Microscale & Miniscale Organic Chemistry Laboratory Experiments [18]	–	42	–	–	–	65	42
Experiments in Biochemistry: A Hands-On Approach [19]	–	8	–	–	–	13	8
Total	12	191	21	5	0	445	229

3.2 *Evaluation of Laboratory Curriculum*

This rubric can also be used to evaluate a course or departmental curriculum. In the USA accrediting agencies have brought pressure on universities, college, programs, and departments to engage in authentic evaluation practices. A by-product of these actions has been for departments to evaluate courses and programs with respect to departmental goals. If a department wishes students to develop skills associated with inquiry, then it is reasonable to evaluate the types of experiences and laboratories that are part of the curriculum using this rubric. Once the results are known, data-driven decisions can be made about the trajectories of inquiry across a course, program, or department.

3.3 *Modification of Laboratories*

If circumstances arise where an increase in inquiry in a course or program is desired, the practitioner-based literature provides examples of modified laboratories that incorporate more inquiry through greater degrees of student independence [19–25]. Space does not allow us to provide a full written example of each type of laboratory. However, it is illustrative to use a single laboratory activity, such as the determination of vitamin C in orange juice, to demonstrate how guidance to the student changes as the level of inquiry increases. Brief examples are shown in Table 5.

A Level 0 confirmation laboratory confirms a principle or outcome that the students know in advance. Note that the brief one-paragraph statement in Table 5 discloses the results of the investigation in advance – the vitamin C concentration should decrease. A full student laboratory handout would provide the question, background, procedures, analysis, communication, and conclusion. The laboratory activity provides little, if any, student independence.

A Level ½ or 1 (structured or guided inquiry) laboratory can be developed from the Level 0 laboratory through simple modifications, although the students continue

Table 5 Examples of increasing levels of inquiry in the determination of vitamin C in orange juice

Level of inquiry	Description of lab in one paragraph.
0	Confirmation – *Confirm a principle or outcome that the students know in advance.* Today in laboratory you will confirm the decrease in vitamin C concentration via oxidation between fresh and 7-day old orange juice. The amount of ascorbic acid (vitamin C) will be determined via a redox titration using a standardized solution of iodine. You will first titrate an ascorbic acid solution to hone your skills, and then titrate two orange juice solutions, one that is fresh, and one that is a week old. Record the results as indicated, complete the analysis, and demonstrate that the concentration of vitamin C decreases.
1/2 or 1	Structured or guided inquiry – *Conduct an investigation using a provided procedure. The students may be provided with the methods of analysis (Level ½) or be required to generate it (Level 1).* Today in laboratory you will determine the concentration of vitamin C in orange juice. The amount of ascorbic acid (vitamin C) will be determined via a redox titration using a standardized solution of iodine. You will first titrate an ascorbic acid solution to hone your skills, and then titrate two orange juice solutions, one that is fresh, and one that is a week old. Record the results as indicated, analyze the data, and effectively communicate your findings.
2	Open inquiry – *Conduct an investigation using a procedure adapted from the literature, analyze the data appropriately, interpret it, and present your findings.* Design an experiment to explore the factors that affect the oxidation of vitamin C in orange juice. Locate an appropriate procedure in the literature, adapt it to the resources available in our classroom, and have it approved by your instructor. Record the data, analyze it appropriately, and effectively communicate the results of your investigation.

using a prescribed procedure. First, the results of the lab are removed from the problem statement. Second, the method of communicating the results of the laboratory is left in the hands of the student, and finally the analysis can be provided (Level ½) or removed (Level 1) depending on the level of student independence desired.

Level 2 guided-inquiry laboratory can be produced by requiring the students to locate an appropriate procedure rather than providing it. Faculty approval of the procedure allows for dialogue about the methods, resources available, data collection, and analysis. This approach incorporates greater degrees of student independence by requiring the students to develop the methods, analysis, interpretation, and communication of results.

4 Conclusion

The rubric for characterizing the level of inquiry in undergraduate chemistry laboratory may be used for research and practice-oriented purposes. We used the rubric to characterize the laboratories in 13 texts. The majority of the laboratories we analyzed were highly structured fostering low levels of student independence. Only one laboratory manual contained Level 2 open-inquiry activities.

The rubric may also assist faculty in developing laboratories with a greater degree of student independence. It can be used to evaluate the trajectories of inquiry across a course, program, or department. The path traced out by the laboratory curriculum may be modified to address departmental or programmatic goals. Thus, the rubric provides a route to data-driven practices.

Acknowledgments This research is supported by the National Science Foundation under Grants No. 0737784 (CCLI) and No. 0536776. Any opinions, findings, and conclusions or recommendations expressed in this material are those of the authors and do not necessarily reflect the views of the National Science Foundation.

References

1. Schwab J J (1962) The teaching of science as enquiry. In: Schwab J J and Brandwein P F (ed) The teaching of science, Harvard University Press, Cambridge, MA.
2. Herron M D (1971) The nature of scientific enquiry. Sch Rev 79: 171–212.
3. Fuhrman M, Lunetta V N, Novick S, Tamir P (1978) The laboratory structure and task analysis inventory (LAI): A user's handbook. Science Education Center, University of Iowa: Iowa City.
4. Tamir P, Lunetta V N (1978) An analysis of laboratory inquiries in the BSCS yellow version. Am Biol Teach 40(6): 353–357.
5. Germann P J, Haskins S, and Auls S (1996) Analysis of nine high school biology laboratory manuals: Promoting scientific inquiry. J Res Sci Teach 33(5): 475–499.
6. Chinn C A, Malhotra B A (2002) Epistemologically authentic inquiry in schools: A theoretical framework for evaluating inquiry tasks. Sci Educ 86(2): 175–218.
7. Bauer R D, Birk J P, Sawyer D J (2005) Laboratory inquiry in chemistry 2nd ed. Thompson Brooks/Cole.
8. Cooper M M (2006) Cooperative chemistry laboratory manual 3rd ed. McGraw Hill: New York.
9. Wink D J, Gislason S F, Kuehn J E (2005) Working with chemistry: A laboratory inquiry program 2nd ed. WH Freeman: New York.
10. Abraham M R, Pavelich M J (1999) Inquiries into chemistry 3rd ed. Waveland Press: Long Grove, IL.
11. Peller J R (2004) Exploring chemistry: Laboratory experiments in general, organic, and biological chemistry 2nd ed. Pearson/Prentice Hall: Upper Saddle River, NJ.
12. Timberlake K C (2007) Laboratory manual for general, organic, and biological chemistry. Pearson/Benjamin Cummings: Boston.
13. Zare R N, Spencer B H, Springer D S, Jacobson M P (1995) LASER experiments for beginners. University Science Books: Sausalito, CA.
14. Purdue University Department of Chemistry (2006) CHM 115 Laboratory manual fall 2006. Hayden McNeil: Plymouth, MA.
15. Mohrig J R, Hammond C N, Schatz P F, Morrill T C (2003) Modern projects and experiments in organic chemistry: Miniscale and standard taper microscale 2nd ed. WH Freeman: New York.
16. Doxsee K M, Hutchison J E (2006) Green organic chemistry: Strategies, tools, and laboratory experiments. Brooks/Cole.

17. Bell C E Jr, Taber D F, Clark A K (2001) Organic chemistry laboratory with qualitative analysis: Standard and microscale experiments 3rd ed. Thomson Brooks/Cole.
18. Schoffstall A M, Gaddis B A, Druelinger M L (2004) Microscale and miniscale organic chemistry laboratory experiments 2nd ed. McGraw Hill: New York.
19. Farrell S O, Taylor L E (2006) Experiments in biochemistry: A hands-on approach. 2nd ed. Thomson Brooks/Cole.
20. Farrell J J, Moog R S, Spencer J N (1999) A guided inquiry general chemistry course. J Chem Educ 76(4): 570–574.
21. Huber R A, Moore C J (2001) A model for extending hands-on science to be inquiry based. Sch Sci Math 101(1): 32–41.
22. Mohrig J R, Hammond C N, Colby D A (2007) On the successful use of inquiry-driven experiments in the organic chemistry laboratory. J Chem Educ 84(6): 992–998.
23. Oliver-Hoyo, M, Allen D, Anderson M (2004) Inquiry guided instruction: Practical issues of implementation. J Coll Sci Teach 33(6): 20–24.
24. Pavelich M J, Abraham M R (1977) Guided inquiry laboratories for general chemistry students. J Coll Sci Teach 7(1): 23–26.
25. Uno, G E (1990) Inquiry in the classroom. Biosci 40(11): 841–843.

Investigating the Effectiveness of Computer Simulations in the Teaching of "Atomic Structure and Bonding"

S. Abdoolatiff and F. B. Narod

Abstract The purpose of this study was to investigate the effectiveness of using computer simulations on students' understanding of "Atomic Structure and Bonding" over the traditionally designed instruction. The impact of computer simulations on students' attitude and motivation was also studied during the research. The study was carried out in a girls' State Secondary School in Mauritius. The sample consisted of 44 students in two Form IV (students of age 14–15 years) Chemistry classes. One class ($n = 21$) represented the treatment group in which computer simulations were used to teach "Atomic Structure and Bonding" using the software "Atoms, Bonding and Structure" (an interactive Chemistry tutor). The other class ($n = 23$) comprised the control group in which the same topic was taught by the traditional expository instruction without the use of computer simulations.

Our findings have revealed an improved performance of students in the treatment group as compared to students in the control group. Furthermore, the study has also shown that students in the treatment group were very enthusiastic and highly motivated when computer simulations were used. Our results thus indicate that computer simulations have not only enhanced students' understanding of "Atomic Structure and Bonding," but also increased their motivation during the lessons.

1 Introduction

It is an undeniable fact that the structure of matter is one of the most important concepts in Chemistry, and that the shape and structure of particles, including the location and bonding of atoms in molecules and the location of electrons, are

S. Abdoolatiff
Lady Sushil Ramgoolam State Secondary School, Mauritius

F.B. Narod (✉)
Department of Science Education, Mauritius Institute of Education, Mauritius
e-mail: zia373@eudoramail.com

M. Gupta-Bhowon et al. (eds.), *Chemistry Education in the ICT Age*,
DOI 10.1007/978-1-4020-9732-4_10, © Springer Science+Business Media B.V. 2009

determining factors in understanding chemical reactions and the reactivity of chemical substances. Thus, conceptual understanding in Chemistry includes the ability to represent and translate chemical problems using macroscopic (observable), molecular (particulate), and symbolic forms of representation [1]. Students are required to think at the molecular level and explain changes at macroscopic level in terms of interactions between individual atoms and molecules [2]. Furthermore, an understanding of the concepts of atoms and molecules is fundamental to the learning of Chemistry; any misconceptions and alternate conceptions that students harbor about these microscopic particles will impede further learning of Chemistry [3].

However, it is equally true that one of the most difficult challenges in the teaching and learning of Chemistry at secondary level involves conveying to students the three-dimensional structure and dynamic interactions of atoms and molecules. Because of students' difficulties in visualizing atoms and molecules in three dimensions, many of them are not able to understand the molecular basis for chemical phenomena. Indeed, Gabel et al. [4] have reported that many students are not able to understand the meaning of chemical symbols used to represent macroscopic and microscopic systems. Other authors have also stated that students have difficulties in explaining the chemical phenomena at the molecular level when prompted to think in terms of atoms and molecules [5–7]. Furthermore, research has also shown that many students have trouble in understanding concepts like elements, mixtures, and compounds, and in understanding the difference between atoms, molecules, and ions, as well as in visualizing the dynamic interactions of chemical events [8]. In accordance with the above findings, Tsaparlis [9] attempted to explain students' difficulties and misconceptions about atomic and molecular structure based on the Piagetian Developmental Perspective [10], the Ausubelian Theory of Meaningful Learning [11], and the Alternative Conceptions Movement [12]. Based on the findings of Tsaparlis [9] and on the above-mentioned studies, students' difficulties and misconceptions about "Atomic Structure and Bonding" may be attributed to one or more of the following factors:

- Inability to visualize atoms and molecules in three dimensions.
- Failure to employ formal operations.
- Lack of association or anchorage for learning about these topics since concepts related to atomic structure and bonding are usually built on entirely new grounds.
- The way the topic has been traditionally taught in classroom situation.

On the other hand, computer-based technologies are able to provide powerful means for fostering molecular understanding as they can represent the multilevel thought in chemistry. Indeed, computers have the capacity to support molecular-level animations of chemical phenomena that are not directly perceivable by other means. According to Williamson and Abraham [13], the dynamic quality of computer animations enables deeper coding and more expert-like mental models of the particulate nature of matter, in contrast to static visuals such as text book pictures and chalk diagrams. Furthermore, computer technology not only allows modeling of expert thinking at the molecular level, but also permits simultaneous representation of molecular and macroscopic views of the same phenomena [14]. Research has also shown that computer technology can engage students in challenging and authentic learning [15], as it raises their level of motivation and can make a positive difference in classroom teaching.

More recently, a study by Ardac and Akaygun [16] has revealed that chemistry students who received computer-based instruction that enabled simultaneous display of molecular representations corresponding to macroscopic observations, showed enhanced performance, as compared to students who received only traditional instruction on the same concepts. Similarly, Stieff [17] has reported that computer-aided visualizations can offer important benefits to chemistry students by connecting together the molecular, macroscopic, and symbolic levels of chemistry. The same author also highlighted that computer simulations can offer support to students by encouraging them to appreciate how macro-level chemistry concepts emerge from the molecular-level interactions of chemical species, thus providing students with additional footholds to enhance learning.

In view of the above discussions, it has been decided in the present study to incorporate the use of computer simulations in the teaching of "Atomic Structure and Bonding," in an attempt to improve acquisition of scientific conceptions related to these topics, as well as to arouse sustainable interest among students during the Chemistry lessons.

2 Methodology

2.1 *Sample Description*

The sample consisted of 44 students (girls of age 14–15 years) enrolled in two Form IV classes in a girls' State Secondary School in a village located in the North West of Mauritius. The study was carried out during eight lessons of 40 min each in 2005, and the lesson objectives were in line with the University of Cambridge Local Examinations Syndicate (UCLES) "O" – level Chemistry syllabus, as the sample students were in fact preparing to sit for this examination in the following year.

One Form IV class (n = 21) represented the treatment group in which the students were taught the topic "Atomic Structure and Bonding" by making use of computer simulations. The lessons for the treatment group were held both in the Chemistry Laboratory and the Computer Laboratory with due approval from the Rector and the Head of the Computer Education department of the school. All students in the treatment group were computer literate and had the basic computer skills required for working with the selected software. The other Form IV class (n = 23) represented the control group in which the same concepts (with same lesson objectives) were taught by the regular expository instruction without making use of computer simulations. The instructional plan used for the control group was similar to the plan used for the treatment group, except for the fact that no ICT resources was used, and the lessons were held only in the Chemistry Laboratory. Students in both treatment and control groups were of mixed ability.

2.2 Data Collection

The following techniques were used to collect data during the study:

1. Observation checklists

During each lesson in the treatment group, an observation checklist was used and filled by the teacher to gather information about students' motivation and interest when computer simulations were used. The criteria for observation included time management, students' motivation, students' participation and interest, as well as the effectiveness of the use of computer simulations in the teaching of "Atomic Structure and Bonding."

2. Students' questionnaire

After the study, a questionnaire was administered to students of the treatment group to gather information about their attitudes, perceptions, and personal opinions regarding the use of computer simulations during the teaching of "Atomic Structure and Bonding." The questionnaire was designed in such a way that it consisted only of multiple-choice questions and students had to select the answer they found most appropriate. In addition, the final part of the questionnaire included a blank space in which students were free to make any relevant comments to express themselves about the use of computer simulations in classroom teaching.

3. Student's achievement tests

Two paper-and-pencil tests, Test 1 and Test 2, were administered to both treatment and control groups at different stages of the study to assess students' understanding of the concepts taught during the study. Students were allocated 40 min to complete each test, which was administered simultaneously to both groups. Test 1 and Test 2 were designed to measure and compare the achievement and performance of students in the treatment and control groups. The maximum scores for both tests were of 20 marks.

2.3 Description of the Software Used During the Study

The software used in the present study "Atoms, Bonding and Structure," designed by Ray Le Couteur, is an interactive Chemistry tutor intended for school use at "O"-Level, and includes eight units as illustrated in Fig. 1. Each unit consists of a number of sequential pages that present the concepts and test students' understanding as they learn. The software also comprises three Interactive Simulations, which are as follows:

1. The "Atom Builder" simulation allows students to "build" any atom of the first 20 elements of the Periodic Table on screen (Fig. 2).

2. The "Ionic Bonding" simulation enables students to investigate the bonding of dozens of ionic compounds interactively on screen (Fig. 3).
3. The "Covalent Bonding" simulation enables students to model the bonding in 12 common covalent compounds on screen (Fig. 4).

In addition, the software also includes an option, "Structures" from the menu bar, which allows students to select from a range of common structures which ones they would like to investigate. One very important feature of the software is that most examples are accompanied by questions that help to probe students' understanding of the concept under study.

Other important and interesting aspects of the "Atoms, Bonding and Structure" software include:

- Interactive simulations to teach the structure of the atom, ionic bonding, and covalent bonding.
- Web extensions activities for each unit.
- Sounds are used (optionally) to indicate correct and wrong answers.
- Answers to the questions are optionally displayed after three wrong attempts.
- Teaching of skills through familiar examples to reinforce general chemistry knowledge.
- Inclusion of an interactive Periodic Table.

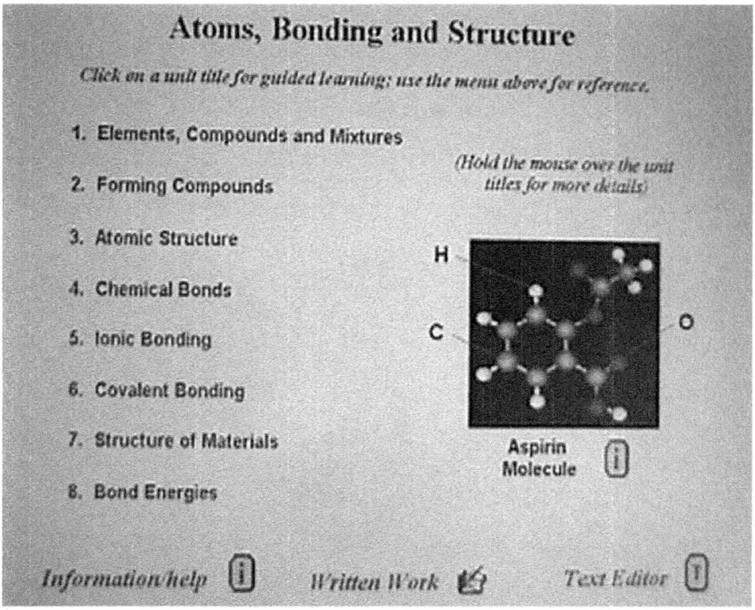

Fig. 1 Units included in the "Atoms, Bonding and Structure"' software

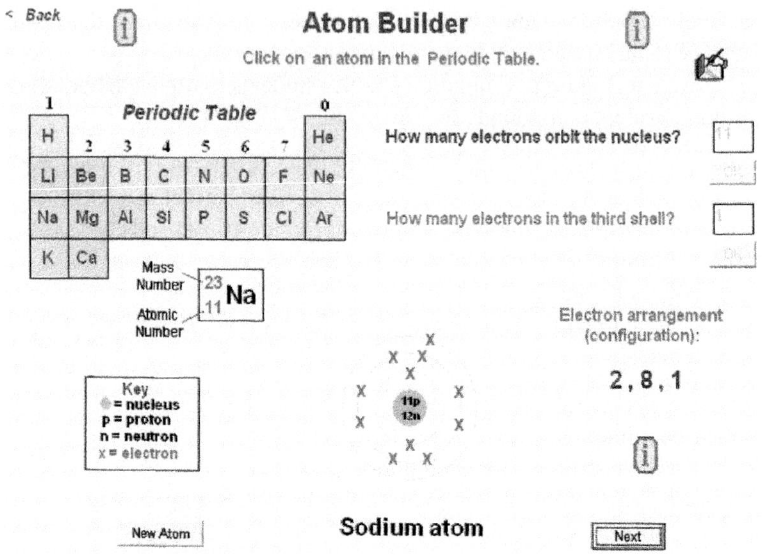

Fig. 2 The "Atom Builder" simulation

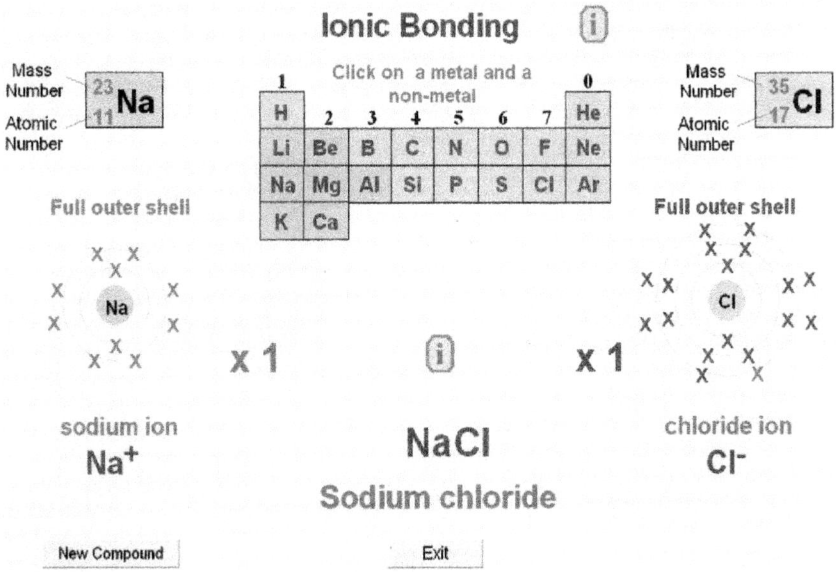

Fig. 3 The "Ionic Bonding" simulation

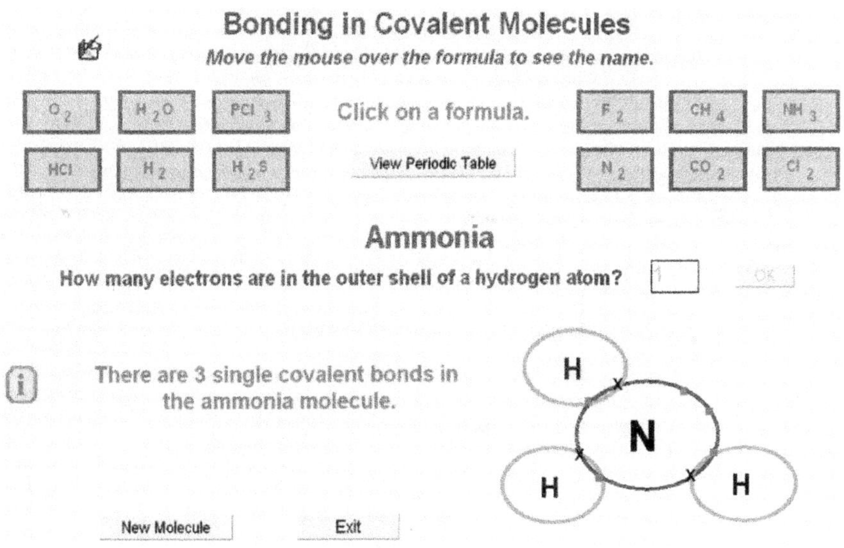

Fig. 4 The "Covalent Bonding" simulation

In the present study, the software was used to teach "Atomic Structure and Bonding" to the treatment group as follows:

- The interactive simulation, "The Atom Builder" (Fig. 2) was used to explain students how to draw the electronic arrangement in an atom and how to write down the electronic configuration of the atom.
- The software was used to introduce students to Chemical Bonding (Fig. 5), and to highlight that atoms can be joined chemically by two main types of bonding, namely ionic and covalent bonding.
- The "Ionic Bonding" and the "Covalent Bonding" simulations (Figs. 3 and 4) were used to explain the major features of ionic and covalent bonds and their formation.
- The software was also used to provide formative assessment tasks for testing students' understanding of the concepts being taught during the lessons. Students were required to interact with the software to "build up" the atomic structure of selected atoms and to simulate the formation of ionic and covalent bonds between selected elements.
- Students were also given the opportunity to use the software to investigate about the bonding in different molecules, and about the formation of ionic bonds in selected compounds.

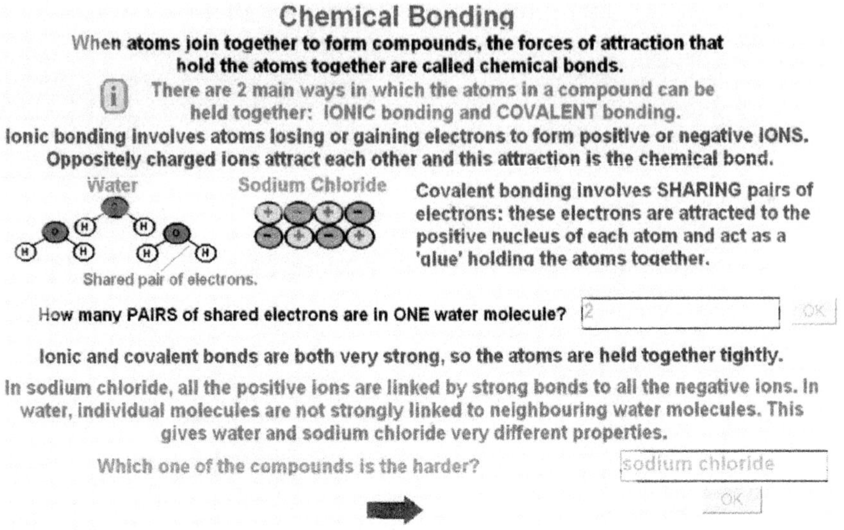

Fig. 5 Introducing students to "Chemical Bonding" using the software

In the control group, the students were taught about the same concepts, that is, the atomic structure and the formation of ionic and covalent bonds by the regular instruction method without using computer simulations.

3 Results and Discussions

3.1 Data Collected from Observation Checklists and Questionnaire

Findings from the present study have revealed that when computer simulations were first introduced to students in the treatment group using a notebook and projector, they were more enthusiastic and more motivated. They were found to be more interested and more attentive during the lessons, and followed the instructions given in the software program with much awareness. Furthermore, it was also evident from students' facial expressions, that they were highly interested in the lessons. The different windows shown on screen were carefully observed and students were quite keen to interact with the software. In subsequent lessons of the study, students in the treatment group were enthusiastic about working in the Computer Laboratory, and were particularly thrilled to be interacting with the computer simulations on their own. This finding is further confirmed by data collected from students' questionnaire, in which all students of the treatment group stated that the lessons involving computer simulations were more interesting and that they were more motivated when simulations were used; this clearly points out that computer simulations have helped to arouse their interest in the lesson. This finding is in accordance with Means and Olson [15] who noted that technology engaged students in challenging and authentic learning while being of help to students as it raised their level of motivation.

Data collected through observations in the treatment group have also shown that students were very enthusiastic to work in pairs on the computer simulations program and they were kept very much on task during the lesson. They were seen to participate actively by answering questions set by the teacher and by the software; furthermore, it was also observed that some students were trying to help their friends during the computer simulations sessions. This indicates that when computer simulations were used along with group work, students were able to interact among themselves and provide help and support to their peers; indeed, use of ICT can help to create an interactive environment in classrooms, and can engage students in the learning process. In addition, it was also evident from students' responses in the questionnaire that a higher percentage of them (57%) preferred to work in groups during the computer simulations sessions rather than individually (43%) as illustrated in Fig. 6. This finding further supports observational data during the lessons where it was noted that students were very enthusiastic to work with their peers and were discussing among themselves to clear misunderstanding and to help each other. This can be explained by the fact that students feel more comfortable when they are in a group as they can ask for help from their friends. According to Vygotsky [18], social interaction is important for learning because higher mental processes such as reasoning, comprehension, critical thinking, and problem solving skills find their origins in social interactions.

Observations in the treatment group have also indicated that computer simulations were effective in enhancing acquisition of concepts during the teaching process; the students were able to complete the formative assessment tasks set by the teacher

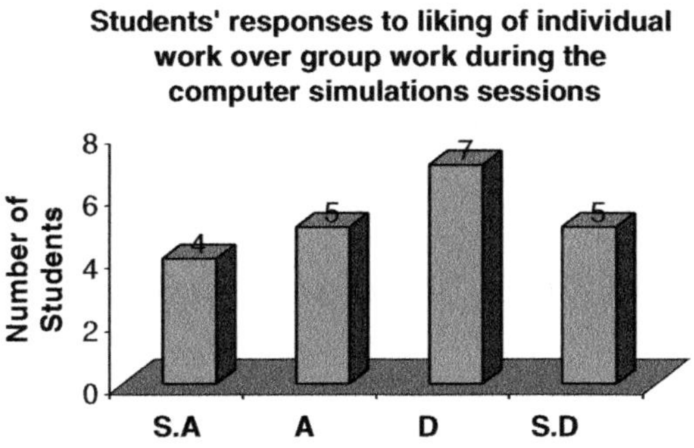

Fig. 6 Students' responses to liking of individual work over group work during the computer simulations sessions (S.A: strongly agree; A: Agree; D: Disagree and S.D: Strongly disagree)

and by the software. This result is further supported by students' responses in the questionnaire where all students in the treatment group have stated that they were better able to understand the concepts being taught when computer simulations were used. This clearly shows that computer simulations can enhance conceptual understanding as earlier reported by Sanger [19]. In the same line, our results have also revealed that all students in the treatment group affirmed that the computer simulations have helped to clear out any misconception they had in the topic. Thus, our findings indicate that use of computer simulations can help to avoid and to clear out misconceptions as earlier reported by Russel et al. [14]. Indeed, the latter authors did report a marked decrease in "misconceptual" statements and an increase in students' understanding in their study on the use of prototype software.

Data collected during the lessons in the treatment group have also indicated that the computer simulations were better appreciated than regular instruction by the students. Besides, all students of the treatment group have stated in the questionnaire that they have liked the visual effects provided by computer simulations; indeed, Somekh [20] has earlier reported that computer technology can provide strong visual elements that would be readily welcomed by teachers and students. In addition, Papert [21] has also stated that early exposure to science through visualization promotes independent learning by using the child's intrinsic curiosity, and by integrating science into his understanding of the world.

In the questionnaire, when students in the treatment group were asked whether sufficient exercises were given for practice by the teacher and the software, all of them answered positively. As in all teaching/learning situations, it was deemed important to check students' conceptual understanding at different stages in the treatment group. Small exercises were therefore given, as diagnostic and/or formative assessment by the teacher; furthermore students' conceptual understanding was also tested by questions in the "Atoms, Bonding and Structure" software after presentation of different concepts.

It was also interesting to note that all students in the treatment group asserted that they were able to manipulate the computer simulations software and did not have any difficulty to use it. This finding points out to the fact that the software was indeed appropriate to the level of the students and was user friendly. At this stage, it needs to be highlighted that when selecting software or any other ICT resources for teaching purposes, teachers must ascertain themselves that its manipulation and use would suit the students' abilities and needs; else, students would not be able to benefit from the resources and would be put off in the learning process.

3.2 Data Collected from Students' Achievement Tests

As mentioned in the Methodology, two achievement tests (Tests 1 and 2) based on the concepts taught during the study were developed and administered to both treatment and control groups to measure and compare the performances of students in both groups. The tests were set to both groups concurrently at two different stages of the research.

Students' scores in the two tests are shown in Tables 1 and 2. Students' names have not been revealed for ethical reasons; the letters C1 to C23 represent the coded names of students in the control group, while T1 to T21 represent the students in the treatment group. The scores were converted to percentages, and the performances of the two groups in the two tests were analyzed and compared (Figs. 7–10).

From Table 1, it can be observed that only 2 students (representing 9%) out of 23 in the control group scored 100% in Test 1. The lowest percentage obtained in Test 1 in the control group is 60%. It is also evident from Table 1 that no student obtained 100% in Test 2 in the control group, while the majority (87%) of the students scored between 60% and 80% in Test 2. As far as the treatment group is concerned, Table 2 indicates that 6 students (representing 29%) out of 21 obtained 100% in Test 1 and no student scored less than 75%; the majority (76%) of the students scored 90% or more in Test 1. Table 2 also shows that in Test 2, 8 students (representing 38%) in the treatment group scored 100%, and no one obtained less than 80%; the majority of the students of the treatment group, 17 out of 21 (representing about 81%) scored 90–100% in Test 2.

On comparing the test scores obtained by students in Test 1 for the control and treatment groups (Figs. 7 and 8), it can be seen that overall, students in the treatment

Table 1 Students' scores and percentage scores obtained in Tests 1 and 2 (Control Group)

	TEST 1		TEST 2	
Student	Score (out of 20 marks)	Percentage	Score (out of 20 marks)	Percentage
C1	17	85	16	80
C2	14	70	13	65
C3	15	75	14	70
C4	20	100	19	95
C5	14	70	15	75
C6	14	70	16	80
C7	16	80	15	75
C8	18	90	17	85
C9	20	100	18	90
C10	15	75	14	70
C11	17	85	15	75
C12	15	75	16	80
C13	16	80	13	65
C14	13	65	12	60
C15	16	80	15	75
C16	12	60	14	70
C17	14	70	15	75
C18	17	85	16	80
C19	15	75	15	75
C20	14	70	15	75
C21	15	75	16	80
C22	13	65	12	60
C23	18	90	16	80

Table 2 Students' scores and percentage scores obtained in Tests 1 and 2 (Treatment Group)

	TEST 1		TEST 2	
Student	Score (out of 20 marks)	Percentage	Score (out of 20 marks)	Percentage
T1	19	95	19	95
T2	16	80	17	85
T3	20	100	20	100
T4	18	90	19	95
T5	20	100	20	100
T6	16	80	17	85
T7	17	85	18	90
T8	20	100	20	100
T9	19	95	20	100
T10	17	85	18	90
T11	19	95	19	95
T12	20	100	20	100
T13	18	90	19	95
T14	16	80	17	85
T15	15	75	16	80
T16	18	90	18	90
T17	20	100	20	100
T18	19	95	20	100
T19	17	85	18	90
T20	18	90	19	95
T21	20	100	20	100

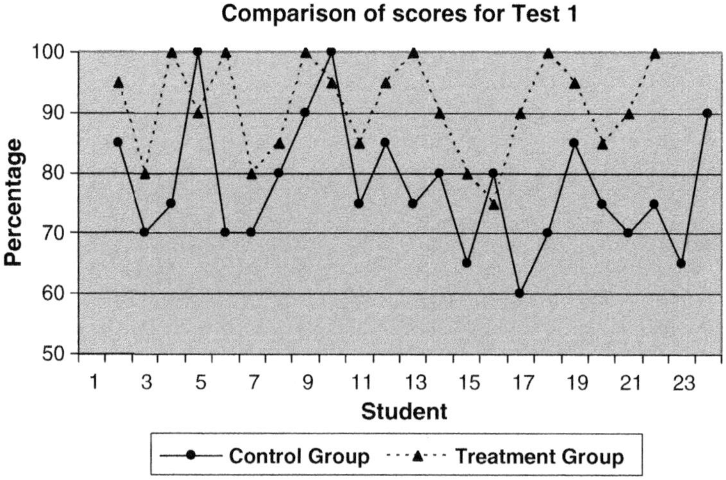

Fig. 7 Comparison of the performance of students in the Control and Treatment Groups for Test 1

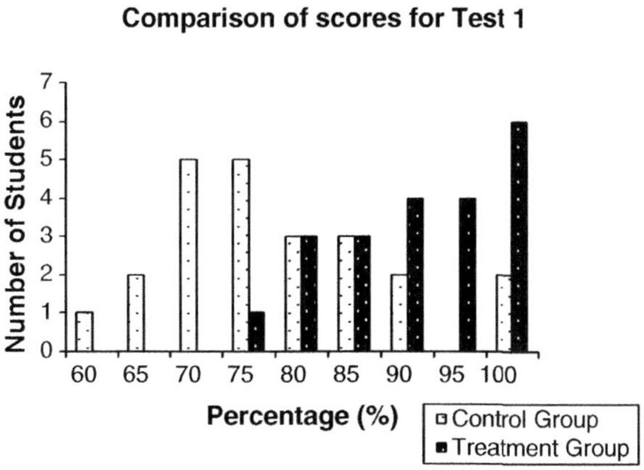

Fig. 8 Comparison of percentage scores obtained by the Control and Treatment Groups for Test 1

group obtained better scores than those in the control group. This can be attributed to the incorporation of computer simulations in lessons in the treatment group. Moreover, the second test (Test 2) was used to assess achievement of learning outcomes covered throughout the study, and the results obtained showed that students from the control group did not perform as well as in Test 1. On the other hand, students in the treatment group showed a marked improvement in their results from Test 1 to Test 2; all students in the treatment group scored a better mark in Test 2 compared to Test 1. This further indicates enhanced understanding of the concepts being taught when using computer simulations. A comparison of scores for the two groups in Test 2 (Figs. 9 and 10) also reveals that students in the treatment group have demarcated themselves well away from those of the control group, indicating that they were much more motivated and confident in Test 2 as opposed to the control group.

Thus, data collected from the achievement tests clearly indicate that students in the treatment group have performed better in the two tests than those of the control group. Our findings undoubtedly reveal that the use of computer simulations in the teaching of "Atomic Structure and Chemical Bonding" has enhanced conceptual understanding of the treatment group leading to a positive impact on their performance. This is in line with findings from Rieber [22] who reported that computer simulations enhance recall of facts and concepts, and from Sanger [19] who highlighted that computer simulations help to improve conceptual understanding of science. More recently, Condie and Munro [23] have indeed highlighted that computer simulations are particularly effective in supporting understanding of abstract and microscopic concepts and processes in science.

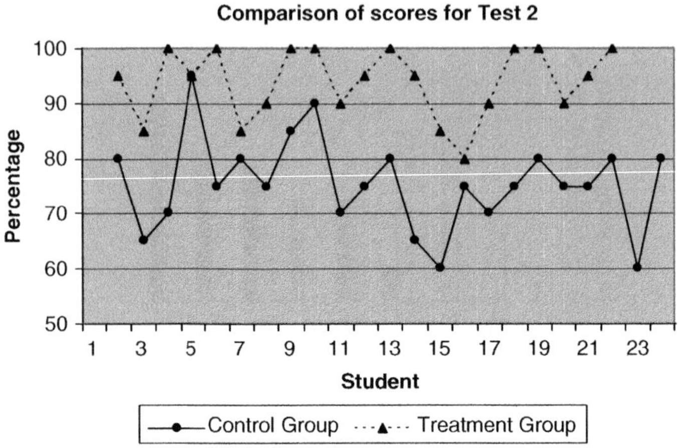

Fig. 9 Comparison of the performance of students in the Control and Treatment Groups for Test 2

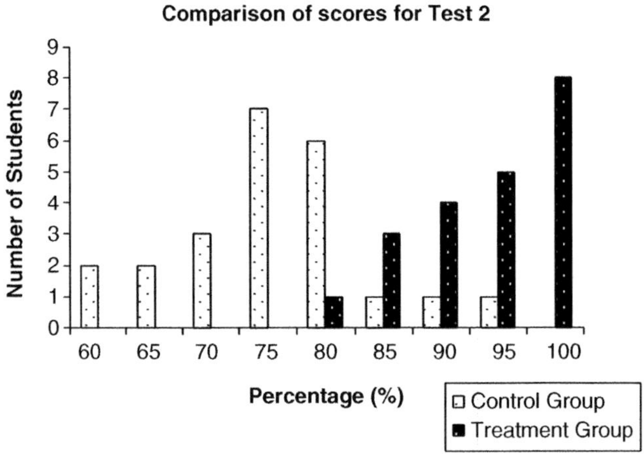

Fig. 10 Comparison of percentage scores obtained by the Control and Treatment Groups for Test 2

4 Conclusions

In conclusion, findings from the present study have clearly shown that the use of computer simulations in the teaching of "Atomic Structure and Bonding" had a positive impact on students, both in terms of their performance and their attitudes, as well as on their motivation during the lessons. Results from the achievement tests have undoubtedly indicated that students in the treatment group outperformed their friends in the control group. Thus, from the present research, it can be concluded that computer simulations were effective in enhancing acquisition of concepts involved in “Atomic Structure and Bonding” in the treatment group. Williamson and Abraham [13] have indeed reported that students who have been exposed to computer simulations and animations achieved higher test scores than those who had viewed still images of the same molecular-level animations. Our findings are also in accordance with Rieber [22] who postulated that animations enhance recall of facts, concepts, and principles, and with Sanger [19] who reported that computer simulations helped to improve understanding of science concepts. Results from our study also support findings from Condie and Munro [23] who have indeed highlighted that computer simulations can help to enhance understanding of abstract and microscopic science concepts and processes

Furthermore, our research has also demonstrated that students showed signs of greater motivation, interest and engagement during the lessons involving use of computer simulations. This fact is supported by Means and Olson [15] who noted that technology engaged students in challenging and authentic learning. Students from the treatment group have also asserted that they preferred computer simulations-based instruction over regular instruction; this finding lends support to Somekh [20] who has indeed highlighted that computer technology can provide strong visual elements that would be readily welcomed by teachers and students.

In view of the present results, it is worth noting that our traditional teaching practices that were developed well before the particulate state of matter was characterized are thus no longer sufficient to prepare students for the microscopic world of modern chemistry. Students need to become familiar with molecular-level structures and phenomena for a proper understanding of chemistry concepts. Consequently, the integration of technology is very important in the classroom; by providing students with different outlets to discover and use technology in the learning process, teachers would allow them to become more creative and more engaged in the lessons leading to meaningful and lifelong learning.

Acknowledgments The authors kindly acknowledge and thank Ray Le Couteur, the designer of the “Atoms, Bonding and Structure” software used. Abdoolatiff also thanks Ray Le Couteur for allowing her to use the software for classroom teaching.

References

1. Bowen CW (1998) Item design considerations for computer-based testing of student learning in chemistry. J Chem Educ 75:1172–1175
2. Ben-Zvi R, Eylon BS, Silberstein J (1986) Is an atom of copper malleable? J Chem Educ 63:64–66
3. Griffiths A, Preston K (1992) Grade 12 students’ misconceptions relating to fundamental characteristics of atoms and molecules. J Res Sci Teach 29:611–628
4. Gabel DL, Samuel KV, Hunn D (1987) Understanding the particulate nature of matter. J Chem Educ 64:695–697
5. Haider AH, Abraham MR (1991) A comparison of applied and theoretical knowledge of concepts based on particulate nature of matter. J Res Sci Teach 28:919–938

6. Abraham MR, Grybowski EB, Renner JW, Marek EA (1992) Understanding and misunderstanding of eighth graders of five chemistry concepts found in textbooks. J Res Sci Teach 29:105–120
7. Abraham MR, Williamson VM, Westbrook SL (1994) A cross-age study of understanding five chemistry concepts. J Res Sci Teach 31:147–165
8. Nakhleh MB (1992) Why some students don't learn chemistry. J Chem Educ 69:191–196
9. Tsaparlis G (1997) Atomic and molecular structure in chemical education: A critical analysis from various perspectives of science education. J Chem Educ 74(8):923–925
10. Piaget J, Inhelder B (1969) The psychology of the child. Basic Books, New York
11. Ausubel DP (1968) Educational psychology – a cognitive view. Holt, Rinehart & Winston, New York
12. Novak JD (1984) Application of advances in learning theory and philosophy of science to the improvement of chemistry teaching. J Chem Educ 61(7):607–612
13. Williamson VM, Abraham MR (1995) The effects of computer animation on the particulate mental models of college chemistry students. J Res Sci Teach 32:521–534
14. Russel JW, Kozma RB, Jones T, Wykoff J, Marx N, Davis J (1997) Use of simultaneous, synchronized macroscopic, microscopic and symbolic representations to enhance the teaching and learning of chemical concepts. J Chem Educ 74:330–334
15. Means B, Olson K (1995) Technology's role in education reform. http://www.ncrel.org/tplan/cbtl/execsum.htm. Accessed 17 March 2006
16. Ardac D, Akaygun S (2004) Effectiveness of multimedia-based instruction that emphasizes molecular representations on students' understanding of chemical change. J Res Sci Teach 41(4):317–337
17. Stieff M (2005) Connected chemistry – A novel modeling environment for the chemistry classroom. J Chem Educ 82(3):489–493
18. Vygotsky LS (1978) Mind in society. Harvard University Press, London
19. Sanger MJ (2000) Using particulate drawings to determine and improve students' conceptions of pure substances and mixtures. J Chem Educ 77:762–766
20. Somekh B (2001) The role of evaluation in ensuring excellence in communication and information technology initiatives. Edu Comm Info 1(1):75–101
21. Papert S (1980) Mindstorms, children computers and powerful ideas. Basic Books, New York
22. Rieber L (1990) Using computer animated graphics in science instruction with children. J Educ Psychol 82:135–140
23. Condie R, Munro B (2007) The Impact of ICT in Schools – A landscape review. Becta Research, Coventry

Networked Learning for an Online Coastal Zone Management Module

R. T. Ramessur

Abstract Networked learning for the online coastal zone management (CZM) module during the period 2004–2007 was interactive, flexible, promoting active engagement and social interaction and providing new opportunities for group working for students. Eighty percent of the CZM students actively participated in the asynchronous forums and regular contributions to online discussions were integral to the determination that a student was keeping pace with the learner-centered activities and achieving the outcomes of the module. About 25% of the messages were made by the course instructor when the asynchronous CZM discussion forums were operational during the 3-year period. The remaining 75% posts were from the students on the two asynchronous forums used for CZM. The asynchronous online forums used provided a good collaborative environment for multiple interactions. Most of the students following the CZM module had not come across the online mode of learning and were experiencing this change in the teaching and learning process for the first time. Over 75% postings in CZM forums were new questions with less than 20% postings of higher type questions for the CZM module during the 3-year period. As part of the quality assurance for the online course, students were requested to complete an evaluation questionnaire of the online course. Most students found that the online CZM forums were a good learning experience and that they benefited from it.

1 Introduction

The Declaration of Principles stated by the World Summit on the Information Society (WSIS, 2003) indicated the necessity to "share information and knowledge." It thereby places emphasis on the particular importance of Information and

---

R.T. Ramessur (✉)
Department of Chemistry, Faculty of Science, University of Mauritius, Réduit
e-mail: ramessur@uom.ac.mu

M. Gupta-Bhowon et al. (eds.), *Chemistry Education in the ICT Age*,
DOI 10.1007/978-1-4020-9732-4_11, © Springer Science + Business Media B.V. 2009

Communication Technologies (ICT) within the context of knowledge for development. But previous experience has demonstrated that the introduction of ICT-supported environmental education is either likely to fail or that the benefits will be limited if the introduction of the new technologies is not combined with the development of important skills to operate and maintain the new technologies. It has become necessary to develop the appropriate organizational infrastructures in which the newly acquired skills become embedded.

E-learning Africa has provided the venue for the creation of a new network of decision makers, helping to link representatives from government and public administration with universities, schools, governmental and private training providers, industry, and important partners in development cooperation in recent years. Flexible computer platforms enable learners to conduct experiments with other team members, over computer networks, and to participate in interactive simulations, all in a context of distance learning. In one of Saliah-Hassane's [1] research projects on electronic measurement and telemanipulation over computer networks, a procedure showing that ICT within a relatively short time frame can be used to create virtual laboratories with collaboration and data sharing for conducting experimental simulations with digital models and accessing the facilities of a technical or teaching assistance resource centre [1–2]. Kinshuk and Lin [3] further explored how to improve the learning process by adapting course content presentation to student learning styles in multi-platform environments such as PC and PDA. The asynchronous online forum described in this paper for the online coastal zone management (CZM) module provided a very good collaborative environment for multiple interactions for tutor–students and students–students unlike traditional classroom settings. This paper also examines student interaction in asynchronous discussion forums for 3 consecutive years and the instructional environment that need to be arranged so that students can critically reflect, discuss, generate, and exchange information and ideas during CZM networked learning as compared to studies done previously which focused on forums for a semester.

2 The Online CZM Module

The online coastal zone management CZM module which is equivalent to three credits (45 h) covers six CZM topics and runs over 15 weeks for final year undergraduate students. Assessment was made on three topic assignments including an EIA case desk study. Students were able to interact with one another and their tutor in an asynchronous mode through forums online on *i*-learn. Students were also assessed for their participation and contribution on online forums. The evaluation was based on three main criteria – reading of messages, confirming peers opinions, or adding value to any discussion threads.

For every activity of the CZM module at the University of Mauritius there was a forum at the disposal of the students. There was also a forum that addressed technical issues concerning the module. The web site of the virtual platform *i*-learn was

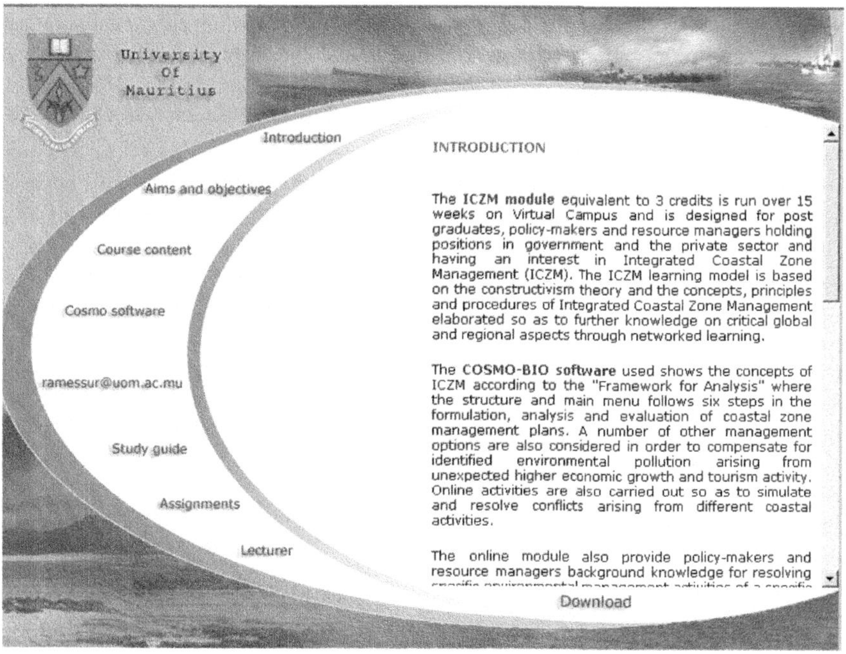

Fig. 1 Online CZM module on *i*-learn at the University of Mauritius

put at the disposal of the students for them to access the study guide and related information about the module (Fig. 1). The highly interactive electronic teleconference media, especially personal computers and audioconference media, allowed a more intensive, personal, individual, dynamic dialogue that can be achieved in using recorded medium and more effectively bridge the transactional distance [6].

A number of activities facilitated the CZM learning process after reading each topic including short questions for self-assessment and for later use in working the assignments. Self-assessment questions were also monitored by the system and feedback on students' participation and performance was directly available to the tutor as well as to the lecturer from the *i*-learn platform. A photo gallery was available online and linked to the various CZM topics in the module.

3 Networked Learning in CZM – Evaluation of Interactivity in CZM Forums

The separation of students and tutor in space and time during the CZM online module involved an asynchronous system where messages were posted on online forums for use by learners at different time periods. E-learning of CZM was flexible and adaptable so that learners could study anywhere and anytime. Networked learning during the CZM online module delivery at the University of Mauritius was interactive, flexible, promoting active engagement and social interaction and providing new opportunities for group working [4–8]. The online forums used in the CZM module provided a very good collaborative environment for multiple interactions unlike traditional classroom settings. The difference was that through the forums, students took the lead and viewed the lecturer as a resource and a guide. The lecturer monitored the processes of group and collaborative work easier since the contribution of each student was made more obvious when they worked online. An evaluation criterion was devised based on three aspects of online participation in the forums run over the 3 years – reading messages, replying (confirming or rejecting peer views), and adding value to the messages that obviously had a higher weightage. The average CZM online class during the period 2004–2007 consisted of about 20–30 students with more than 80% participating in the discussion forums. The tutor read the students' postings, provided answers to queries, added pedagogical comments, and provided advice.

4 Results and Discussion

During the 15 weeks that the asynchronous discussion forums were operational from 2004 to 2007, an average of about 60–75 posts from each cohort were made with about 25% made by the course instructor. The remaining 75% posts were from the students on the two asynchronous forums used for CZM. Distinguishing the postings that are first-degree replies from those that build on the other students' messages helped to assess what proportion of the postings were built on another students' message. A first-type contribution replied directly to the question posted in the discussion forums while a second-type contribution was a message that responded to another students reply (or first-type contribution) to the question posted online. A third-type contribution was a message that responded to students' second-type contribution and so on [8]. Over 75% postings during the 3 years were new questions with no replies and it was not surprising to see so many postings were new questions and first-type and less than 20% postings of higher type.

The notion of flexibility and autonomy was encouraged with the CZM online module and created independence among distance learners. The main defining feature of online delivery was the separation of tutor and learner, usually in both time and space. This separation fostered noncontiguous communication (communication that occurred between the learner and teacher from a distance), which had to be mediated. Consequently, mediated communication became the second defining feature of e-learning as mediated communication is an important feature in both e-learning and traditional distance education [9].

Interactions between the instructor and students and interactions among students with the CZM asynchronous online forum provided opportunities for an educational transaction. Without interaction, teaching becomes simply "passing on content as if it were dogmatic truth," and the cycle of knowledge acquisition-critical evaluation-knowledge validation is nonexistent. The use of new technologies during the CZM online delivery permitted live interaction and allowed for immediate feedback and interaction between teacher and student(s). Successful interaction in the mediated educational transaction was highly dependent upon how comfortable the learner felt in working with the delivery medium. The inability to achieve learner–interface interaction successfully can be a significant problem to those comfortable with technology yet unfamiliar with the specific communication protocols. Successful learner–interface interaction required the learner to operate from a paradigm that included understanding not only the procedures of working with the interface, but also the reasons why these procedures obtain results. An orientation session was important as a means of "climate setting" in order to introduce learners to the technology during the CZM module. The session which separated learning to use the delivery technology from the course content had several benefits: it did not compete with the course content, it helped ensure uniform minimal proficiency, and it removed the stress of graded performance.

5 Conclusions

Active and focused participation was an expectation of the online CZM module as regular contributions to online discussions were integral to the determination that a student was keeping pace with the learner-centered activities and achieving the outcomes of the module. Most CZM students make use of the Internet as it increases flexibility in regard to when, where, and how to learn, whereas mobile-learning is now able to support new trends in workplace learning and in fieldwork – particularly informal learning – by supporting the delivery of short pieces of content on demand and at the point of need. In order to monitor the very impact of e-learning/blended learning activities in CZM, a set of criteria has to be developed and implemented whereas electronic conferences can be designed to facilitate discussion among students. Videoconference is still highly inaccessible because of the cost of telecommunications involved in the process, while with the limited Internet bandwidth available in mainly developing countries and with reference to Mauritius, videoconferencing is still not a widely used feature [7]. The implied costs for 3G services are still too high for the wide adoption of this technology among the students. The problem that exists with the emergent technologies nowadays is that it seems imperative that there is a feeling that it is compelling to integrate every new technology in the teaching and learning process [6]. The availability of advanced mobile technologies, such as high bandwidth infrastructure, wireless technologies (WiFi, Bluetooth) and small handheld mobile devices (iPods, PDAs), has however started to extend e-learning towards m-learning [10]. Problems such as the structure of remote laboratory and the design of the graphical interfaces still need to be upgraded for dynamic object real-time control during field visits and lab experimentation in CZM. These objects can represent distributed laboratory measuring instruments and can be used in the context of various pedagogical scenarios.

Generic distributed graphical interfaces need to be designed and used together with Web services. Some functionalities from popular software in e-science such as Matlab or LabVIEW for real-time data acquisition with measurement instruments can also provide effective solutions while meeting the requirements of a collaborative remote laboratory platform in CZM where research can be planned to extend the platform in a grid as discussed by Saliah-Hassane [11]

References

1. Saliah-Hassane H, Saad M, Nerguizian C and Kodjo D (1997) "Environnements virtuels de formation et d'apprentissage sur le réseau Internet, du rêve à la réalité" Initiatives 97: Inforoutes et technologies de l'information, une chance pour le développement dans l'espace francophone. (AUPELF-UREF), Hanoï.
2. Saliah-Hassane H, Dumont-Burnett P and Loizeau C (2001) "Design of a Web-based virtual laboratory instrument measurement interface", Chronicle of Worldwide Innovations, iNEER, Preston King Station, Arlington, VA.
3. Kinshuk and Lin T (2004) Application of learning styles adaptivity in mobile learning environments. Third Pan Commonwealth Forum on Open Learning, 4–8 July 2004, Dunedin, New Zealand.
4. Ramessur R T and Santally M I (2003) Design and online mode delivery of a coastal zone management module using a constructivist approach. ICOOL Proceedings.
5. Ramessur R T and Santally M I (2005) Asynchronous Computer Mediated Conferencing for an Online ICZM Module. ICOOL Proceedings. WCCE 2005.
6. Ramessur R T (2006) Pedagogical aspects of computer mediated conferencing for an online CZM module. Arpege online Proceedings.
7. Ramessur R T and Santally M I (2007) Computer mediated communications for effective teaching-learning of coastal zone management module. International Journal of Education and Development 3(1).
8. Ramessur R T, Santally M I and Ah Fock S (2007) Comparison of asynchronous computer mediated communications for two online modules – CMCP and CZM. ICOOL Conference Proceedings.
9. McConnel D (1994) Implementing computer supported cooperative learning, Kogan Page, London.
10. Sharples M (2000) The design of personal mobile technologies for lifelong learning, Computers and Education 34:177–193.
11. Saliah-Hassane H, Fattouh B and Yan Y (2004) Grid enable technologies and web services for online laboratories. PROLEARN Online Experiments Workshop 2004: "Integrating European research in online experiments".

Moodle in Teacher Training

M. Miranowicz

1 Introduction

Continuous learning has become a natural element of every European's everyday life. It addresses the needs and expectations of contemporary labor market. Rapidly changing technology and continuously developing knowledge make continuous education a must in today's world.

The European Union believes that continuous learning constitutes an essential manner of building knowledge-based economy. This was clearly defined in the Lisbon Strategy as well as in the document on continuous development of education by 2009 issued by Polish Ministry of Education (MENiS) [1].

2 I*Teach Innovative Teacher

The project called *I*Teach (Innovative Teacher)* was introduced in the Fall of 2005 in order to meet the concrete needs of gaining skills and competencies which are essential to knowledge-based activities as well as independent continuous learning. The project aims at designing a set of practical methodologies, approaches, and tools to be used on a daily basis by trainers and teachers. The project answers the assumption of the Leonardo da Vinci project for the years 2005–2006 which is "continuous education of trainers and teachers" [2].

*I*Teach Innovative Teacher* is implemented within the pilot projects of Leonardo da Vinci program [3]. It is a common effort of academics and teachers from seven European countries, i.e., Bulgaria, Germany, Holland, Italy, Lithuania, Romania, and Poland. *I*Teach Innovative Teacher* is a research project into the application of IT tools in the process of learning/teaching in order to expand emotional skills related to cooperation, analytical, logical, and critical thinking, introducing

M. Miranowicz (✉)
Department of Chemical Education, Adam Mickiewicz University, Poznań, Poland
e-mail: groch@amu.edu.pl

M. Gupta-Bhowon et al. (eds.), *Chemistry Education in the ICT Age*,
DOI 10.1007/978-1-4020-9732-4_12,

teachers to e-learning as well as applying this form of teaching into the system of intramural and extramural education.

*I*Teach Innovative Teacher* is an educational pilot project. The aim of such projects is to improve the quality of learning and vocational education as well as career advising. It promotes designing educational products employing innovative methods of teaching applying the latest technological advancements. Particular support is offered to projects whose aim is to improve the quality of educational systems as well as vocational education, counteracting discrimination, increasing the awareness of cross-cultural differences as well as those projects which address the needs of employees moving from one country to another. All important information about the *I*Teach Innovative Teacher* project are published on portal http://i-teach.fmi.uni-sofia.bg/ (Fig. 1, Table 1).

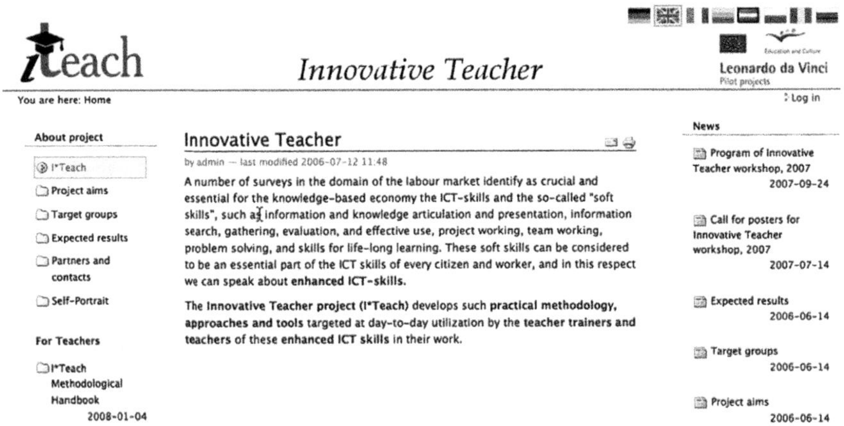

Fig. 1 I*Teach Innovative Teacher portal http://i-teach.fmi.uni-sofia.bg/

Table 1 Addresses of VTC of I*Teach Innovative Teacher project

Country	VTC address
Bulgaria	http://e-learning.fmi.uni-sofia.bg/moodle
Holland	http://www.utwente.nl/elan/iteach/
Lithuania	http://distance.ktu.lt/moodle/
Germany	http://nats-www.informatik.uni-hamburg.de:8080/iteach/moodle/
Poland	http://zdch.amu.edu.pl/moodle
Romania	http://i-teach.info.uaic.ro/moodle/
Italy	https://i-teach.unige.it/

*I*Teach Innovative Teacher* is a multistage and multilevel project. One of the levels covers training academic teachers, secondary school teachers, as well as trainers working for institutions dealing with teacher training (intramural and extramural education) which aim to apply e-learning in their everyday work.

*I*Teach Innovative Teacher* is a research project which serves to verify whether IT tools applied in the educational process can influence the expansion of *emotional* skills (analytical, logical, and critical thinking), *social* skills (communicating and negotiating with the other team members, teamwork, presentation skills, project work, problem solving), as well as *information-related* skills.

The project is designed for:

- Teacher trainers from universities and other institutions dealing with teacher training (intramural and extramural)
- Trainers and teachers of IT-related subjects (both intramural and extramural) working in secondary schools with certain educational profile (Mathematics profiled junior high schools) as well as those working for organizations and faculties teaching future educators
- The end users of this project, i.e., the students of the above secondary schools who take part in IT courses or study Information Technology
- Potential users of this project (trainers and teachers of all subjects) who intend to apply Information Technology in their work as well as people who might potentially use the project, i.e., students and participants of training sessions

The aim of the project is:

- To design a manual of practical methods, methodological tools as well as software for trainers and teachers to help their students acquire IT skills and competencies
- To create a teacher training curriculum which uses the methodologies presented in the manual as well as teaching advanced IT skills
- To design a multi-language online repository of contents which would constitute the source of essential and methodological knowledge for teachers
- To create Virtual Training Centers in project partner countries in order to offer continuous support to teachers and to train them to apply e-learning in intramural and extramural education [3]

3 Virtual Training Centers

In all of the project partner countries, Moodle-based Virtual Training Centers have been founded. The aim of these centers was to create a community of people interested in the project as well as provide a space where courses for teachers could be carried out. On one hand the courses allow teachers to get familiar with the methodology of the project and the manner of implementation of didactic research and on the other to present teachers with the operation of the Moodle and with the methodology of distance learning as well as to train them to use this form of learning in order to improve their own qualifications.

Virtual Training Centers do not have a homogenous character related to their form and contents. It was agreed that the local centers should address the needs of local teachers and educators and that all of the materials should be presented in local languages.

In Poland, on the distance learning platform E-education ZDCH (Fig. 2) set up by the Department of Chemical Education an e-learning course "Moodle Step by Step" has been published within the I*Teach Innovative Teacher project (http://zdch.amu.edu.pl/moodle/ Fig. 3.).

Its aim was to prepare teachers to create e-learning courses as well as to practically apply distance learning tools and methods in education. The course covers the basics of designing and administering course materials published on the Moodle, the methodology of e-teaching as well as preparing the final project. It takes 11 weeks for the participants of the course to learn the basics of Moodle work using the tools offered to design their own course which at the same time constitutes their final project (Figs. 4 and 5).

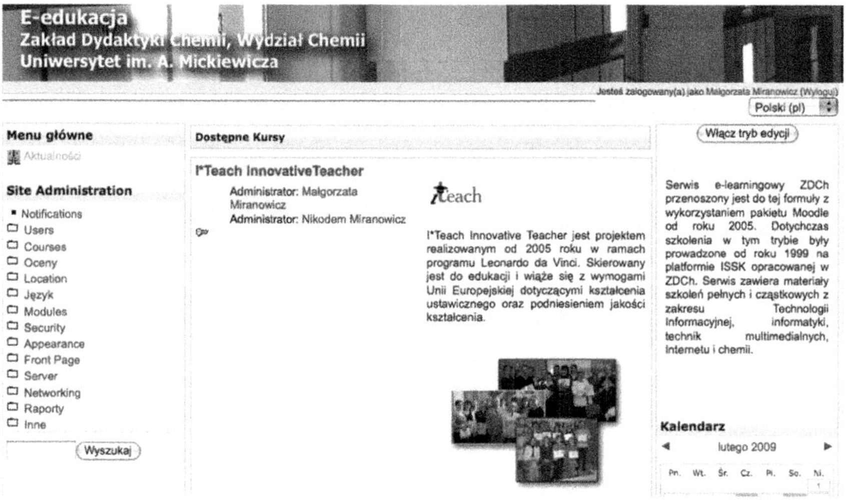

Fig. 2 E-education platform of ZDCH UAM Poznań

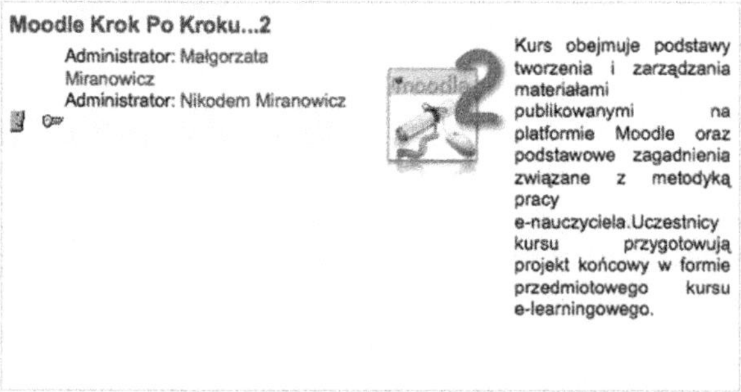

Fig. 3 Moodle Step by Step published on the ZDCh UAM Poznań platform

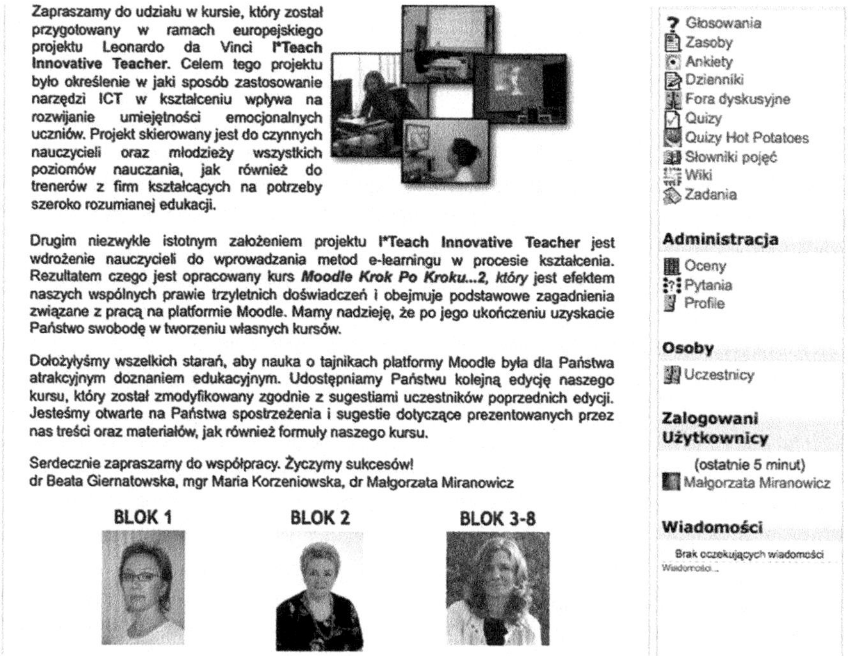

Fig. 4 Inside Moodle Step by Step course

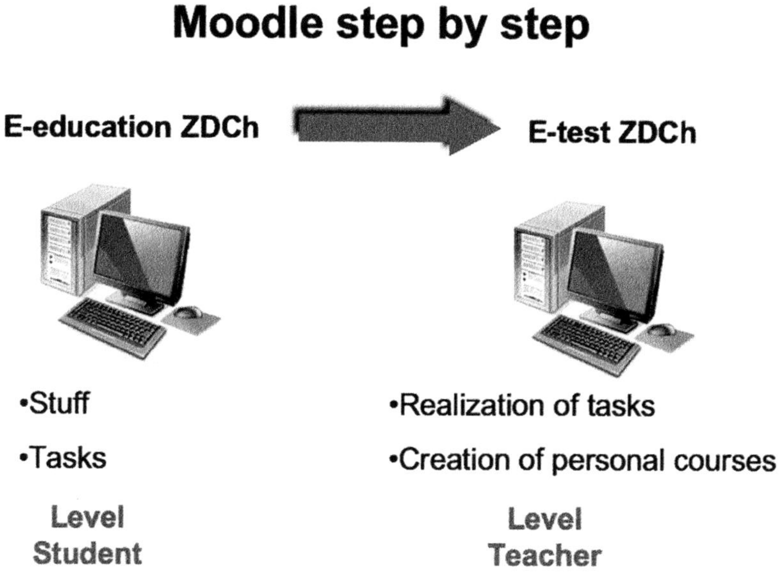

Fig. 5 Different access level on two different platforms

4 Conclusion

One of the main aims of any educational system is to prepare students for living in the contemporary world. Any changes in education are concerned with adjusting it to current conditions and the development of civilization where Information Technology occupies its special place. The students' ability to use communication and information technology in effect means that they will be able to use it in their adult and independent lives. These abilities, however, should be further developed and updated. The ability to learn continuously is extremely important in the context of information society and knowledge-based economy. Continuous learning constitutes an essential element of knowledge-based economy which has been stressed in the Lisbon Strategy as well as in the strategy of continuous development of learning until 2009 prepared by the Ministry of Education [4].

The use of Moodle within teacher training course proved to be extremely fruitful and yielded excellent results. For many participants it was the first opportunity to use this type of tool which resulted in creating their independent school platforms. Within the course very interesting materials were prepared which are currently used by teachers in their school work. E-education and E-test platforms provided by the Department of Chemical Education is at the participants' disposal. Almost 200 teachers have finished e-learning course Moodle step by step in Poland. Here are a few comments cited on the discussion forum of the course (Fig. 6):

> Thank you so much for the past weeks we spent in cyberspace☺!!!
> I have not learned with so much determination and pleasure for such a long time. I must admit that I got to know the world which had been completely unknown to me. So far

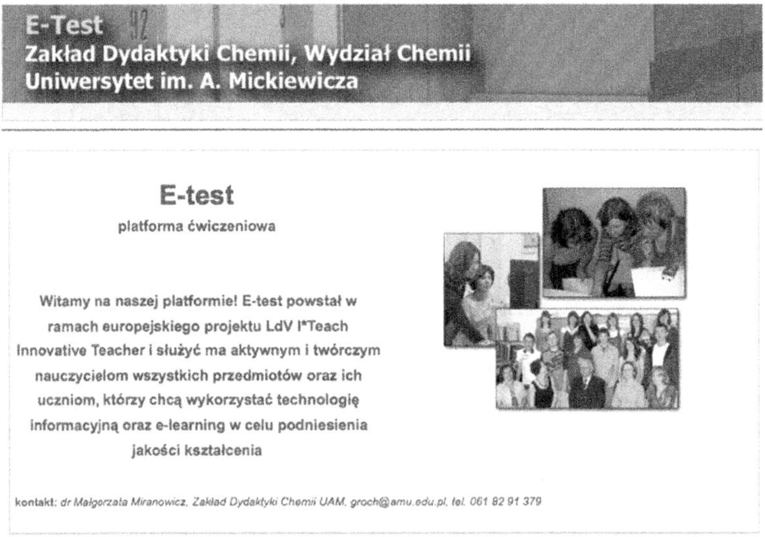

Fig. 6 E-test platform of ZDCh UAM Poznań

I have only been treating the computer as a tool for creating documents which made my work easier. Now I can do really great things here, the things that my students will definitely enjoy. This fascinating world has opened my eyes to a totally different perception of the Internet. I know that I will pop in here and use our resources and I might even let my students have a glimpse of that and I do hope that they will thoroughly enjoy it.

... [M]any teachers are afraid of the new claiming that this type of teaching does not suit their subject. I do think, however, that getting to know the new can lead to a better understanding. If there are three or four teachers who use e-learning then it will be enough for the students to encourage other teachers in their school to use it too. There are so many teachers (not all of them young), who will willingly apply this new teaching method – what they really need is to get to know it. And our administrators should see to it that there are more courses like ours.☺

Whilst preparing my students for Matura Exam I frequently use e-mail. My students send their work or type in their answers in Word documents which I prepare. I believe that the course we are trying to create here will be an excellent addition to the optional classes students participate in before their most important exam. During regular classes we could either solve typical tasks or those which pose most difficulties. Additional knowledge might be acquired in the system of e-learning. This would allow each student to work at their own pace and most convenient time. When in doubt, students can talk about it on the forum or take part in a chat. And the issues that seem difficult for most students could be explained in class again. In addition to this, the method is great for those graduate students who want to improve their previous year results as it is impossible for them to participate in regular lessons at school.

References

1. Delors J., *Learning. The Treasure Within* Report to UNESCO of the International Commission on Education for the 21st Century, UNESCO Publishing, Paris (1996)
2. http://ec.europa.eu/education/programmes/leonardo/objectives_en.html
3. I*Teach Innovative Teacher portal, http://i-teach.fmi.uni-sofia.bg/
4. Rodrigues M. J., European Policies for a Knowledge Economy, Edward Elgar, Cheltenham. 2003

SIA@SFU (Science in Action at Simon Fraser University)

S. Lavieri and J. P. Canal

Abstract SIA@SFU is an innovative, interdisciplinary outreach program that offers high school students a Science Immersion Day in subjects ranging from chemistry to biology to physics. High school students who may have limited exposure to the sciences spend the day interacting with scientists and performing experiment in a laboratory setting. This paper will illustrate our experiences, the benefits and the challenges of developing and starting a multidisciplinary outreach program.

1 Introduction

Science in Action at Simon Fraser University (SIA@SFU) is an innovative science outreach program that is being developed by volunteers in the Faculty of Science at SFU. The program consists of a science immersion day with tailored hands-on activities in chemistry, biology and physics for students in grades 8 through 12 and is provided at no cost to the schools. Although SIA@SFU primarily targets high school students who have limited exposure to the sciences, such as students in inner-city and aboriginal schools or those in outlying communities that do not have ready access to a university environment, the program is open to all schools. Students have the opportunity to carry out hands-on experiments designed to stimulate their interest and encourage them to take science courses in high school and university and ultimately pursue a science-related career.

SIA@SFU started over 2 years ago as CIA@SFU (Chemistry in Action at Simon Fraser University). As per the request of numerous high school teachers, it is developing into a collaborative program between different departments in the Faculty of Science at SFU. Described is the present status of our program, which is evolving to better serve the needs of the students.

S. Lavieri and J.P. Canal (✉)
Department of Chemistry, Simon Fraser University, Canada
E-mail: jcanal@sfu.ca

M. Gupta-Bhowon et al. (eds.), *Chemistry Education in the ICT Age*,
DOI 10.1007/978-1-4020-9732-4_13,

2 Methodology

2.1 Typical Day

The SIA@SFU program is tailored to match the curriculum of the participating school. The students arrive at SFU at 9:30 a.m. and are welcomed by a faculty member. After an introduction with short demonstrations, students perform science experiments in a laboratory (2 h), then complete and hand in a report sheet. The students are assisted by faculty members, graduate and undergraduate students at a student teacher ratio of 10:1, with as many as 150 students participating in one session.

A supervised lunch is provided to the students and they assist in making their own ice cream [1], as an example of everyday science. After lunch, the students have a guided tour of the science departments and other places of interest at SFU. The students have the opportunity to interact with graduate students and find out what it is like to do graduate studies. Following this, the students attend a 1 h session of chemistry demonstrations conducted by a faculty member in a lecture hall. Before leaving, the students receive a sponsors' science-related memento.

The experiments performed by the students include both hands-on and observational activities. These activities are adapted to the educational level of the class and what they are currently discussing in their science lessons. The experiments and presentations engage the students in the scientific method by creating hypotheses, testing their predictions and presenting their findings to the class. The response to this format has been overwhelmingly positive and over 2,000 students who have already participated have been very enthusiastic about their experiences [2].

Experiments, such as qualitative analysis of ions via precipitation with a cation or anion [3], the isolation of banana (or strawberry) DNA [4], creating a DNA alias [5], coagulation of milk [6], examining reactions for exothermic or endothermic properties [7], Tollens test for aldehydes [8] and the conversion of copper pennies to 'silver' [9] have been completed by the participating students.

3 Results and Discussion

3.1 Target Students

Schools that are not frequently exposed to these types of experiences, in particular due to economic status and location, are specifically being approached to participate in this program. While many students in schools from larger urban areas have access to laboratory facilities, many of the less wealthy, inner city, aboriginal and outlying small community schools have no such facilities. It is thus possible for a student in science to graduate from high school without ever having conducted an experiment in a wet chemistry, biology or physics laboratory. The strain on the teachers' time and resources makes it difficult for teachers in these areas to show their students that science is interesting and accessible. Many students come to the conclusion that science is 'just too difficult' for them, without ever having the experience of actually trying it. Because jobs for scientists are rare in smaller communities, many young Canadians never get to meet a 'real' scientist; thus the stereotypes of the mad scientist, the absent-minded professor and the science geek persist, which may discourage some students from continuing in science beyond the most basic requirements for their high school degree [9]. It becomes even more difficult to recruit students into science programs when they come with pre-existing notions of science being inaccessible and irrelevant to them.

The SIA@SFU program strives to counter this. Through the experiments students get real hands-on experience and are able to interact with undergraduate and graduate students, as well as the post-doctoral researchers and faculty members. Given the diverse background of the volunteers, it is our goal that students become aware that the sciences are available to all and that various levels of training and many different areas of expertise are required in the scientific field.

4 Experiences

To date, the feedback from the students and teachers indicates that the SIA@SFU program is well received and greatly impacts the students [2]. Several schools, which previously attended one session, are already scheduled to return in the following academic years, thus our program has become part of their curriculum. It is too early to tell whether the SIA@SFU program will increase the number of students enrolled in university science degree programs, but teachers at the high school level have noticed an increased interest in the sciences of their students after their participation.

SIA@SFU will also provide ongoing support for students and teachers through the SIA@SFU web site (www.siasfu.ca), which will grow to contain links to interesting and age-appropriate web sites for chemistry, biology and physics, including virtual experiments. The web site will contain helpful information for science teachers and experiments that can be conducted at home or the classroom with commonly available materials.

The development and implementation of the SIA@SFU program has required a great deal of effort. If not for the willingness of the members of the science departments to volunteer their time and the support of the departments' chairperson, as well as the Dean of Science to donate the space and resources, this program could not have run. Many undergraduate and graduate students now consider this program as an important volunteering opportunity.

Time must also be taken to reassess the program to better serve the students. Through the experience of running many sessions, feedback from the teachers and the suggestions received, the program is able to grow.

The need for outreach programs clearly exists. Solely by word of mouth, the number of schools requesting to participate has ballooned. To date, no school has been turned away. As we are striving to reach students in less wealthy communities, we do not charge the school to participate. The need for funding clearly exists and we are actively pursuing external grants. As the program grows and its needs become more evident, further sources of funding will be required.

5 Conclusion

SIA@SFU is a science outreach program that has the opportunity to impact groups of students who would not ordinarily come into contact with a scientist in their daily lives. By engaging these students through a Science Immersion Day, the SIA@SFU program strives to stimulate their interest in the sciences, encourage them to take science courses and possible pursue careers in this field.

Acknowledgements A very special thank you to all the volunteers (faculty, staff, graduate and undergraduate students) of the SIA@SFU program, who freely donate of their time and expertise. We would also like to thank the National Science and Engineering Research Council of Canada (NSERC) Pacific, NSERC PromoScience, the Faculty of Science at Simon Fraser University (SFU) and the Departments of Biology, Chemistry and Physics at SFU for funding.

References

1. Thomas Jefferson National Accelerator Facility – Office of Science Education (2008) http://education.jlab.org/beamsactivity/other/ln2icecream/index.html. Accessed 21 April 2008
2. http://www.siasfu.ca. Accessed 21 April 2008
3. Chemistry 121 Laboratory Manual Version 1.7. Simon Fraser University, Burnaby (2006/2007)
4. http://www.rhfleet.org/site/education/BananaDNAExtraction.pdf. Accessed 21 April 2008
5. DNA Detectives What is your DNA alias? http://www.nature.ca/genome/05/051/0511/0511_m205_e.cfm. Accessed 21 April 2008
6. Van Cleave J (1999) Icy, freezing, frosty, cool & wild experiments. Wiley, Toronto
7. Holmquist D, Volz D (1997) Chemistry with computers. Vernier Software, Portland
8. Kemp M (1981) Silver Mirror. J Chem Ed 58:655–656
9. Szczepankiewicz SH, Bieron JF, Kozik M (1995) The "golden penny" demonstration. J Chem Ed 72:386–388
10. Schwarcz J (2008) An image problem. Can Chem News 60:12–13

Context and Chemistry Going Dutch? The Development of a Context-Based Curriculum in the Netherlands

J. H. Apotheker

Abstract In the Netherlands a new context-based curriculum has been designed to cope with the problem of the diminishing attention for chemistry as a subject. The curriculum consists of a number of modules in which a context is introduced. Based on scientific questions related to the context chemical concepts are discussed with the students. An experiment preliminary to implementation is being carried out in 20 schools. The design of the curriculum and the first results of this experiment are presented in this paper. A number of problems have been encountered. It seems possible to attain the ultimate goal of nationwide implementation in 2012.

1 Introduction

1.1 Not Enough Science Students in Europe

There has been a large decline in the number of students interested in taking up a study of sciences or mathematics. In a recent report [1] produced for the European Communities the authors indicate that action is needed to reverse that trend. They propose a renewal of the pedagogical practices used in the schools. A change from the normally used deductive method to an inquiry-based method. They also assume that teachers play a key role in the renewal of science education: "*Networks of teachers stimulate teacher's professional development ... and stimulate morale and motivation.*"

In a report to the Nuffield foundation [2], the authors formulate a number of recommendations about the scope of science education in secondary schools. Apart from a function in addressing the problems of low student motivation science education

J.H. Apotheker (✉)
Department of Chemistry Education, University of Groningen, The Netherlands
Member Steering Committee, 'nieuwe scheikunde', The Netherlands
e-mail: j.h.apotheker@rug.nl

M. Gupta-Bhowon et al. (eds.), *Chemistry Education in the ICT Age*,
DOI 10.1007/978-1-4020-9732-4_14, © Springer Science + Business Media B.V. 2009

should "*educate students both about the major explanations of the material world that science offers and about the way science works.*"

1.2 Dutch Analysis

In the Netherlands, Dutch chemistry education was analyzed by a committee [3] that reported a number of problems similar to those elsewhere in Europe [1]:

- The image of chemistry is negative.
- There is no relationship between the content of chemistry in secondary education and chemistry in research and industry.
- The current chemistry curriculum leaves teachers and students not enough time to make chemistry education more challenging and interesting.

1.3 Dutch Proposal

This committee advised the installation of a second committee that should propose a renewal of chemistry education in the Netherlands. This committee reported in 2003 [4] a context – concept-based approach for chemistry education. Contexts are to be used as a bridge between the material world and chemical concepts. The committee stressed the role of the teachers as designers of education. The new curriculum is to be developed in small networks of teachers. The committee advised the installation of a steering committee aimed at the development of a new chemistry curriculum.

1.4 New Dutch Curriculum

The steering committee was installed in 2003 and has been working with seven teacher networks in the development of the new curriculum. Chemistry is taught in the last year of the first phase of secondary education. In the second phase it is an optional subject for students. In Fig. 1 the years in which chemistry is taught are indicated.

The steering committee has formulated a number of basic principles, to be used for the development of the new curriculum: It should be

- Based on a context concept approach
- Based on achievement of scientific literacy
- Developed by networks of teachers coached by professionals

In September 2007 enough material was developed to be able to start an experiment in 20 schools. In these schools the new curriculum is to be tried out both in the pre-vocational as well as the pre-university stream of secondary education. In this paper the results of the first year of this experiment are presented.

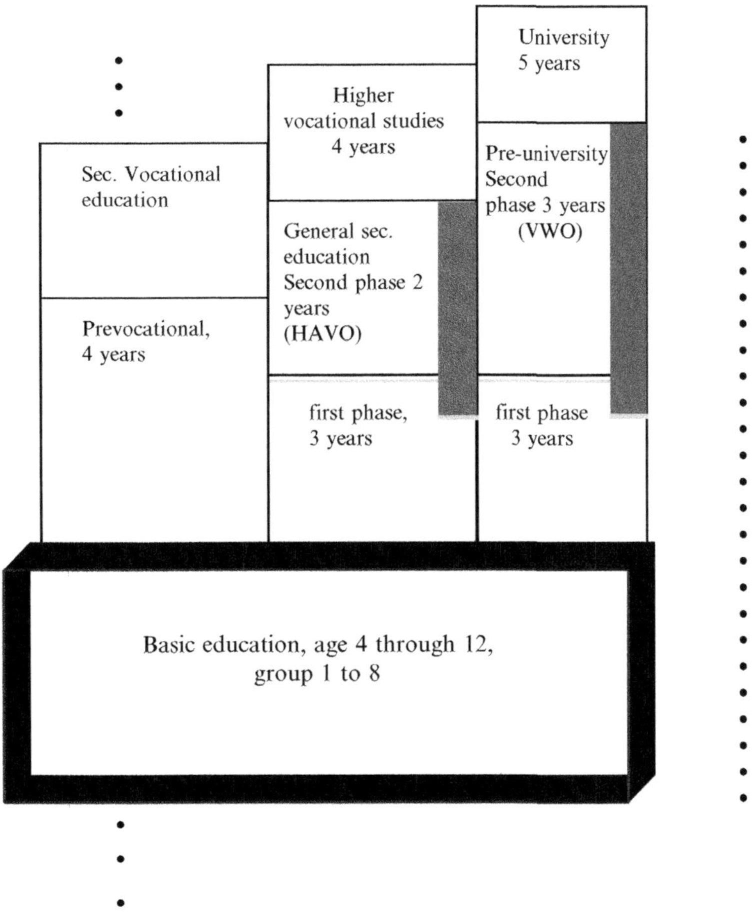

Fig. 1 The Dutch educational system. The years chemistry is taught are indicated in grey (last year of first phase and second phase)

2 Methodology

2.1 Design New Curriculum

Goodlad et al. [5] have analyzed the steps involved in the introduction of a new curriculum. In Fig. 2 these steps have been visualized. Usually the ideals for a curriculum are formulated, this is then formalized in a set of rules, by the relevant authorities. These rules are then interpreted by the teachers and used to design the lessons in the classroom. In the classroom, students experience the curriculum. The learning outcome finally is the attained curriculum. All these steps involve interpretations by the individuals working within each set. The difference between the ideal curriculum and the attained curriculum therefore can be quite notable.

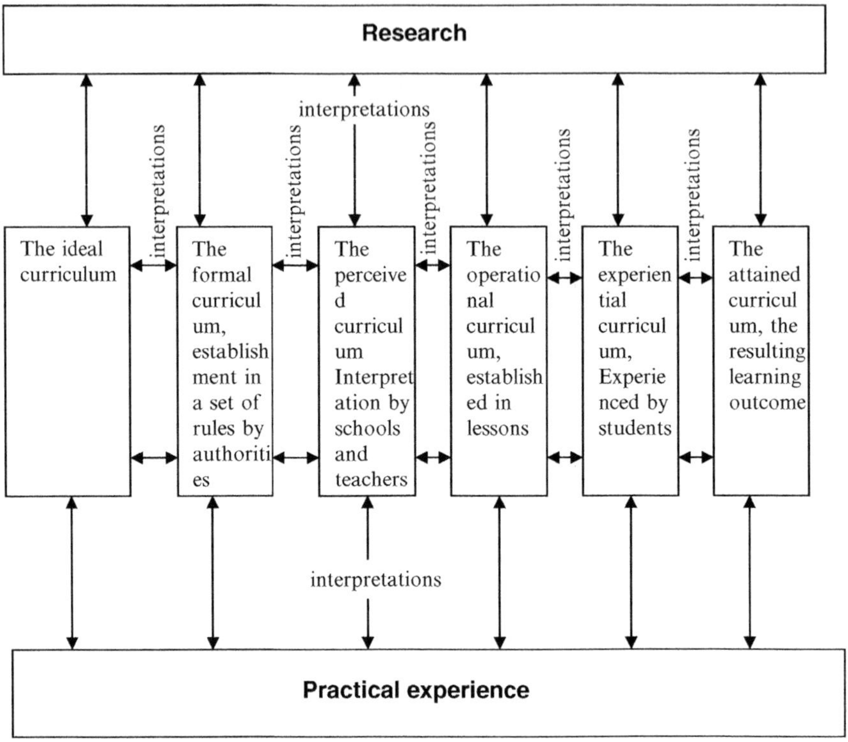

Fig. 2 Curriculum development based on Goodlad [5]

In the design of the curriculum the aim was to make the difference between the ideal curriculum and the operational curriculum as small as possible. A number of basic principles were formulated, after which networks of teachers were asked to develop material for the new curriculum. This way a direct link between the ideal and the operational curriculum was established. This material was to be tested in class, so that the attained curriculum could be determined. The material was then adapted.

2.2 Principles of the Designed Material

The material produced was based on the work done for "Chemie in Kontext" [6]. The structure chosen there is comparable to that of the 5 E model as formulated by Bybee [7].

Schematically the model has the following phases:

- Engage

  Teacher engages the students by introducing the context.

- Explore

Teachers and students explore the scientific questions students have concerning the context and plan on how to answer them.

- Explain

Students explain the answer to the questions by acquiring the scientific knowledge needed to answer the questions posed.

- Elaborate

Teacher and students elaborate the knowledge the students have by using the knowledge in a different context.

- Evaluate

Teacher and students evaluate how knowledge was obtained and whether the answers and the procedure were satisfactory.

This scheme was adapted as indicated in Table 1.

2.3 *Basic Structure of Learning Lines*

For the curriculum a basic structure for the learning line was developed in which two modules, each about 12 lessons, are followed by 3–6 lessons called a bridge. This gives teachers the opportunity and freedom to decide for themselves the sequence of the modules. The bridge is used to discuss the knowledge gained by the students. Students also have the opportunity to further develop skills acquired in the modules. Finally the bridge can be used to demonstrate that the knowledge gained can be used in a different context. At the end the bridge is used as a step-up to the next two modules.

In Table 2 a sequence of modules used in this experiment is given.

Table 1 Phases in a module

Phases in module		
Phase 1: Engage	Introduce context	Formulate context question
Phase 2: Explore	Derive relevant scientific questions	Define needed knowledge
Phase 3: Explain	1. Collect knowledge, acquire skills 2. Exchange and scaffold knowledge 3. Answer context question	
Phase 4: Explore	Can other context questions be answered?	connection to following module

Table 2 Example of a sequence of modules used in the second phase of secondary education (years 4, 5, and 6)

Period	Year 4 prevocational	Year 5 prevocational	Year 4 pre-university	Year 5 pre-university	Year 6 pre-university
1	Unbreakable mugs, ceramic materials	Antibiotics, synthesis in industry	Perfume (esters, alcohols, carboxylic acids)	ECO-travel (stoichiometry, biotechnology)	Chemistry in the mouth (buffers, receptors, polymers in dentistry)
2	Superslurpers (polymers, special side groups)	Green chemistry	Growing (salts, fertilizer and pesticides)	Energy to take away (redox)	Gasification (technology, alternative fuels, electro. technology)
3	ECO-travel (stoichiometry, biotechnology)	Chemistry of life (biochemistry)	Swallow or shoot, The route of Medicine (influence of pH on organic substances)	Smart materials	Own research project
4	Growing (salts, fertilizer and pesticides)	Training for central examination	Artificial sweeteners, stereochemistry, peptides and biotechnology	Moving chemistry, (polymers energy in the cell)	Exam training
5	Nobel prize, (the atom and the periodic table)		Nobel prize, (the atom and the periodic table)	Green chemistry	

2.4 *Experiment*

In September 2007, 20 schools started with the new curriculum in the last years of secondary education. Schools were grouped in three more or less geographically oriented groups. A professional coaches each group. Each group was able to choose its own learning line. The group coached by the author chose the line presented in Table 2.

During the run of the experiment coaches have the task to survey the problems encountered. Together with the teachers and other coaches solutions were found.

This curriculum is supposed to be introduced nationally in 2012. The ultimate goal of this experiment is to formulate a curriculum that can be used by most teachers in the Netherlands. Up to now teachers used two or three modules in a school year. In the experiment only modules and bridge lessons are used.

Specific questions in this experiment to be answered are:

- Implementation of the curriculum
  - In what way are teachers and students coping with the new curriculum?
  - Which problems arise and how are they solved?
  - How does the curriculum fit in the school organization?
  - Are the basic principles observed?
- Results of the new curriculum
  - Does it result in a coherent curriculum?
  - Is the program suitable for all teachers?
  - Does the program induce more students to choose chemistry as a subject?
  - How do teachers and students evaluate the material?

These questions were discussed with the teachers in meetings with the coach and during meetings in which all 20 schools met together. From these meetings a number of conclusions can be drawn.

3 Results and Discussion

3.1 *Implementation of the Curriculum*

3.1.1 In What Way Are Teachers and Students Coping with the New Curriculum?

Both teachers and students need time to adapt to this way of working. One comment that was received mentioned the difficulty in helping weak students during extra lessons.

A number of teachers were not used to working with groups and had no experience with cooperative learning methods that were used. During the coach meetings they received additional training. Both students and teachers have little experience working this way.

Students enjoyed working with the material. They were interested in the subject and worked hard as a rule in class. Outside the classroom it was difficult to get them to do homework.

Teachers indicate that students learn a lot in a short time. One of the problems signaled is that because of this, students don't practice their skills enough.

Students take more responsibility in class. Time differentiation and level differentiation is possible. It takes time for the students to get used to this way of learning.

Coaches report that teachers have problems at the start of the experiment. Quite a number of things needed to be adapted. This took more time and effort than the teachers expected. The teachers felt that the material was not always good enough. It took an extra effort to adapt the material to the needs of the teachers. Most often this concerned the texts that were produced to give information to the students. In another case the experiments were deemed to be too much cook book and not open enough.

3.1.2 Which Problems Arise and How Are They Solved?

The material that was given to the teachers was still fresh. It still needed editing. Teachers felt they had enough opportunity to adapt the material to their own wishes. This took an extra investment of time.

Even though students were enthusiastic about the subjects they scored low on tests. During the evaluation teachers suggested this was caused by a lack of practice in the skills needed to solve problems. For example in the module "growing" precipitation reactions are introduced. Students do write a number of equations but they are not practiced as much as they would have otherwise.

Coaches felt most problems were solvable. They expressed the opinion that the problems signaled will lead to a better product and ultimately to a workable curriculum.

3.1.3 How Does the Curriculum Fit in the School Organization?

The organization of the lessons took more time. Because not all students do the same experiment at the same time, the organization is more complicated. The organization of different experiments did not lead to major problems however.

The number of students in each class is large (between 25 and 30). In some cases six or seven classes were doing the same module. This factor did result in organizational problems in the laboratory.

3.2 *Are the Basic Principles Observed?*

3.2.1 Teacher as Designer of Education

Teachers indicate they have enough room to adapt the material, or choose their own module. They indicate it is important to be consequent in the pedagogy chosen.

3.2.2 Need to Know

The "need to know" is adequately taken up by the students. They get the idea that the knowledge they acquire helps answer the context questions posed.

3.2.3 Macro Micro

Connecting the properties on the macro level and the properties on the micro level is also achieved. The focus of the first two modules in HAVO is specifically aimed at this aspect. In unbreakable mugs students learn about the relationship between properties on the macro level (breaking) and the micro/meso scale (meso structure of clay particles). In the second structure in which polymers used to absorb water in diapers is discussed, students learn about the molecular structure of the polymers.

3.3 *Results of the New Curriculum*

3.3.1 Coherent Curriculum

Teachers report that students get a much better idea of the role of chemistry in society. Contexts are effective in engaging the students. Students are confronted with a lot of knowledge in a short time. They learn a lot. There are still problems however. Using the knowledge gained in other contexts needs to be established more firmly. There is a need for more practice with skills like calculations in stoichiometry, writing equations in precipitation reactions or acid–base reactions.

Teachers and coaches have the idea that a coherent program is achieved. Students are able to connect their chemical knowledge to societal issues. They learn to communicate about chemistry. They obtain an adequate basic knowledge of chemical methodology and chemical principles to be able to continue in further education.

3.3.2 Is The Curriculum Suited for All Teachers?

Van Berkel [8] discusses different pedagogies used by teachers. Teachers have been used to what he calls: "correct explanation and solid foundation." Teachers build up a set of knowledge that can be used to explain the material world. The concepts play a central role. After enough knowledge is acquired the material world is introduced. In the new curriculum "Science Technology and Society" and "History and Philosophy of Science" are used. Science and Technology uses a context as an introduction and reason to find new knowledge to be able to understand the scientific aspects of the context. In "Philosophy and History of Science" the development of concepts in history is used as a context. The development of concepts follows the historical development closely. Contexts are used as an introduction for concepts.

This changeover from one pedagogy to another leads to problems for the teachers. They are not always familiar with collaborative groups. The experiment also showed teachers need some extra training before they start working with the modules. In this experiment the training had a positive result on the attitude of the teachers.

4 More Students

After 1 year it is not yet possible to conclude that this program yields more students. Students are much more enthusiastic about chemistry than before. In one particular school 90% of the students chose chemistry as a subject. It is however too early to claim a significant increase in the number of students.

5 Evaluation

In the meetings teachers reported a number of positive and negative points. In the planning for the next school year however they decided to plan the same program, with minor changes. This indicates they are satisfied with the way things went in the past year. The material still needs a lot of work. Texts that were used need to be edited. A number of the practicals need revision so they can become more open.

6 Conclusion

The start of the experiment demonstrated a number of problems. The material that was offered to the teacher and students did not have the quality it needed to have. After this was changed both students and teachers were more satisfied. Especially in HAVO, material needed to be more structured.

The experiment also showed teachers need some extra training before they start working with the new program. A professionalization program is needed when the new curriculum is implemented. In this experiment the training had a positive result on the attitude of the teachers.

Contexts are a good way to engage students. They get a good impression of the role of chemistry in the material world. Students learn a lot in a short time. More time however is needed for practicing skills. This has partly been found in the bridge lessons.

As expected a number of problems were encountered with the start of the curriculum. Most of the problems have been solved or are expected to be solved. A separate research group will carry out the evaluation of this curriculum. This evaluation will be used to determine the final version of the new curriculum.

Acknowledgments This paper is based on a multitude of material produced in the past year in connection with the experiment by a large group of people

References

1. Rocard, M. and Csermely, P. et al. (2007). *Science Education Now: a renewed pedagogy for the future of Europe*. Luxembourg, Office for official publications of the European Communities.
2. Osborne, J. and Dillon, J. (2008). *Science Education in Europe: Critical Reflections*. London. The Nuffield Foundation.
3. Koten, G. v., De Kruijff, B., Driessen, H. P. W., Kerkstra A. and Meinema H. A. (2002). *Bouwen aan scheikunde*. Enschede, SLO. http://www.slo.nl/themas/00051/Map9/Map1/. Accessed 2 July 2008.
4. Driessen, H. P. W. and Meinema, H. A. (2003). *Chemie tussen context en concept*. Enschede, SLO. http://www.slo.nl/themas/00051/Map9/Map1/. Accessed 2 July 2008
5. Goodlad, J. L. and Associates. (1979). *Curriculum inquiry: The study of curriculum practice*. New York, McGraw-Hill.
6. Parchmann, I. and Ralle, B. (1998). Chemie in kontext- ein konzept zur verbesserung der akzeptanz von chemieunterricht? In A. Kornetzt (Ed.), *Chemieunterricht im spannungsfeld gesellschaft-chemie-umwelt* 12–24. Berlin, Cornelesen Verlag.
7. Bybee, R. W. (2002). *Learning science and the science of learning*. Arlington, VA, NSTA Press.
8. Berkel, B. van. (2005). *The structure of school chemistry*, PhD thesis, University of Utrecht.

Using New Technologies in Teaching Chemistry

H. Gulińska

Abstract A well-designed combination of various elements of imaging and visualization, high level of interactivity, as well as creating associations between the learners' emotions and impressions is key to effective Chemistry teaching. The above assumptions are all implemented in the manual *Absorbing Chemistry* whose integral part are CD-ROMs. The handbook is to be used by teachers of junior high schools whose students are 13–15 years of age (Fig. 1). Each multimedia unit found on the CD contains the following elements: animation and simulations of chemical processes and phenomena; dynamic 3D models of chemical compounds; video presenting, among others, the course of chemistry experiments; tests and problem-solving tasks.

1 Multimedia Handbooks in Teaching Chemistry

A good example of introducing new technologies to teaching chemistry in Poland can be found in *The Interesting Chemistry* handbook (Fig. 2), which is supplemented with integral CDs. The aim of the handbook is to facilitate teaching junior high school students in Poland, aged 13–15 [1]. It was prepared according to the latest scientific and didactic trends; therefore, teaching was based primarily on experimental work of teachers and students. The fundamental assumption of *The Interesting Chemistry* cycle is to make the most of junior high school students' inherent curiosity, in order to stir their interest in chemistry and its significance in everyday life. Inspiration for practical activities can be found in the films included in the CDs, which illustrate the course of many experiments and present students with animation of the observed phenomena.

Each chapter of *The Interesting Chemistry* opens with a question which students, guided by the teacher, try to answer, using both printed and multimedia components of the handbook. Students can find additional aids in color illustrations,

H. Gulińska (✉)
Adam Mickiewicz University in Poznan, Faculty of Chemistry, Poland
e-mail: gulinska@amu.edu.pl

M. Gupta-Bhowon et al. (eds.), *Chemistry Education in the ICT Age*,
DOI 10.1007/978-1-4020-9732-4_15,

Fig. 1 The aim of the handbook is to facilitate teaching junior high school students in Poland, aged 13–15

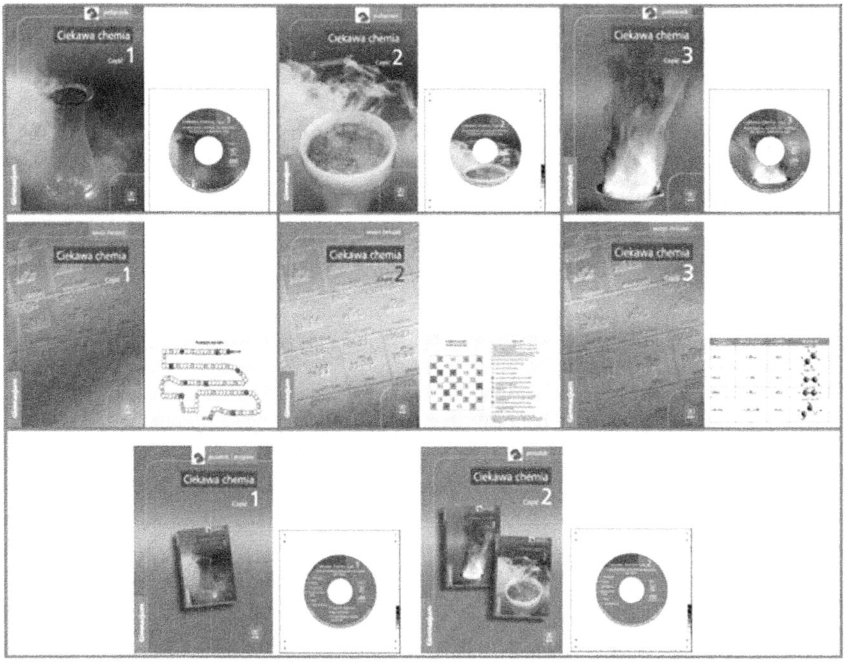

Fig. 2 Polish handbook – *The Interesting Chemistry*

pictures, tables, diagrams, and models, including models of processes occurring in the environment. The proposed tasks and problems are above all practical. Each chapter contains modules *Let's Learn Together* and *Let's Learn Actively* which encourage students to individual, extracurricular learning at home. Such modules as *Key Terms* and *Recapitulation* help students to logically revise newly acquired knowledge, while modules *Solving Tasks and Problems* and tests enable them to assess themselves after such a revision. The handbook contains icons referring students to electronic resources on the CD which they can use according to their own needs. The CD contains diversified teaching aids, such as illustrations with voice-over captions, three-dimensional models of elements and chemical compounds, films, animations of phenomena and processes, arithmetic tasks, tests (e.g., multiple choice tests), a periodic table, a dictionary, as well as various educational games which additionally activate educational processes and – as the research shows, help accomplish educational goals. Exercise books correlated with *The Interesting Chemistry* handbook aim to organize students' work during chemistry classes and at home. The tasks included in the exercise books are related to everyday life situations, thus becoming involving even for those students who demand additional pedagogical assistance. Problem-solving tasks inspire students to use their knowledge extensively. Each thematic section finishes with a subsection *Young Experimenter's Sheet*, in which students are asked to conduct, describe, and sum up a simple experiment. These diversified tasks allow students to revise knowledge conscientiously, in an attractive and stress-free way, using for instance board games. *The Interesting Chemistry* package also includes Teacher's Books which comprise a curricular program and a syllabus for teaching using various educational paths. Additionally, each lesson is supplied with a scenario containing a list of essential didactic means, a lesson plan, remarks on handbook and exercise book tasks, as well as proposals of experiments which can be conducted using simple research methods. In order to individualize teaching processes, additional guidelines were prepared for work with dyslexic students. There are also alternative tasks which can be done during classes, at home, or when preparing students to take part in chemical knowledge competitions. The CDs attached to each volume of the Teacher's Book include a set of materials that largely exceeds teachers' basic pedagogical needs, e.g., films *At the Border of Chemistry*, animations and dynamic models, revision folios, and lesson plans that can be modified by teachers.

Yet another integral element of the handbook comprises Multimedia task sets on CDs (Fig. 3). The CD is designed for students as supplemental and self-assessment materials that enable students to check their knowledge and self-discipline in a stress-free way. The CD can also address teachers, as it is a valuable aid that can be utilized not only when preparing students for competitions and examinations, but also as didactic means facilitating revision after each thematic section. The CD contains arithmetic tasks, multiple choice tasks, tasks with models, chemical reaction puzzles, animation tasks, and elements characteristics. All tasks were constructed in a way which enables the user to check whether the answer to a given question was correct by pressing a key, after which the user can return again to the question chart. Such a solution allows for carrying out a detailed analysis of errors. Each task is awarded points separately, which makes students follow the content of a given task more carefully and think about possible solutions more penetratingly.

The handbook was supplied with Internet help available at the publishing house's web site. The web site collection contains films illustrating the course of

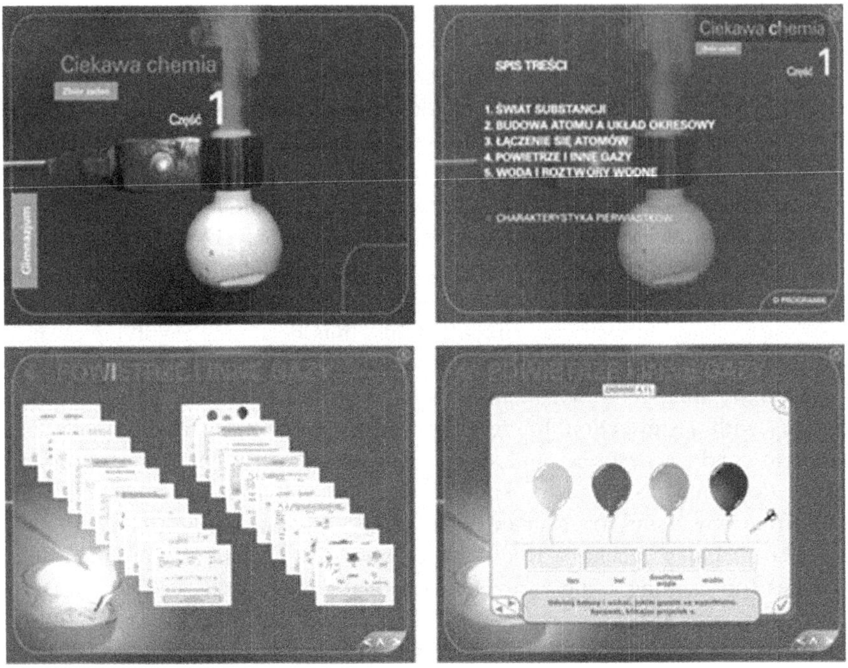

Fig. 3 Integral element of the handbook comprises *Multimedia Task Sets on CDs*

reactions covered by *The Interesting Chemistry* handbook, films illustrating experiments using widely accessible products, as well as pictures of elements and minerals, lesson plans, and additional tasks and tests.

2 Types of Animations in *The Interesting Chemistry* Handbook

The research on the role of didactic means in teaching abstract ideas, conducted at the Department of Chemistry Didactics of the Faculty of Chemistry at Adam Mickiewicz University (AMU) in Poznan, revealed high efficiency of animations and simulations in chemical education, with most beneficial effects observed in the case of students displaying advanced abstract thinking [2, 3]. This and other research was conducted in order to define in which conditions using didactic means which present chemical processes and phenomena in a dynamic and three-dimensional ways can be most beneficial. Another type of research pertained to the functions performed by models and animations in teaching particular subjects, as well as to conditions necessary to utilize these models in teaching processes. These experiments convinced us that many methodological obstacles can be overcome by including animations of simulations and interactive models in properly designed multimedia software, within which they could perform strictly defined didactic functions.

2.1 Preliminary Animations Before Experiments

Some animations included in *The Interesting Chemistry* CDs illustrate laboratory techniques and make it possible to simulate the measuring of solution reactions, the reading of liquid level in burettes, reaction temperature, product properties, as well as to simulate the preparation of particular solutions, and to observe the sequence in frame after frame, including fast forwarding, rewinding, and repetition. These animations also teach students to work neatly and accurately. They can be utilized when preparing students for chemical analyses; some modules can also be used during laboratory classes, helping teachers increase laboratory work efficiency among students (Fig. 4).

2.2 Animations Helpful in Drawing Conclusions from Experiments

By presenting reaction results, controlling analysis conduct, following the course of reactions according to many variables, and using computer algorithms to variously interpret the course of chemical reactions, *The Interesting Chemistry* gains a new type of educational significance. Among the empirical tasks of the user of such software are observations, experiments, and evaluations of experimental data. The result of these actions is empirical cognition, verification, and falsification. The possibility to verify the validity of assumed hypotheses stimulates the generation of original approaches to solving problems. At this stage of the experiment, the users both search for and verify alternative solutions, as well as conduct new experiments and utilize them in subsequent activities. The users assume the role of discoverers within a safer and limited reality. The necessity of detailed analysis of all interdependencies that occur in that reality, paired with the impossibility of achieving the

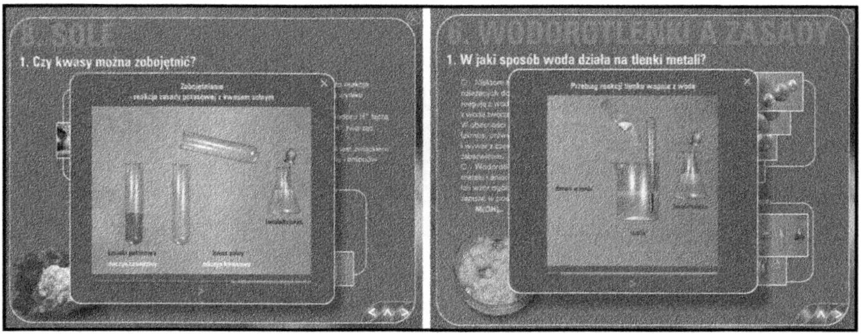

Fig. 4 Preliminary animations included in *The Interesting Chemistry* CD prepare students to conduct experiments

final goal accidentally, contribute to the development of the basic qualities of a good chemist-analyst, such as patience and accuracy (Fig. 5).

2.3 Animations Explaining the Course of Experiments

An interesting kind of animations accompany films which illustrate the course of chemical experiments (Fig. 6). These experiments were recorded in close-up, thus allowing students to take a good look at respective elements of laboratory apparatus, and to follow subsequent experiment steps. It is experiments that require substances unattainable at schools, i.e., those which are expensive, toxic, or hard-to-recycle, that are presented most frequently. Computer animations account on a microscale for the phenomena normally observed on a macroscale, such as dissociation or electrolysis. The teacher's book designed alone *The Interesting Chemistry* handbook contains descriptions of exemplary utilization of respective simulations in class, individually, or in small groups.

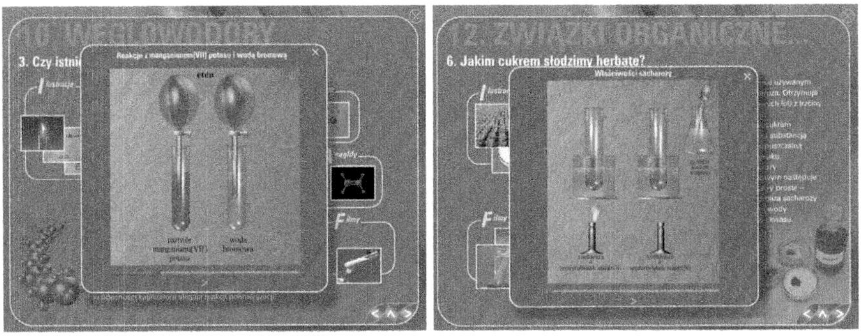

Fig. 5 Animations helpful in drawing conclusions from experiments, included in *The Interesting Chemistry* CDs

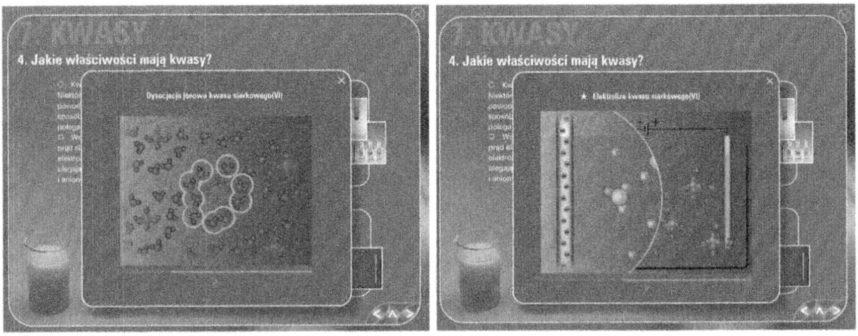

Fig. 6 *The Interesting Chemistry* CD animations explaining the course of chemical processes

2.4 Animations Helpful in Learning Reaction Mechanisms

Animations of reaction mechanisms included in *The Interesting Chemistry* CD were developed in Flash and supplemented with voice-over commentary (Fig. 7). By influencing sight and hearing simultaneously, these animations have a positive impact on learning and memorizing new material, as they illustrate what is normally unnoticeable to the human eye. They make us, for example, see the movement of valence electrons, which is the core of chemical processes. Each new didactic measure should undergo assessment in the course of pedagogical research. The results of the experiment carried out in the 2005/06 school year among students of junior high schools, high schools, and the Faculty of Chemistry at AMU allow for claiming that there is much behind animated visualizations of reaction mechanisms. The group which used animations scored better by 72% in the final test and by 44% in the distant test than the group which used printed materials. The animations positively influence the durability of students' knowledge as well as efficiency of teaching, with particularly good results achieved in typical ad problem-solving situations.

2.5 Animations Aiding Modelling Skills

The Interesting Chemistry CDs open up new didactic opportunities, chiefly thanks to their high potential in terms of interactive presentations, including three-dimensional models. The ability to utilize models and modelling skills is of great importance in imprinting proper mental images among students. The ability to use these images correlates positively with the skill of solving chemical problems. It has been noticed that providing learners with proper information in a visual form facilitates memorizing such terms as atom, particle, particle orbit, or chemical reaction. If students attempt to solve problems that require them to use a particular model in an

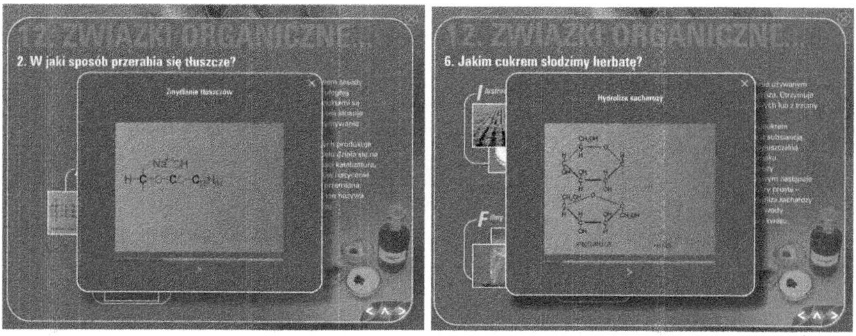

Fig. 7 Animations helpful in learning reaction mechanisms included in *The Interesting Chemistry* CDs

algorithmic way, it means they are unable to come up with images responding to that model, which is an indispensable skill to understand the essence of the problem and to reason properly (Fig. 8).

2.6 Animations Explaining the Problems of Environmental Chemistry

Problems from the border of applied chemistry and environmental chemistry constitute material for many activating methods (Figs. 9 and 10). Among forms of activities of this type, we can distinguish multicode information transfer which is present in films, happenings, scholarly meetings and panels, and multimedia software. Animations included in the CDs attached to *The Interesting Chemistry* handbook encourage carrying out authentic research on environmental parameters using external measuring devices and interfaces transmitting research results to the computer, which makes it possible to create more than a simple teaching aid for a few chemistry lessons, but rather enables users to develop their environmental awareness by activating them. The chief

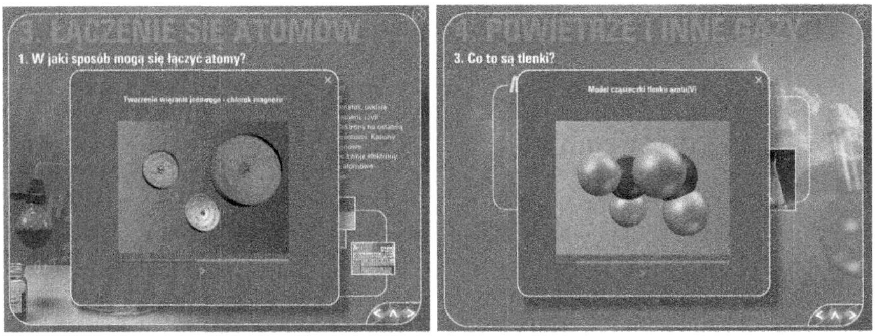

Fig. 8 *The Interesting Chemistry* CD animations explaining compound structures and the course of chemical reactions

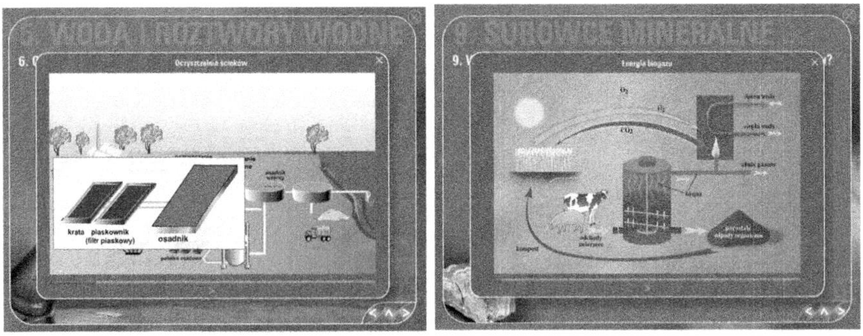

Fig. 9 Animations explaining environmental processes, included in *The Interesting Chemistry* CDs

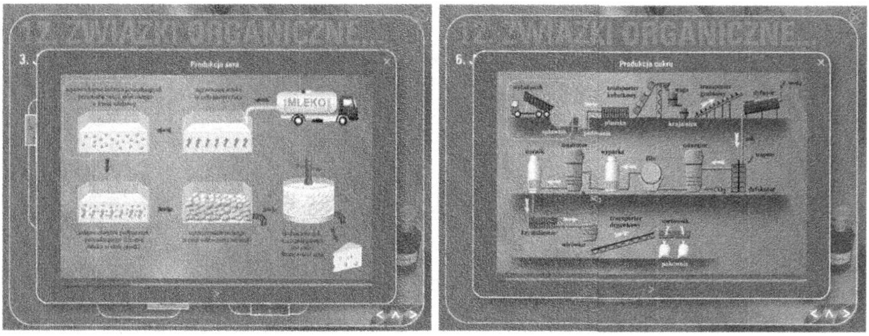

Fig. 10 Animations included in *The Interesting Chemistry* explaining processes occurring in natural environment

goal of education using interactive simulations is to develop one's environmental awareness, paying particular attention to such questions as quantitative balance between respective substances in nature, the harmful impact of certain groups of compounds on the environment, the degradation of natural environment, the types of actions to be taken up for environmental protection, and ways of encouraging students to take part in actions for the environment.

2.7 Animations Explaining Technological Problems

Yet another attractive component among the ones included in *The Interesting Chemistry* CD offer is the set of simulations of technological processes, supported with hypertextual and hypermedia help. In the miniature windows next to working space currently utilized by users, texts and videos can be stored, allowing users to switch between files to find interrelations between them. Using hypertext databases in the learning process facilitates processing information (putting forward hypotheses, confirming or refuting them by checking data), backs the development of communicational skills (deciding which databases are necessary to achieve assumed goals), supports the learning of complex competences (making quick decisions, ordering, analyzing, and interpreting information, synthesizing knowledge, and drawing conclusions), and broadens the number of teaching strategies (increasing autonomy in education).

2.8 Animations Helpful in Knowledge and Abilities Check-Ups

Yet another type of animations facilitates and objectifies the conduct of knowledge and abilities assessment. Simulative tasks included in *The Interesting Chemistry* CDs aim at doing a given activity properly, e.g., titration, weighing, or chemical apparatus assembly. These activities require such qualities as diligence, accuracy,

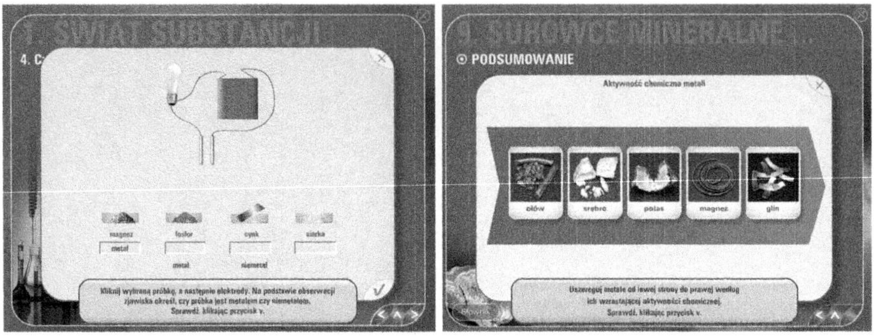

Fig. 11 *The Interesting Chemistry* CD animations helpful in making knowledge and abilities assessment more attractive

and persistence. Being able to solve the problem indicates that the assessed student has reached the required level of competence, done the necessary tasks, found the essential information, and combined these components within the determined time limit (Fig. 11).

2.9 Animations Facilitating Knowledge and Abilities Assessment

One of the crucial functions of animations included in *The Interesting Chemistry* CD is to control the process of preparation for conducting chemical experiments, combined with the assessment of the ability to solve arithmetic tasks (a properly solved task ends with an animation illustrating subsequent activities in the task, e.g., substrate or solution adding). The application of multimedia assessment and self-check methods makes it possible to individualize the pace and content of learning processes, structure and systematize knowledge, supplement it, acquire new abilities, and compensate for knowledge deficiencies. Emotional engagement accompanying work with computers encourages students to work in groups or cooperate with other groups. This is conducive to teaching the youth how to apply zered when assessing students enable teachers to plan further didactic work, while controlling students allows teachers to diagnose individual students in terms of their development and cognitive skills. Attempts to assess objectively seem to follow two directions. While the first direction aims at setting universal assessment criteria, the second reaches toward a system based on statistical interrelations (Fig. 12).

2.10 Educational Games Supporting Self-Checks

Examination methods proposed by *The Interesting Chemistry* often take up the form of strategic and adventure games, or even a virtual laboratory. Each game has a particular didactic goal to achieve, and achieving this goal is only possible

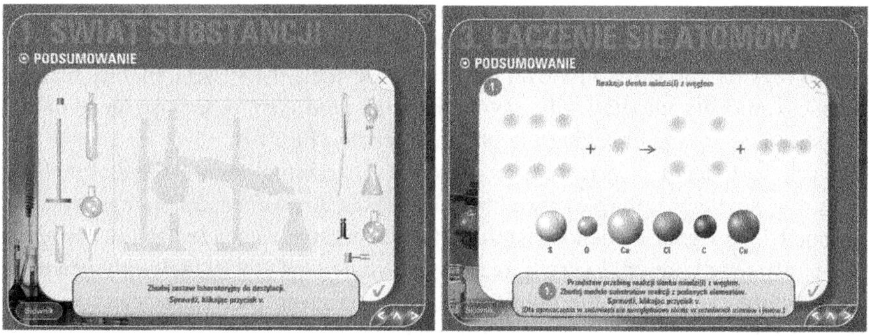

Fig. 12 *The Interesting Chemistry* CD animations facilitating objective assessment of students' progress

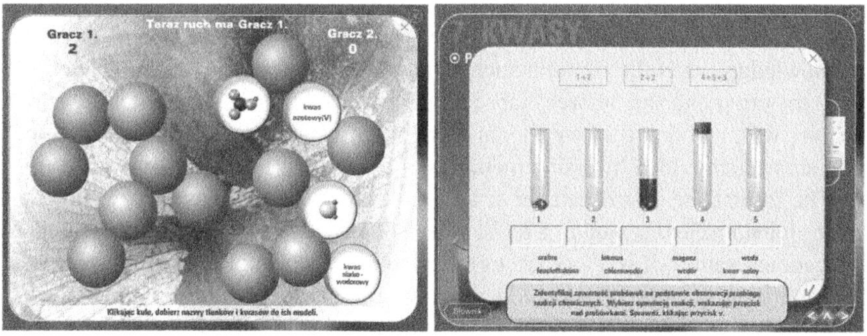

Fig. 13 *The Interesting Chemistry* CD games ensuring objective assessment of students' progress

when consecutive stages of the game are accomplished. This is particularly visible in multilevel adventure games. There are no rigid confines set for each game genre, and many of them have indirect character. The tasks that are apparently different serve in fact the same purpose, i.e., to solidify a given skill (e.g., laboratory apparatus assembly). Many of these simulative games are typically experimental. The possibility to concentrate on respective tasks encourages students to master the required skill in a more diligent manner, which in the long run helps them perform some tested laboratory activities in an actual chemical laboratory (Fig. 13).

3 Information Technology at the Service of School

Multimedia assessment is more and more often supported by systems of interactive tests, polls, and votes. Using wireless remote controls, all respondents can answer questions in real time by pressing a key on the remote responding to the answer they chose. Results are displayed in the form of a chart illustrating answer percentage layout on the computer screen, interactive board, or other large screen. All information concerning obtained answers, the number of attempts, or even answering time, is registered in the system, individually for each participant. Combined with an integrated register and difficulty level calibrating devices, the system enables teachers to assess students automatically, saving them the time spent otherwise on compiling statistics and double-checking the obtained results (Fig. 14).

Each participant of a lecture, conference, or training, answers individually to questions single or multiple) asked by the speaker. Answers are then registered and grouped. The signal is then transmitted to the receiver connected to the computer, using infrared radiation. Every time a given remote sends the signal with data to the receiver, the circle with the number signifying a given participant changes color from green to red, thus informing users the signal has been received. Each remote has its unique ID linked to a given participant's ID on the answers chart. The instructor is only able to display such information on the board as the participant's name, surname, nickname, and the participant's (or remote's) system ID number. It is up to the instructor, which type of data he or she chooses to display on the answers chart. It can either be the answer chosen by the participant, or a mere acknowledgment of the fact that such an answer was provided. After the appointed time for providing the answer elapses, the instructor can indicate what the correct answer was, and demonstrate the answers chart displaying percentages of respective answers provided by group members.

The interactive board can serve as a great tool to create and save interactive educational images, as well as a tool to assess knowledge and skills, and a ground for educational games. It is a modern tool facilitating teachers in organizing their fields of activity and creating distance education systems [4]. The board combines elements of a screen for displaying presentations, an elf-copying board, and a computer. It can be operated using an electronic pen, which allows users to write on it without using ink. Connecting the board with the computer using wire or wireless (infrared) not only enables users to work interactively with multimedia presentations, such as computer animations, film sequences, texts, charts, exercises, and tasks, but it also makes it possible to supplement these presentations with necessary data, (e.g., numerical), interpretations and commentaries by annotating them directly on the presented image. When working with Starboard software, provided by its producer – Hitachi Software

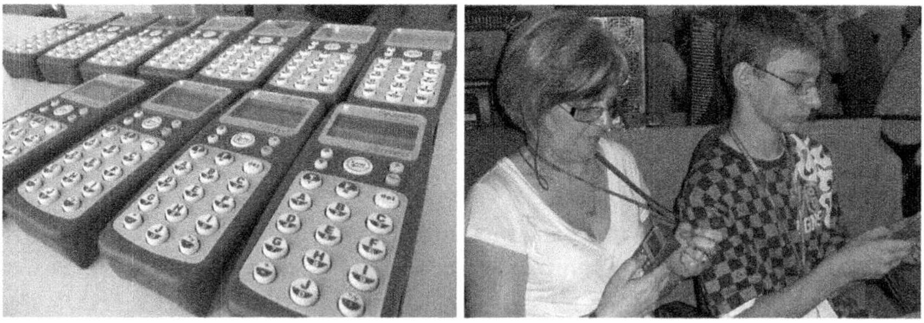

Fig. 14 Using VerdiCT and Santeo systems to examine students' knowledge

– its users can copy respective elements of presentations to Starboard's clipboard, amend them and save on the computer's hard drive (or other data carrier), from which they can be printed and distributed among the participants as handouts. A well thought-out and skillful interactive board use enables one to utilize various media in a reasonable way, so that any possible alterations made during classes can be saved on the computer, thus becoming a basis for further improvements.

The inclusion of interactive board to teaching processes opens new opportunities for:

- Utilizing information technology in (chemical) education in order to make it more efficient
- Activating and developing creative thinking among students to engage them emotionally in learning, and thus improve teaching results
- Effective presentations of films and animations, one that allows users to rewind, pause, and highlight chosen elements, as well as to interact with images (e.g., by annotating them with new information, transforming them or removing selected elements which can be further transformed)
- Teaching problem lessons (generating, registering, and verifying ideas)
- Improving various abilities and skills, e.g., graphic (with a possibility of continuous improvement of projects and their public assessment)
- Using the existing multimedia presentations for teaching new educational aspects
- Utilizing Internet resources in a well thought-out and interactive way by teams working simultaneously (at teleconferences and in distance education)
- Registering a given class and replaying it at any time, which allows teachers to continuously take advantage of the teaching material they worked long to compile.

Using the interactive board can also tempt teachers to reduce direct contact between students and experimental research in favor of impressive multimedia presentations. However, even those teachers who are fascinated by IT should be able to avoid making such a mistake, instead using the board to explain short- or long-lasting, complex processes that are unnoticeable to human eye, and to present those elements of chemical education that demand active participation of the entire student team. Utilizing the board gives its users a chance to bring demonstrative and practical methods closer to each other; therefore, instead of merely watching, we can act, formulate views, solve tasks, design patterns, etc. This is all the more important in situations when experiments turn out particularly unworkable in class surroundings.

One of interactive board's working modes is teleconference. Such an option encourages users to include interactive board in the system of distance education, which, as we think, would minimize the universal flaw of distance education, i.e., the lack of immediate presence of teachers. It would do so by enabling the participants of a given session to cooperate interactively with the teacher. Using a digital felt-tip pen, which performs a similar function to a regular pen, users can both write and draw on the board. This in turn ensures comfort in teaching classes and activates the participants. Another undeniable benefit is the possibility to teach classes for parties interested in specialized subjects, without troublesome commuting to distant locations. Combining two interactive boards provides an opportunity to utilize the available software in an attractive way, to present complex chemical experiments,

Fig. 15 Didactic classes taught using an interactive board (Department of Chemistry Didactics, AMU Faculty of Chemistry)

and to verify didactic ideas jointly. It is the first such didactic solution in Poland. Its innovative character also stems from the fact that, though teleconferences had previously been organized, it was the interactive board that enabled people to undertake interactive actions simultaneously in several aspects, such as the possibilities to simultaneously create joint files, fill in diagrams, take tests, and solve problem tasks (also experimental ones). These actions made it possible to teach classes using *The Interesting Chemistry* at several schools at once, thanks to which students in smaller towns had a chance to see experiments which would therwise be difficult to conduct in their schools (Fig. 15).

References

1. Gulińska H., Smolińska J. (2006), *Polish handbook titled Ciekawa Chemia (Absorbing Chemistry), published with integral CDs*, WSiP, Warszawa
2. Gulińska H. (2007), *Types and functions of simulations in the Absorbing Chemistry handbook*, Modeling and simulation, Coimbatore Institute of Technology, Allied Publishers Pvt. Ltd., Coimbatore, 6
3. Gulińska H. (2006), *Visualization in multimedia handbooks*, Proceedings of the 19th International Conference on Chemical Education "Chemistry and Chemistry Education for Humanity", Seoul, 308
4. Gulińska H., Bartoszewicz M. (2006), *Multimedia software for representation of chemical reaction mechanism – high school and college level*, Journal of Science Education, 7, 1

The Effectiveness of Lecture Demonstrations to Enhance Learning of Chemistry

A. Rade

Abstract Two methods for teaching a 12th grade chemistry course were compared by carrying out a two-group differential investigation in order to evaluate the effectiveness of using lecture demonstrations to improve student learning. The sample was students of four classes at a public girls' high school in Tehran. Initial equivalence between four classes was checked by a standard I.Q.-estimating test and a chemistry knowledge evaluating pretest, the results of which showed no significant difference between averages of four classes. In two classes of experimental group, alongside teaching the chemistry textbook, 11 demonstrations planned in relation with topics to be taught were conducted on appropriate occasions. In the other two classes, the control group, only the textbook was taught. After having taught each topic to both groups they were to answer an identical written quiz on that topic. In all, there were nine quizzes of the kind during the semester and an identical end-of-semester examination of all topics taught in this chemistry course. A statistical analysis of the results of all assessments, using *t-test*, showed with a very high confidence level that students in the experimental group achieved higher scores than students in the control group. Considering the initial equivalence between four classes, identical instructor, and similarity in teaching method and all aspects and conditions of this study, the results suggest that using lecture demonstrations is a much more effective means of instruction than traditional lecturing method.

1 Introduction

Chemistry lecture demonstrations have gained wide acceptance for illustrating fundamental chemical principles and concepts, especially in introductory classes, by promoting discussion around relation between the macroscopic world

---

A. Rade (✉)
Chemistry Teacher, Ministry of Education, Tehran, Iran
e-mail: azra_radeh@yahoo.com

M. Gupta-Bhowon et al. (eds.), *Chemistry Education in the ICT Age*,
DOI 10.1007/978-1-4020-9732-4_16,

of chemistry and the molecular structure [1–3]. Lecture demonstrations have also been used to evaluate conceptual understanding and critical thinking skills of students [4].

In the curriculum for high school currently in use in Iran, no practical activity is provided for 12th grade students' chemistry course [5]. While teaching their first semester chemistry – chemical kinetics and equilibrium – in the current academic year (autumn of 2007), a study was designed and carried out in order to evaluate the effectiveness of using chemistry lecture demonstrations in promoting students' learning in chemistry.

2 Methodology

A two-group differential investigation was used. The sample was 148 students of four 12th grade classes each consisting of 37 students at an average public girls' high school in Tehran, so that it could be considered as a random sample of 12th grade classes at public high schools across this country. The 12th grade students in our high school had been distributed similarly in these four classes according to their achievements at the previous grade, and thus the overall level of students' scientific knowledge and abilities in all four classes was comparable. Nevertheless, equivalence between four classes was checked by a standard I.Q.-estimating test [6, 7], the result of which showed no significant difference between averages of four classes. In addition, their chemistry knowledge achieved in previous grade was pretested. This test also revealed that there was no significant difference between them.

In two classes selected randomly as experimental group, along teaching the chemistry textbook, 11 demonstrations planned in relation with topics to be taught, were conducted in appropriate occasions (Table 1). These demonstrations were selected from available resources [8–14], slightly modified or redesigned to render them more appropriate to our students' level.

In other two classes, the control group, only the textbook was taught using traditional narrative method. But, in order to maintain equivalence in teaching time and conditions for both groups, and also in order that the visual effect of demonstrations could eventually be detected, the chemistry involved in the demonstrations was explained – without performing them – for two classes of the control group as fully as was done for the experimental group.

After having taught each topic to both groups they were to answer an identical written quiz on the basic concepts of the topic. In all, there were nine quizzes of the kind during the semester. In addition, all the sample students had an identical end-of-semester, comprehensive examination of all topics taught in this chemistry course.

Table 1 Demonstrations conducted in this study

Chapter in textbook	Topic in textbook	Demonstration conducted for the topic	Demo. ref.
Kinetics	Effect of physical state on rate of reaction (5, p. 8)	Demo. 1: Two substances (citric acid and sodium hydrogen carbonate), when both in solid state, are mixed; no reaction occurs. But when same substances are dissolved, a vigorous reaction occurs in liquid state.	[12, 13, 14]
	Effect of concentration on rate of reaction (5, p. 9)	Demo. 2: When concentration of sodium thiosulfate in the following reaction is increased, faster production of solid sulfur indicates an increase in the rate of reaction. $S_2O_3{}^{2-}{}_{(aq)} + 2H^+{}_{(aq)} \longrightarrow S_{(s)} + SO_{2(g)} + H_2O_{(l)}$	[12, 13, 14]
	Effect of temperature on rate of reaction (5, p.10)	Demo. 3: In the following reaction, $S_2O_3{}^{2-}{}_{(aq)} + 2H^+{}_{(aq)} \longrightarrow S_{(s)} + SO_{2(g)} + H_2O_{(l)}$ increasing the temperature results in increase in the rate of reaction.	[12, 13, 14]
	Effect of catalyst on rate of reaction (5, p. 19)	Demo. 4: In the following reaction, $2H_2O_{2(aq)} \longrightarrow 2H_2O_{(l)} + O_{2(g)}$ adding potassium iodide, catalyzes the decomposition of hydrogen peroxide.	[10, 12, 14]
Equilibrium	Reversible reactions (5, p. 24)	Demo. 5: The following reaction, $\underset{\text{Yellow- green}}{[CuCl_4]^{2-}{}_{(aq)}} + 4NH_{3\,(g)} \longrightarrow \underset{\text{Very dark blue}}{[Cu(NH_3)_4]^{2+}{}_{(aq)}} + 4Cl^-{}_{(aq)}$ reverses by adding more acid.	[11, 12, 14]
	Effect of concentration on Equilibrium (5, p. 39)	Demo. 6: In the following equilibrium, $\text{Heat} + \underset{\text{Pink}}{[Co(H_2O)_6]Cl_{2\,(aq)}} \rightleftharpoons \underset{\text{Blue}}{[CoCl_2(H_2O)_2]}_{(aq)} + 4H_2O_{(l)}$ changing the concentration of reactants, results in a shift of equilibrium between pink and blue complex ions of cobalt.	[8, 14]
	Equilibrium and Le Chatelier's principle (5, p. 41)	Demo. 7: In the following equilibrium, $Fe^{3+}{}_{(aq)} + SCN^-{}_{(aq)} \rightleftharpoons \underset{\text{Red}}{FeSCN^{2+}{}_{(aq)}}$ changes in the concentration of reactants, result in color changes.	[8, 14]

(continued)

Table 1 (continued)

Chapter in textbook	Topic in textbook	Demonstration conducted for the topic	Demo. ref.
	Effect of pressure on equilibrium (5, p. 41)	Demo. 8: In the following equilibrium, $\underset{\text{Reddish-brown}}{2NO_{2\,(g)}} \rightleftharpoons \underset{\text{colorless}}{N_2O_{4(g)}}$ decreasing (increasing) the volume of the gas, shifts equilibrium to the right (left).	[12, 14]
	Effect of temperature on equilibrium (5, p. 43)	Demo. 9: When the following equilibrium system, $\text{Heat} + \underset{\text{Pink}}{[Co(H_2O)_6]Cl_{2\,(aq)}} \rightleftharpoons \underset{\text{Blue}}{[CoCl_2(H_2O)_2]_{\,(aq)}} + 4H_2O_{\,(l)}$ is heated, a color change from pink to blue indicates a shift of equilibrium to the right. When the solution is cooled, the color change from blue to pink indicates a shift to the left.	[8, 14]
		Demo. 10: In the following equilibrium system, $\text{Heat} + \underset{\text{Blue}}{CuSO_{4(aq)}} + 4KBr_{(aq)} \rightleftharpoons \underset{\text{Green}}{K_2[CuBr_4]_{(aq)}} + K_2SO_{4(aq)}$ changes in the temperature shift the equilibrium.	[8, 14]
		Demo. 11: When the following equilibrium mixture $\underset{\text{Reddish-brown}}{2NO_{2\,(g)}} \rightleftharpoons \underset{\text{colourless}}{N_2O_{4(g)}}$ is heated, it shifts to the left. When cooled, it shifts to the right.	[8, 14]

3 Results and Discussion

A statistical analysis of the results of nine quizzes and the final exam, using *t-test* to compare the experimental and control groups achievement means [15, 16] for each of the measures (Table 2), showed with a confidence level of 95% that in all ten assessments, students in the experimental group achieved higher scores than students in the control group with a statistically significant difference. The *p*-values calculated for each case indicate that this difference is highly significant.

On the basis of these results, it was reasonably concluded that using lecture demonstrations substantially promoted learning of chemistry.

Considering the initial equivalence between four classes, identical instructor and course material, and similarity in teaching method, and all aspects and conditions of this study – with the exception that the demonstrations were conducted only for the experimental group – this teacher suggests that the significant difference revealed between two groups originates from the fact that demonstrations provide attractive interludes for students to relieve the passive listening to a monotonous lecture delivered in a merely downloading procedure and watch some memorable visual images of phenomena containing an element of surprise or intriguing moments. This motivates curiosity of students to reflect and understand these surprise-evoking situations [17–21]. This method of chemistry teaching, therefore, is consistent with the constructivist procedure of learning [22]. Such a cognitive arousal for finding an explanation for their observations, recognition of variables and consistency of concepts, and seeking some rule and relevance between various parts of demonstrations drives students to follow the teacher's lecture more attentively and purposefully. Demonstrations, therefore, make students particularly receptive to education so that they achieve a better comprehension even of abstract and theoretical concepts in chemistry.

Table 2 Results of statistical comparison of sample student groups

	Experimental group		Control group		Standard		
Assessment	Mean	S.D.	Mean	S.D.	error	t	p-value
Quiz 1	16.25	3.72	14.73	3.56	0.60	2.54	< 0.01
Quiz 2	16.62	3.33	15.22	3.19	0.54	2.61	< 0.005
Quiz 3	16.73	3.88	14.75	3.71	0.62	3.17	< 0.001
Quiz 4	16.64	3.64	14.62	3.62	0.60	3.38	< 0.0005
Quiz 5	16.45	3.74	14.58	3.59	0.60	3.10	< 0.001
Quiz 6	16.91	3.84	15.12	3.62	0.61	2.92	< 0.0025
Quiz 7	16.76	3.28	15.24	3.31	0.54	2.81	< 0.0025
Quiz 8	17.21	3.41	15.07	3.08	0.53	4.01	< 0.0005
Quiz 9	17.35	3.45	15.64	3.17	0.54	3.14	< 0.001
Final exam	16.92	3.25	15.50	3.01	0.51	2.76	< 0.005

4 Conclusion

This study showed with a very high confidence level that using lecture demonstrations is a much more effective method in chemistry teaching and learning than traditional narrative method. Conducting a demonstration excites curiosity of students, and they, viewing a demonstration, apply their previous knowledge to understand what is occurring and try to make logical inferences about the chemical processes involved. During this study, this teacher witnessed with contentment that students taught with demonstration, clearly enjoyed learning chemistry. They also exhibited an excited intention for selecting chemistry as their future discipline at university.

On the basis of the results of this study, we shall propose to our Ministry of Education to incorporate the use of demonstrations in the high school chemistry curriculum as an integral part of the teaching–learning process.

References

1. Bowen CW, Phelps AJ (1997) Demonstration-based cooperative testing in general chemistry: a broader assessment-of-learning technique. J Chem Educ 74(6):715–720.
2. Hatcher-Skeers M, Aragon E (2002) Combining active learning with service learning: a student-driven demonstration project. J Chem Educ 79:462–464.
3. Meyer LS, Schmidt S, Nozawa F et al. (2003) Using demonstrations to promote student comprehension in chemistry. J Chem Educ 80:431–435.
4. Deese WC, Ramsey LL, Walczyk J et al. (2000) Using demonstration assessments to improve learning. J Chem Educ 77:1511–1516.
5. Curriculum Development Center (2005) Chemical processes: Chemistry 289/1 for 12th grade in natural and mathematical sciences branches (in Persian). Ministry of education, Tehran
6. Eysenck HJ (1962) Know your own I.Q. Penguin Books, Harmondsworth
7. Eysenck HJ (1966) Check your own I.Q. Penguin Books, Harmondsworth
8. Summerlin LR, Ealy JL Jr. (1988) Chemical demonstrations: a sourcebook for teachers, Vol 1, 2nd edn. American Chemical Society, Washington DC
9. Summerlin LR, Borgford CL, Ealy JB (1988) Chemical demonstrations: a sourcebook for teachers, Vol 2, 2nd edn. American Chemical Society, Washington DC
10. Baker C (2002) Exhibition chemistry: the hydrogen peroxide reaction. Educ Chem 4
11. Baker C (2006) Exhibition chemistry: a spectacular reversible reaction. Educ Chem 3
12. University of Leeds. Chemistry Lecture Demonstrations. http://www.chem.leeds.ac.uk. Accessed 22 May 2008
13. Shakhashiri BZ (1989) Chemical demonstrations: A handbook for teachers of chemistry, Vols 1–4. University of Wisconsin Press, Madison, WI
14. Rade A (2008) A descriptive webliography of chemistry lecture demonstrations. In: Abstract book, The 9th European Conference on Research in Chemical Education. Turkish Chemical Society & EuChems, Istanbul
15. Walpole RE, Myers RH (1978) Probability and statistics for engineers and scientists. Macmillan, New York
16. Miller NM, Miller JC (2000) Statistics and chemometrics for analytical chemistry. Ellis-Harwood
17. Hirst PH, Peters RS (1990) The logic of education. Routledge, London
18. Hilgard ER, Bower GH (1986) Theories of learning. Prentice Hall of India

19. Cohen B (1981) Education and the individual. Allen & Unwin, London
20. Petri HL (1996) Motivation: theory, research, and applications, 4th edn. Brooks/Cole, California
21. Bodner GM (2001) Why lecture demonstrations are 'exocharmic' for both students and their instructors. U Chem Ed 5:31–35.
22. Herron JD, Nurrenburn SC (1999) Chemical education research: improving chemistry learning. J Chem Educ 76:1353–1361.

The Use of Writing-Intensive Learning as a Communication and Learning Tool in an Inorganic Chemistry Laboratory Course

H. Jalali, L. Hanlan, and J. P. Canal

Abstract In the scientific community, effective communication is essential for the illustration and development of ideas, yet this skill is often not specifically taught in the undergraduate curriculum. To address this issue, we have developed and implemented a substantial writing component for a required, second-year inorganic chemistry laboratory course to provide specific instruction in scientific writing. The concept of writing for different audiences is introduced with an emphasis on keeping a laboratory notebook, and writing laboratory reports and a variety of low-stakes (quick writes) and high-stakes (formal laboratory reports) writing assignments.

1 Introduction

At Simon Fraser University (SFU) recent university-wide curriculum revisions require all undergraduate students to take two courses designated as writing-intensive. The goals of this initiative were twofold. Firstly, students would "write to learn in the forms typical of the discipline," that is these courses should use examples, styles, and technical language that are representative of the specific subject. Secondly, feedback and response to students' written work would be provided, aimed at improving the overall quality of writing [1, 2].

Although most chemistry undergraduate curricula are focused on teaching the theory and facts of the field, the use of writing as a learning and communication tool is entering the chemistry curriculum [3–9]. Writing, which is traditionally viewed as an evaluation tool, is increasingly being valued as an active learning technique to engage students and to stimulate discussions [7]. Chemistry programs are increasingly implementing writing components into their courses, in order to teach effective writing styles [10, 11].

H. Jalali, L. Hanlan, and J.P. Canal (✉)
Department of Chemistry, Simon Fraser University, Canada
e-mail: jcanal@sfu.ca

M. Gupta-Bhowon et al. (eds.), *Chemistry Education in the ICT Age*,
DOI 10.1007/978-1-4020-9732-4_17,

This paper will focus on the techniques used to introduce writing as a learning tool and overcome challenges in improving the writing proficiency of students, all in the context of a second-year inorganic laboratory course.

2 Methodology

To meet the new undergraduate curriculum requirements at SFU, the Department of Chemistry recently altered a required second-year inorganic chemistry laboratory course to provide specific instruction in scientific writing. The weekly 4-h laboratory sessions remain, in the traditional way, the time for students to receive hands on experience with basic inorganic chemistry concepts and laboratory techniques. With the addition of weekly 1-h tutorials, detailed instruction could focus on the various aspects of writing typically associated with a laboratory science. Attendance at both the laboratory and tutorial sessions is mandatory.

The writing component of the laboratory course focuses on three types of written assignments: (1) keeping a research style laboratory notebook, (2) completing laboratory report sheets, and (3) writing formal laboratory reports. During the tutorials a variety of instructional methodology is used to establish the standard format and style for each type of assignment. This is done through the use of specific guidelines as well as discussion and analysis of examples taken from various sources; for example, research or technical journal articles, instructor composed examples, or other students' work. "Audience," an important concept in writing, is discussed for each type of assignment and students are encouraged to identify their audience to assist them in determining the level of detail required in their writing [12, 13].

Each assignment type is deconstructed to its basic components and each individual component discussed for its purpose, content, and style. Low stakes writing exercises are employed to provide practice for each of the components with opportunity for instructor feedback or peer review. By starting with the components of the laboratory notebook followed by those of the reports sheets and finally the formal laboratory reports, students progress toward more complex writing styles that require an increase in the understanding of the subject matter.

3 The Laboratory Notebook

In the first tutorial students are introduced to the importance of keeping a proper research style laboratory notebook as proof of scientific discovery [14]. This is stressed by using examples from the literature that detail instances where published articles have been retracted due to poor or nonexistent laboratory notebooks [15, 16], and with a sample of an advertisement for an industrial research position that specifies "maintaining a laboratory notebook" as part of the job description [17].

For all experimental work, students are required to accurately and clearly record their procedures, observations, data, and results in their laboratory notebook on an ongoing basis while in the laboratory. To assist the students with the task of properly completing their notebooks, the audience has been identified as (1) themselves, as they will need the information in the notebook when writing up report sheets or formal reports and (2) other students with similar chemistry background who may want to repeat the experiment.

The basic components of the notebook were identified as: Date, Title, Objective, Data Tables, Equations and Calculations, and Procedure and Observations. Time is spent examining each component. For example, students are taught that the laboratory notebook represents what they *did* and what they *observed* while in the laboratory and therefore the Procedure and Observations section is to be written while they are doing the experiment, not before, and should be in the past tense. With respect to audience, as outlined above, entries are to include enough detail for a student of similar chemistry background to repeat the experiment based on the laboratory notes.

Students are also required to complete some aspects of the notebook before coming to the laboratory session, including the Title, Date, and Objective of the experiment. For each experiment a Data Table is to be prepared, and should include the names and pertinent physical data for all the reagents and solvents to be used in the experiment. Notebooks are regarded as low-stakes writing. They are checked by a Teaching Assistant (TA) at the beginning of each laboratory session and again at the end to evaluate for completeness and to give feedback. No marks are assigned in the first 3 or 4 weeks of the course, giving students the opportunity to improve based on the feedback they received. After that, the weekly laboratory notes are marked for completeness.

4 The Laboratory Report Sheet

Prior to writing formal laboratory reports, students complete a laboratory report sheet for the early experiments. These are essentially template style laboratory reports that require the student to report specific data and results, and answer specific questions about the experiment. The report sheets can be used to build up toward the formal laboratory report as they essentially guide students, with direct questions, to the important aspects of the experiment that are worthy of discussion and, ultimately, to an understanding of the chemical concepts. Report sheets are considered high-stakes writing assignments. They are due 1–2 weeks after completion of the laboratory and are worth a significant portion of the final grade.

5 The Formal Laboratory Report

Formal laboratory reports are an effective communication and learning tool for both the students and the target audience. They provide a means for students to present their work as a polished and complete unit, which therefore requires them to further explore the experiment and gain an in-depth understanding of the chemistry involved. Unlike the guided question/answer format of the report sheets, formal reports are more open-ended and students must consider for themselves what aspects of the chemistry are important to discuss. Students are expected to use the previous assignments as guidelines to compile the contents required in a formal laboratory report. To aid in determining the level of detail required in a formal laboratory report, we identified the audience to be someone with a firm understanding of the fundamentals of basic chemistry but not the particulars of the reaction being reported.

In the tutorial, a scientific paper is deconstructed into its components (Introduction, Experimental, Results and Discussion, Conclusions, References and Abstract). Each component is examined and decoded to identify the key elements and features that are typically included. For example, when writing an Introduction to a scientific paper, we recommend the students address the following questions: (1) What has been done before? (2) Why are we doing this experiment? (3) What are we going to do? (4) How are we going to do it?

Low stakes, quick writes are used for additional practice in writing the various components of a formal report. Students might be asked to take their laboratory notes from an experiment and re-write them as an Experimental section for a formal report or to write a Conclusion for a previously completed laboratory. This work is assessed in peer review groups and fellow students provide each other with constructive feedback. Higher stakes writing exercises are also assigned, to be handed in at the next tutorial and subsequently graded by the TA or instructor.

Once the students have progressed through all the components of a scientific paper in the tutorial setting, the initial report sheet assignments in the laboratory are replaced with the higher stakes formal laboratory reports. Three formal laboratory reports are required in the course. For the first one the students are given considerable guidelines and suggestions as to what should be included in the discussion. The amount of guidance is reduced for the second formal report and for the third one it is left entirely for the individual student to decide what should be included.

The first formal laboratory report is evaluated both for the writing and the chemistry contents but the writing portion of the grade is not recorded. Each student is given a detailed feedback form describing what was done correctly and where they need improvement (Table 1). This feedback form is provided to students as a guide for improving subsequent formal reports. The writing aspect of the later formal laboratory reports do contribute to the final grade.

6 Other Writing Assignments

In addition to the three main types of assignments detailed above, a variety of other assignments have been used to enhance the "writing to learn" experience. Various examples of safety data sheets were examined in tutorial and students were then assigned the task of compiling their own safety data sheet for a specific chemical.

Table 1 Formal Report Feedback Sheet

Writing Criteria	√	X	Comments	Writing Criteria	√	X	Comments
Title page				***Discussion***			
• Title				• Clarity			
• Name/Station#				• Logical connections			
• Date				• Grammar, paragraph and sentence style			
• Course information				• Communicates an understanding of the chemistry			
Introduction				• Use of relevant references to support discussion			
• What is known				***Conclusion***			
• What will we do; why				• Reflects the objective			
• How will we do it				• Summarizes the results without replicating discussion			
• Clear statement of objective				• Short and concise			
• Paragraph/sentence style				***References***			
• Communicates an understanding of chemistry				• Appropriate Reference section			
• Use of relevant references and background information				• Citations follow the ACS style			
Experimental				• Information referenced completely			
• Paragraph/sentence style				• All references cited			
• Past tense, passive voice				***Concerns***			
• Procedure is reproducible				• Plagiarism/ "cut and paste writing"			
• Includes important observations				• Point form/ bullet style			
• Appropriate detail				• Spelling/ grammar check			
Presentation of Data/ Results				• Readability			
• Tables							
• Structures							
• Equations							
• Calculations							
• ***General Comments***							

One tutorial is dedicated to reference material: how to search for information, how to properly reference the literature, identifying, and avoiding plagiarism. Students were given a journal article with citations removed, and were asked to identify places where they felt a reference was required.

7 Results and Discussion

Several challenges were encountered in introducing the writing-intensive component into a laboratory course. It was clear early on that trying to give writing instruction in the laboratory setting did not work. There were too many distractions and students were always keen to get on with the experimental work. It was important to have separate classroom time, so the addition of the tutorial hour was critical. Pre-course surveys indicated that many students were skeptical about "learning to write" in chemistry. Some initially felt that "writing intensive" meant a lot more work writing and less time spent on chemistry. They did not have a lot of confidence in their ability to write and some gave that as a reason they were taking science courses, not arts courses. In addition, a significant proportion of our students have English as a second language; their writing skills are often very weak.

Although the writing intensive aspects that we have adopted for our laboratory course focus on "learning to write in the discipline" to promote effective communication of scientific discovery, we believe we have done this in the greater context of using writing as a tool to learn chemistry. When keeping a laboratory notebook the student must determine what information is important to include, and what is unnecessary detail. This requires an understanding and appreciation of the techniques being used and the chemistry observed. They must read and understand background chemistry on the topic in order to summarize and write an effective introduction. The discussion section of a formal report will require an in-depth understanding of the chemical concepts in order to discuss the significance of the experimental results.

In a writing intensive course, students receive appropriate feedback on their work, directed to improving the quality of writing. This often involves several drafts and revisions of each assignment. Since the writing assignments for the laboratory course must coordinate with the experimental work being done, multiple revisions of any one assignment are not done. Instead we chose to use a progression of assignments where the feedback from an assignment based on one experiment, is used to improve upon the writing for a similar assignment based on the next experiment. As the students work through their seven or eight different experiments in the laboratory they have multiple opportunities for writing and feedback for each type of assignment that we have detailed above. The nature of the feedback varies depending on the complexity of the writing assignment. Quick notations and verbal feedback are used for the laboratory notebooks and peer review is used to evaluate the quick write in tutorials. The report sheets receive in-depth feedback on the style and the appropriateness of the audience, with the formal laboratory reports receiving a more structured, high stakes feedback in the form of a detailed feedback sheet (see Table 1). As the requirements and demands of the format increase the students receive more complete feedback from the instructor and teaching assistants.

Our second year laboratory course has now been offered five times in this new writing-intensive format. Each time we have noted a general improvement in the quality of student writing as the semester progressed, as seen in the marks assigned to written work. There were students whose writing immediately improved while others did so more slowly. A few students remained "stuck" on previously adopted writing styles from other disciplines and did not feel that they needed to make adjustments. Several former students have commented that the writing instruction they received in our course was a definite help to them in other laboratory courses that required written work. This is reflected in the experience of instructors in third and fourth year laboratory courses as well, who have noticed an improvement in the written work submitted by their students.

At this point we have not been able to correlate whether the focus on writing in a chemistry laboratory course has led directly to an improvement in the understanding of the chemical concepts. We hope to be able to track such improvements over the next several semesters that the course is offered.

8 Conclusion

A required second year inorganic chemistry laboratory course was recently altered to include a writing-intensive component. It was designed to improve the writing abilities of our students in the Department of Chemistry. Although, students initially have reservations of the importance of this program, the quality of writing by our students within the department has shown signs of improvement. As we develop the program and tailor the lessons to maximize the learning outcome, our initial goals will be further met. Although instructors have noticed improvement, the overall influence of this course on students' ability to better understand and learn chemical concepts is to be determined in years to come.

Acknowledgements We would like to thank the Writing Intensive Learning Office at SFU (Kathryn Alexander and Wendy Strachan) for initially helping develop this program and the Department of Chemistry at SFU for funding. Thank you to all the Teaching Assistants and students who have taken this course, whose suggestions have helped the program develop.

References

1. Simon Fraser University (2008) http://www2.sfu.ca/ugcr/For_Faculty/WQB_Cr-iteria_and_Definitions/Writing. Accessed 10 May 2008
2. Beall H (1991) A tool for increasing comprehension and communication. J Chem Ed 68:148–149
3. Ablin L (2008) Student perceptions of the benefits of a learner-based writing assignment in organic chemistry. J Chem Ed 85:237–239
4. Robinson M S, Stoller F L, Jones J K (2008) Using the ACS journal search to validate assumptions about writing in chemistry and improve chemistry writing instruction. J Chem Ed 85:650–654
5. Whelan R J, Zare R N (2003) Teaching effective communication in a writing-intensive analytical chemistry course. J Chem Ed 80:904–906
6. Kovac J, Sherwood D W (1999) Writing in chemistry: An effective learning tool. J Chem Ed 76:1399–1403
7. Bressette A R, Bretson G W (2001) Using writing to enhance the undergraduate research experience. J Chem Ed 78:1626–1627
8. Tilstra L (2001) Using journal articles to teach writing skills for laboratory reports in general chemistry. J Chem Ed 78:762–764
9. Kovac J, Sherwood D W (2001) Writing across the chemistry curriculum. An instructors handbook. Prentice Hall, Upper Saddle River, NJ
10. Schepmann H G, Hughes L A (2006) Chemical research writing: A preparatory course for student capstone research. J Chem Ed 83:1024–1028

11. Bailey R A, Geisler C (1991) An approach to improving communication skills in a laboratory setting. J Chem Ed 64:996–998
12. Werner T C (1986) Reflections on the emphasis of communication skills I the undergraduate chemistry curriculum. J Chem Ed 63:140–141
13. Rosenthal L C (1987) Writing across the curriculum: Chemistry lab reports. J Chem Ed 64:996–998
14. Eisenberg A (1982) Keeping a laboratory notebook. J Chem Ed 59:1045–1046
15. Kawaskai H, Taira K (2006) Retraction: Induction of DNA methylatin and gene silencing by short interfering RNAs in human cells. Nature 441:1176
16. Wadman M (2006) Journal reveals plans to fight fraud. Nature 444:658
17. Xenon Pharmaceuticals Inc. (2006) Research associate-medical chemistry http://www.bcbiotech.ca/Employment/job07250605.asp. Accessed 25 July 2006

A Study on the Use of Concept Maps in the Teaching of 'Chemical Periodicity' at the Upper Secondary Level

A. Douquia and F. B. Narod

Abstract Concept maps are special forms of web diagrams for exploring knowledge and for gathering and sharing information; they illustrate meaningful relationships between concepts. The purpose of this study was to use concept maps in the teaching of 'Chemical Periodicity' in view of investigating their effectiveness on students' motivation, interest and conceptual understanding. The study was carried out through an action research over three cycles in a boys' State Secondary School in Mauritius. Data was collected through observation during the lessons, and through achievement tests administered after each cycle. Furthermore, at the end of the action research, a questionnaire was administered to all students in the sample, and a group interview was conducted to gather information about their attitudes and views on the use of concept maps in the teaching of 'Chemical Periodicity'. Our findings have revealed that the use of concept maps in the teaching of 'Chemical Periodicity' has led to enhanced conceptual understanding, as evidenced by an improvement in students' performance in the tests in the second and third cycles. The results have also indicated that the use of concept maps has aroused and maintained a high level of motivation and interest among the students during the lessons, and has promoted interactions between teacher and students.

1 Introduction

According to Ausubel [1], meaningful learning can only take place when the learner is able to relate new knowledge to existing knowledge, and to perceive a relationship among the different concepts to be learnt. Based on Ausubel's learning principles, Novak has developed the use of concept maps as a strategy to represent meaningful relationships between concepts [2, 3]. A concept map is a special form of web diagram

A. Douquia
John Kennedy College, Mauritius

F.B. Narod (✉)
Department of Science Education, Mauritius Institute of Education, Mauritius
e-mail: zia373@eudoramail.com

M. Gupta-Bhowon et al. (eds.), *Chemistry Education in the ICT Age*,
DOI 10.1007/978-1-4020-9732-4_18,

for exploring knowledge and for gathering and sharing information [4]. Furthermore, Hyerle [5] (cited in Costa, 2001) stated that teachers should make use of visual tools such as webs, organizers and thinking process maps that would encourage students to think in patterns rather than organizing information linearly; there is a cognitive dissonance between the linear form in which knowledge is presented and the non-linear formats in which the brains of students process this knowledge. Indeed, concept maps are reported to be an invaluable tool that can be used to support students in mapping the patterns of knowledge for meaningful learning (GRAPHIC.ORG).

On the other hand, 'Chemical Periodicity' is a concept-rich topic at the University of Cambridge Local Examinations Syndicate (UCLES) Advanced ('A')-level; students are required to handle, interpret, analyze and evaluate a large amount of information to identify patterns, report trends and draw inferences. However, it has been observed that learners have trouble in constructing a holistic view of the topic; they are confused by the vast amount of facts they need to assimilate and are often not able to organize the content of this topic in a systematic way to facilitate retrieval when required. Furthermore, students tend to learn 'Chemical Periodicity' by rote and thus, they fail to answer questions requiring higher-order thinking skills. Indeed, examiners have reported that many students seem to learn Inorganic Chemistry by rote and are unable to apply the concepts to novel situations [6]. Besides, examiners have also highlighted that many students are unable to select the relevant data to answer a particular question and they tend to write at considerable length about irrelevant points thereby losing time and missing key ideas [6].

In view of the above discussions, it was decided in the present study to use concept maps to teach 'Chemical Periodicity' in an attempt to help students organize the large amount of information they need to handle and to enhance conceptual understanding. Concept maps have indeed been reported to encourage students to adopt a deep and holistic approach to learning [7]. Furthermore, concept maps, by their structure, can also help students to process complex and new information [8]. It is hoped that this would make learning of the topic more meaningful and discourage passive rote-learning.

2 Methodology

2.1 Sample Description

The study was conducted in an urban Boys' State Secondary School in Mauritius over a period of 6 weeks, and consisted of 18 periods (each of 40 min) in August and September 2006. The sample comprised 24 students (boys of age 15–17 years) who were preparing to sit for the UCLES 'A'-level examinations in November 2007.

2.2 Research Design

Since the study involved the implementation of a new strategy, namely use of concept maps, in classroom situation, it was decided to use the action research model, which is reported to be the most suitable method when a new approach is to be grafted onto an existing system [9]. The present study was indeed based on an action research that was carried out over three cycles to teach 'Chemical Periodicity'. In the first cycle, the lessons were carried out by the regular instruction method, while in the second and third cycles, use of concept maps was incorporated in the teaching of the topic. Details of the three cycles are given in Table 1.

Table 1 Details of the action research cycles and the use of concept maps in the teaching of 'Chemical Periodicity' in the present study

CYCLE 1	CONTENT	STRATEGY
Lesson 1	-Introduction to the Periodic Table.	Regular instruction – expository, questioning and group discussion.
Lesson 2	-Variations in atomic and ionic radii across Period 3.	
Lesson 3	-Variations in melting points and electrical conductivities across Period 3	
CYCLE 2	CONTENT	STRATEGY
Lesson 1	-Introducing students to the structure of a concept map. -Variation in first ionization energies of elements of Period 3.	-Use of Figs. 1 and 2 to explain structure of a concept map. -Brainstorming followed by construction of a concept map (Fig. 3). -Use of a flowchart concept map (Fig. 4) to summarize lesson.
Lesson 2	-Reaction of elements of Period 3 with oxygen. -Variation in melting points of oxides of Period 3 elements.	-Use of a spider concept map (Fig. 5) to test students' prior knowledge. -Use of a concept map (Fig. 6) for formative assessment. -Use of a spider concept map (Fig. 7) for summarizing lesson and as an evaluative tool.
Lesson 3	- Properties of oxides and hydroxides of Period 3 elements.	-Use of a hierarchy concept map (Fig. 8) to test students' prior learning about oxides of different elements. -Use of a spider concept map (Fig. 9) to explain reactions of the oxides with water. -Use of a spider concept map (Fig. 10) to explain the properties of some metal hydroxides. -Use of three hierarchy concept maps (Figs. 11, 12 and 13) for formative assessment.
CYCLE 3	CONTENT	STRATEGY
Lesson 1	- Chlorides of Period 3 elements.	-Students were required to construct relevant concept maps during the lesson using the pre-structured formats given in Figs. 14 and 15.
Lesson 2	-Reactions of chlorides of sodium, magnesium, aluminium and silicon with water.	-Use of a spider concept map (Fig. 16) for testing students' prior knowledge about the chlorides of Period 3 elements. -Students were required to construct relevant concept maps to describe the reactions of the elements with water.
Lesson 3	-Reactions of chlorides of phosphorus with water.	-Students were required to construct a concept map about the reactions of chlorides of phosphorus with water through self-study.

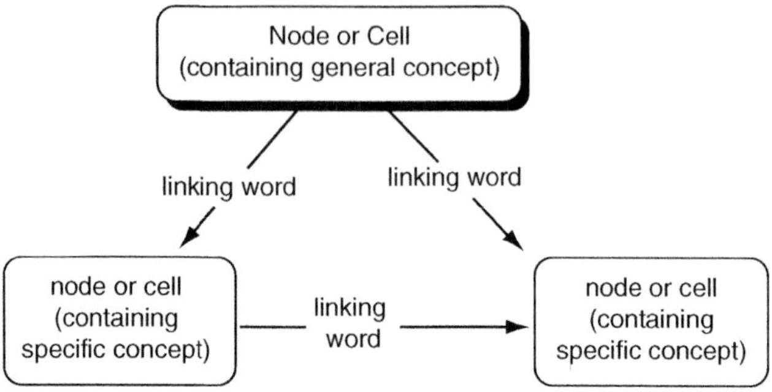

Fig. 1 Introducing students to the structure of a concept map

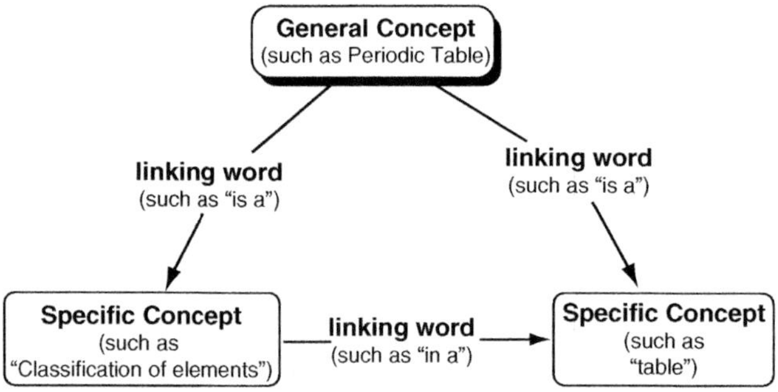

Fig. 2 Concept map used as an example

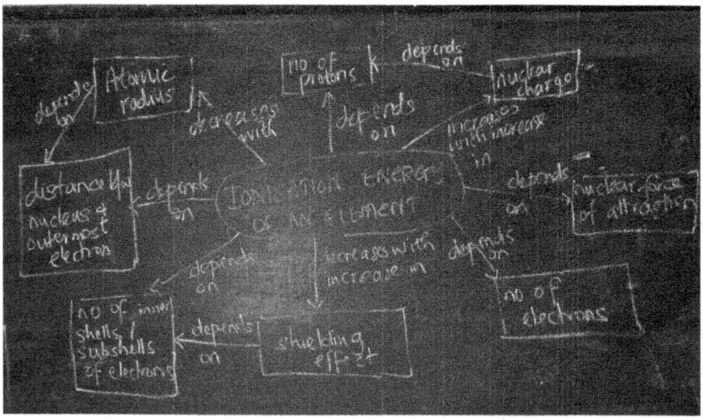

Fig. 3 Concept map constructed on factors affecting the ionization energy of elements

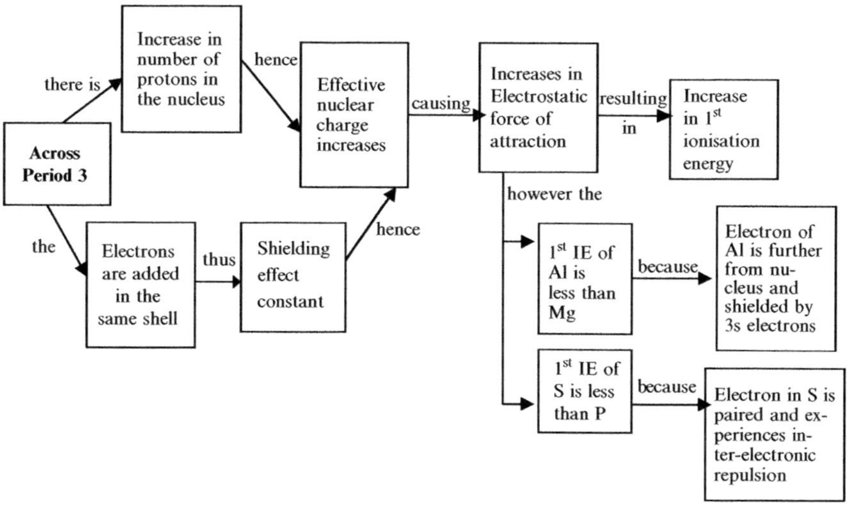

Fig. 4 Flowchart concept map showing trend in first ionization energy across Period 3

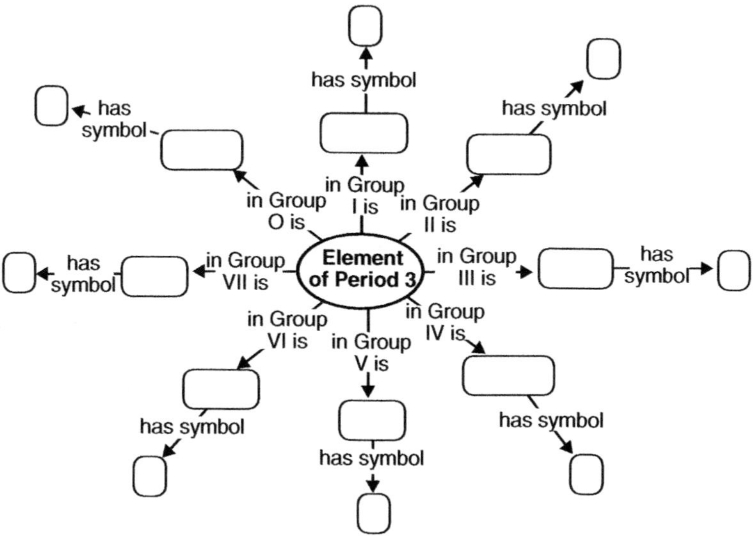

Fig. 5 Spider concept map used for testing students' prior knowledge about Period 3 elements

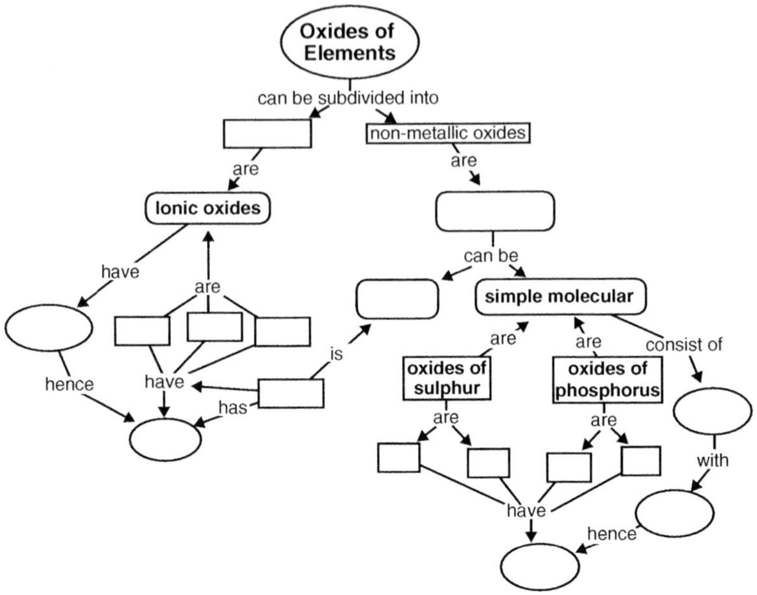

Fig. 6 Concept map used for assessing students' understanding of the oxides of Period 3 elements

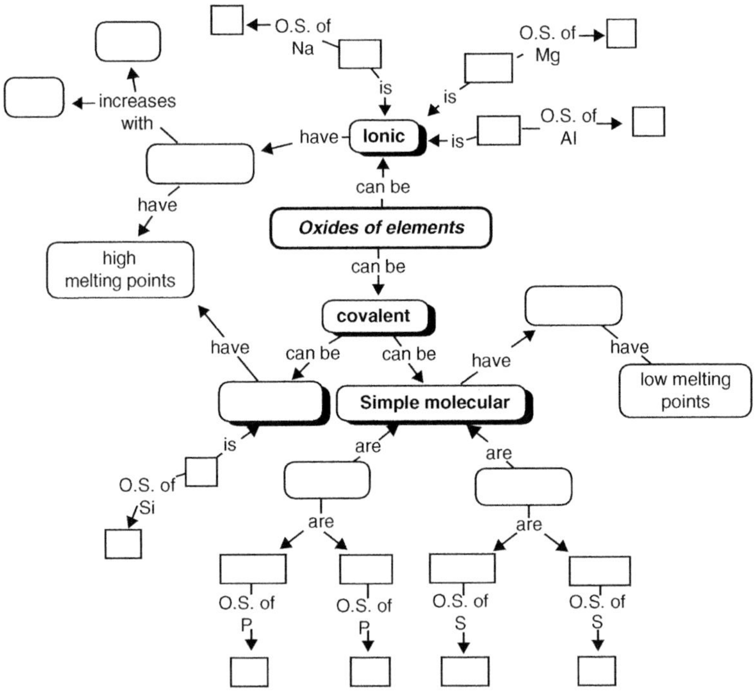

Fig. 7 Spider concept map used for summarizing information about the oxides of Period 3 elements (completed by students during the lesson)

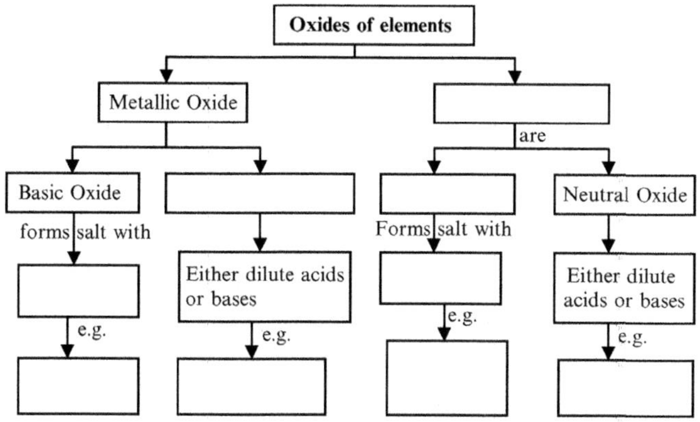

List of Concept words:

Non-metallic oxide	sulphur dioxide	lead (II) oxide
Amphoteric oxide	sulphur trioxide	magnesium oxide
Basic oxide	sodium oxide	phosphorus (III) oxide
Aluminium oxide	silicon dioxide	phosphorus (V) oxide
Barium oxide	nitrogen dioxide	water
Carbon dioxide	carbon monoxide	zinc oxide

List of Linking words: are, can be classified as, for example, with bases only, with acids only, forms salts with, does not form salt with

Fig. 8 Hierarchy concept map used to test students' prior knowledge about oxides of elements. (Students were provided with a list of concepts and linking words from which they could select.)

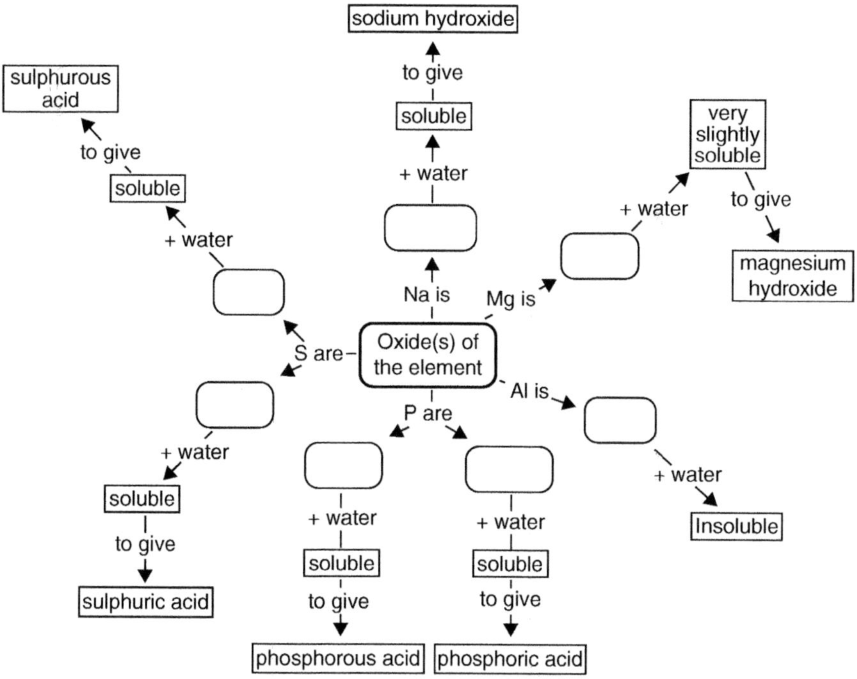

Fig. 9 Spider concept map used to explain the reactions of the oxides of Period 3 elements with water

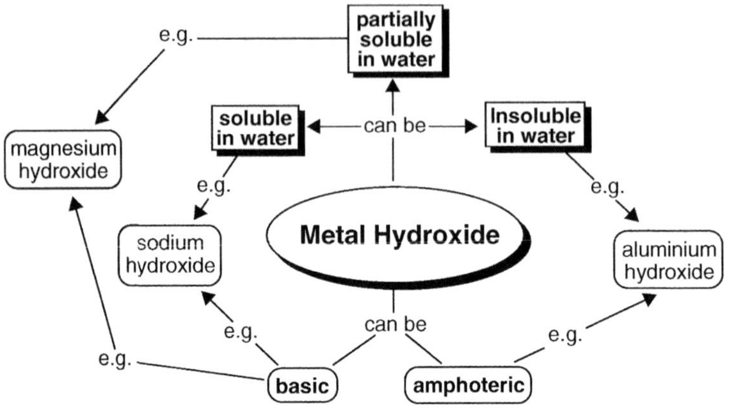

Fig. 10 Spider concept map illustrating the properties of hydroxides of sodium, magnesium and aluminium

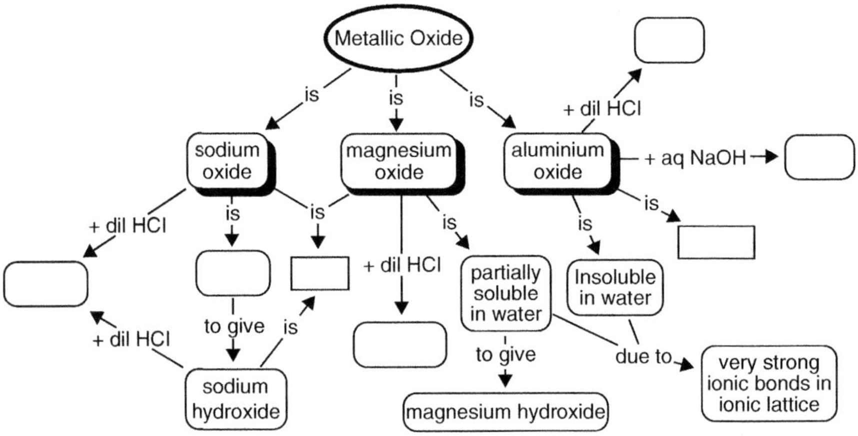

Fig. 11 Hierarchy concept map used for formative assessment (properties of metallic oxides)

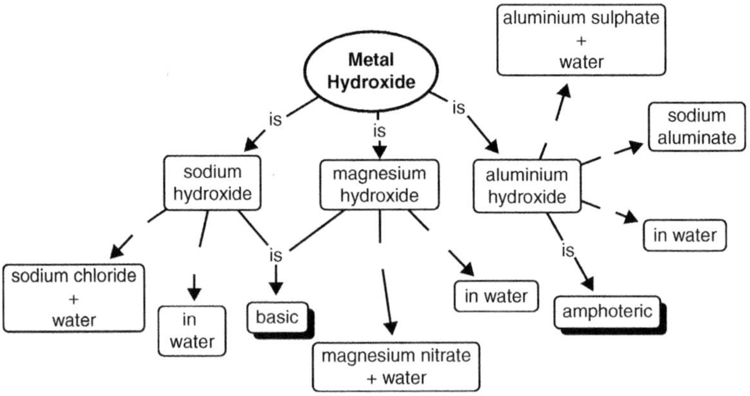

Fig. 12 Hierarchy concept map used for formative assessment (properties of metallic hydroxides)

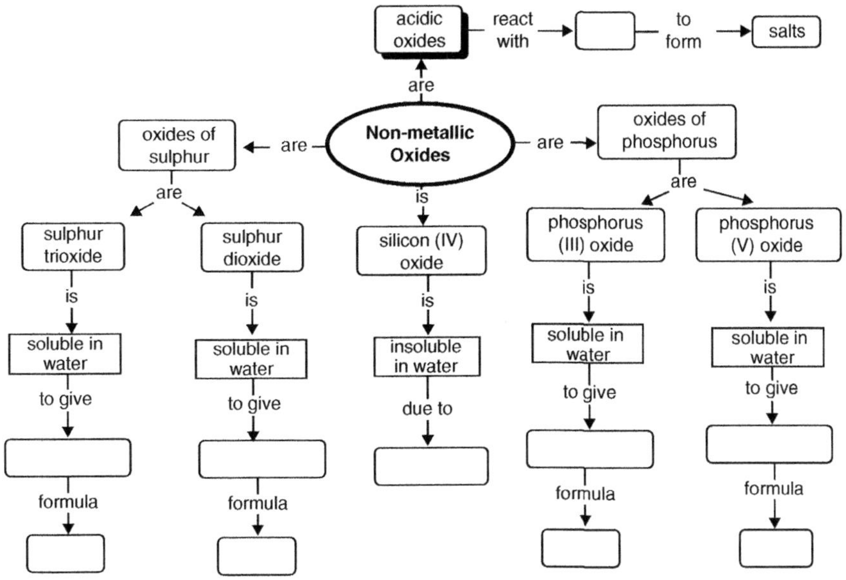

Fig. 13 Spider concept map used for formative assessment (properties of non-metallic oxides)

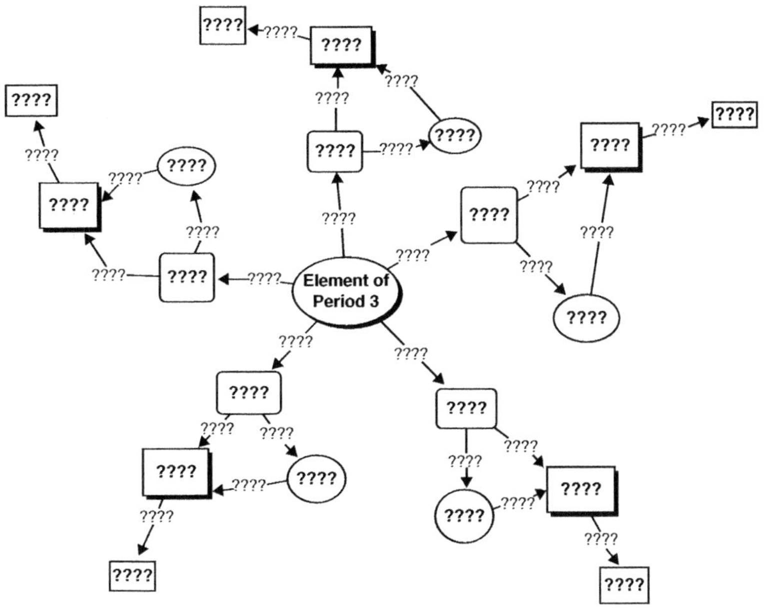

Fig. 14 Pre-structured format used by students to construct concept maps in Cycle 3 on the chlorides of Period 3 elements

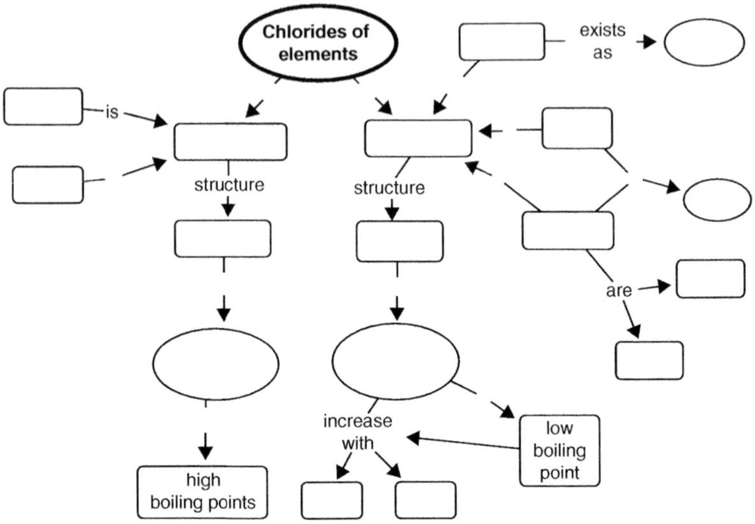

Fig. 15 Spider concept map used in Cycle 3. (Students were required to complete the concept map to relate boiling points to bonding and structure of the chlorides.)

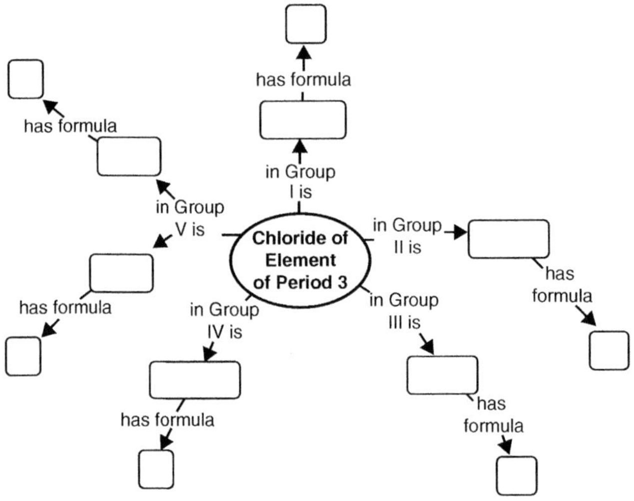

Fig. 16 Spider concept map on the chlorides of Period 3 elements (completed by students during the lesson)

2.3 Data Collection Methods

A multi-method approach, known as *triangulation*, was used in the present study to collect data in order to ensure validity and reliability [10]. In fact, four data-collecting techniques were used in our present research; namely observation checklists, students' achievement tests, questionnaire and an informal interview.

2.3.1 Observation Checklists

The observer is a 'key and active participant' in action research since the findings of the research depend a lot on 'observation and behavioural data' [11, 12]). Throughout the research, the students were observed at work to get a holistic picture of classroom events by making use of observation checklists. Detailed observation checklists which outlined the criteria to be observed were prepared for each lesson. The criteria to be observed included students' motivation and interest, and their participation in the lessons, as well their responses during the different steps of the lessons. Besides the above criteria, for the lessons including use of concept maps, additional criteria were included that relate to the use of concept maps, students' understanding of information presented in the form of concept maps, as well as their abilities to complete pre-structured concept maps and to draw concept maps from raw data.

In addition, a space was provided in the observation checklists to record any other relevant observation that was not included in the list of criteria.

2.3.2 Students' Achievement Tests

At the end of each cycle of the present action research, a students' achievement test was administered to assess students' understanding of the concepts taught during each cycle. Three tests were set, namely Tests 1, 2 and 3 which were administered at the end of cycles 1, 2 and 3, respectively. The achievement tests, designed by the teacher, were criterion-referenced since they aimed to test whether students had achieved specific objectives through the lessons [9]. The criteria for the three tests were standardized aiming at measuring learning at different levels according to Bloom's taxonomy [13, 14] . All the tests had the same structure consisting of two sections; Section A consisted of ten multiple choice questions while Section B consisted of a set of structured questions.

2.3.3 Students' Questionnaire

In order to gather information about the attitudes and opinions of students involved in the research, a questionnaire was administered to them. The questionnaire was carefully designed so that the questions did not present any degree of threat or sensitivity to the students [9]; it included mainly structured and semi-structured questions which aimed at obtaining feedback on their attitudes and reactions on the use of concept maps in classroom teaching. After piloting the questionnaire to two students of the sample, it was administered to all the participants at the end of the action research.

2.3.4 Interview

A semi-structured group interview was conducted in conjunction with the other data collecting methods to find out facts which could not be observed or obtained from the questionnaire and also to triangulate the information collected. A list of questions related to issues of the present study was drawn focusing on the perceptions and attitudes of students towards the use of concept maps in the teaching of 'Chemical Periodicity'. These questions were set to students and the answers provided by students were jotted down in note form paying attention to non-verbal messages (smiles, nodes, hesitations) for subsequent reconstruction of the events. In addition, questions were also improvised based on the interviewees' responses and noted accordingly. The interview was carried out after the third cycle; students were given the freedom to answer in either English or Creole (mother tongue) to allow them to express themselves fully and clearly.

3 Results and Discussions

In the present study, data have been collected through observation checklists, achievement tests, questionnaire and an informal interview; thus, the data presented are both qualitative and quantitative. The data collected from the different data-collecting tools are presented below.

3.1 Data Gathered from Observation Checklists

In the first cycle of the action research, it was observed that many students did not appreciate the bulk of notes that was directed to them through dictation or overhead projector. Some students were copying the notes without even trying to understand their meaning since they were not motivated. They were also apprehensive about the lengthy notes they were required to take down. It was also observed that some students found it difficult to follow the pace of the class discussions due to low linguistic intelligence [15]; they were not able to cope with the situation since the method employed did not match their learning styles [16].

In the second cycle of our research, concept mapping was introduced to the students. Rye [17] stressed that the key in beginning to develop concept mapping skills in students is to use subject matter where they can apply existing knowledge to think of valid relationships. Thus, the first exposure to concept mapping was based on the construction of a concept map on 'factors influencing the ionization energy of elements', an already studied concept relevant to the topic 'Chemical Periodicity'. The brainstorming activity, which preceded the introduction of concept mapping, resulted in the board being filled with disorganized information. The mapping technique produced a neat diagram (Fig. 3) that showed the ideas more clearly, and highlighted the relationships between the various concepts. Students were able to appreciate that concept maps present an effective way to arrange a large amount of information in an organized manner, and involved use of minimum text [18]. It was observed that students experienced some difficulties in drawing the concept maps in their note books. According to Rye [17], students develop the competence in concept mapping only if the early exposure to the approach is enjoyable. Hence, in the following lesson, pre-constructed concept maps were distributed on worksheets, and students were required to fill in blank spaces with appropriate concepts or linking words in an attempt to make the integration of concept mapping easier and smooth to the students. During the second cycle, it was also observed that some students experienced difficulties in reading the propositions, and in finding the appropriate words to complete the concept maps provided. Subsequently, students were provided with a list of words from which they could select to complete pre-structured concept maps. It was indeed found that students' understanding was enhanced when they were provided with a list of words to explain the relationships between concepts in pre-structured concept maps. This is in accordance with Ujoodha's claim [25] that teachers should vary their techniques of using concept maps with respect to the needs of their students. Findings from observation have also revealed that students succeeded to make neat diagrams when the concept maps were simple and did not contain too many concepts. Moreover, it was also noticed in the second cycle that students who had mastered the concept-mapping skills were helping their friends, that is, the concept-mapping technique was promoting student–student interactions and development of social skills as reported by Ujoodha [25].

In the third cycle of our study, students were required to construct their own concept maps on selected subtopics of 'Chemical Periodicity' during the lessons. Students were provided with information in small paragraphs, from which they had to extract the propositions, identify concepts and linking words to build up relevant concept maps through individual as well as group work. Our findings have shown that students had more difficulties in constructing concept maps through individual work than through group activity; thus, it can be concluded that group work facilitates the construction of concept maps since students have the opportunity to discuss with their peers to clear misconceptions. Findings from observation have also revealed that during the construction of concept maps, many students tend to omit linking words, and found it difficult to select appropriate concept words and linking words to construct good propositions. Indeed, Novak and Cañas [19] have recently advocated that the most challenging and difficult aspect of constructing a concept map relates to the construction of propositions where linking words that clearly depict the relationship between concepts have to be determined. They also reported that the process of identifying the most prominent linking words involves high levels of cognitive performance like evaluation and synthesis of knowledge. Moreover, Striebel [20] reported that students are engaged in metacognitive thinking during the construction of concept maps, and this process promotes meaningful learning. Data collected through observation have also revealed that the concept mapping skills of students had significantly improved compared to cycle 2 of the action research. The concept maps prepared by students were neat; the students were found to be very creative in drawing their concept maps.

Most importantly, data collected through observation in the second and third cycles have revealed that the use of concept maps resulted in a high degree of motivation among the students; they were fully receptive and engaged during the lessons. Concept mapping has proved to be a good technique to motivate students and encourage hard work. Our findings have also shown that there was enhanced teacher–students interaction during the lessons, since the frequency of students asking for help and clarification from teacher has significantly increased from cycle 1 to cycle 3. This is in line with Prnka [21] who reported that concept mapping strategy improves the communication between teacher and students, in view of the fact that it encourages the latter to express confidence and ownership in their learning. Our results have also shown that use of concept maps as a strategy allowed teacher to provide quick feedback on any misconception that might occur during the teaching process as earlier stated by Zimmaro and Cawley [22].

3.2 Data Gathered from Students' Achievement Tests

An achievement test was administered to the students after each cycle, namely Test 1 after cycle 1, Test 2 after cycle 2 and Test 3 after cycle 3, in order to assess the impact of the use of concept maps on their performance. Table 2 shows the marks scored by each student represented by a code number in the three tests; names of students have not been displayed for ethical reasons. On the other hand, Fig. 17 illustrates the variation in the number of students scoring marks in specific percentage ranges. Table 3 and Fig. 18 illustrate a comparison of the minimum, maximum and average marks obtained by students in the three tests.

In Test 1, the overall results revealed that the performance of students in the achievement test was not very satisfactory. As indicated in Fig. 17, four students (representing 16.7%) did not obtain pass marks (above 40% marks) and no one managed to obtain a score higher than 90% marks. Scripts of students revealed that questions testing knowledge and comprehension were generally well answered. However, questions based on higher-order thinking proved to be somewhat difficult and less scoring. Moreover, many students wrote at length giving irrelevant information and missing on important points. This shows that students have learnt the facts by rote and have been unable to mould their knowledge according to the requirements of the questions. The major shortcoming of the students in Test 1 was their inability to apply their knowledge to answer specific questions. They appeared to have learnt the content of the topic by rote without proper conceptual understanding; they stated the facts as answers to questions without giving any importance to what is asked in the question.

Findings from Test 2 have revealed that there was a slight but definite improvement in the performance of the students in Test 2 compared to Test 1 (Table 2). The maximum mark has increased from 30 to 33, the minimum mark from 7 to 11 and

Table 2 Students' scores in Tests 1, 2 and 3

Student	Test 1 (Score out of 35)	Test 2 (Score out of 35)	Test 3 (Score out of 35)
A1	17	20	25
A2	25	26	28
A3	24	27	29
A4	9	11	20
A5	15	20	22
A6	30	29	32
A7	24	28	29
A8	25	30	28
A9	29	32	32
A10	23	25	26
A11	17	19	21
A12	24	26	29
A13	14	20	21
A14	29	33	25
A15	22	23	26
A16	20	24	24
A17	23	23	29
A18	17	15	21
A19	24	29	30
A20	18	26	24
A21	16	19	20
A22	19	22	21
A23	11	15	17
A24	7	12	20

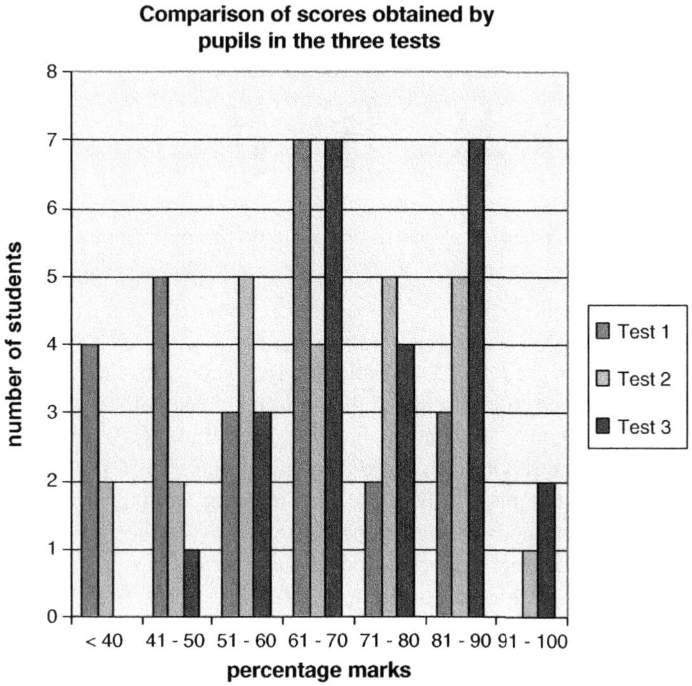

Fig. 17 Comparison of students' performance in Tests 1, 2 and 3 in terms of percentage marks

Table 3 Comparing the maximum, minimum and average marks in Tests 1, 2 and 3

	Test 1	Test 2	Test 3
Maximum mark	30	33	32
Minimum mark	7	11	17
Mean	20.1	23.1	25
Median	21	23.5	25
Standard deviation	6.2	6	4.3

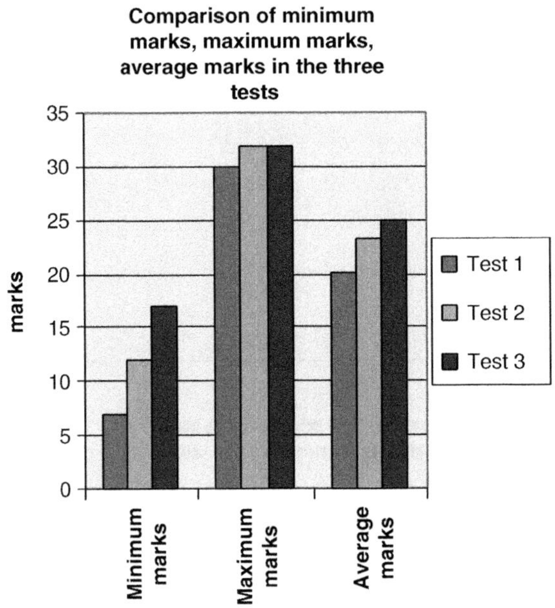

Fig. 18 Comparison of minimum, maximum and average marks obtained by students in Tests 1, 2 and 3

the mean mark from 20.1 to 23.1 (Table 3). This shows that the use of concept maps in the teaching of 'Chemical Periodicity' has helped to improve the performance of students. This finding is in line with findings from several authors [7, 23–26], who have reported that use of concept maps has a positive impact on the performance of students. Furthermore, data obtained from the scripts of students in Test 2 have also revealed that the degree of irrelevancy had decreased considerably and the answers were more concise and precise. Students were more critical in answering questions and the regurgitation of facts learned by rote had decreased. This clearly shows that use of concept maps in the second cycle of the action research has encouraged the learners to think about the relationships between different concepts, and has encouraged critical thinking and enhanced conceptual understanding.

In addition, data gathered from Test 3 have clearly shown that there was further improvement in the performance of students as compared to Test 2 (Table 2). The minimum mark has increased from 11 to 17 and the mean mark has increased from 23.1 to 25.0 (Table 3). It is interesting to note that no student obtained less than 40% marks in Test 3, while two students scored more than 90% marks as indicated in Fig. 17. These results clearly indicate that the construction of concept maps by students in cycle 3 has further enhanced conceptual understanding in 'Chemical Periodicity'. Findings from the students' scripts of Test 3 have undoubtedly shown that students were better able to use and apply relevant concepts while answering questions; in addition, the answers were structured in a more sequential and logical manner. Novak and Wandersee [27] have indeed reported that the construction of concept maps by students leads to meaningful learning. Moreover, during the construction of concept maps, a powerful knowledge framework is created, which permits utilization of the knowledge in new contexts, as well as the retention of the knowledge for longer periods. Thus, it can be seen that allowing the students to construct their own concept maps during cycle 3 has further enhanced acquisition of concepts and has resulted in students performing better and achieving higher marks in Test 3.

It is evident from Fig. 18 that there has been an improvement in the performance of the students in Tests 2 and 3 as compared to Test 1. The minimum, average and maximum marks scored by students have increased considerably, showing that concept mapping, as a strategy, has enhanced conceptual understanding resulting in improved performances. Moreover, it can be seen that the number of students obtaining low percentage marks (between 41% and 50%) has decreased from Test 1 to Test 3, whereas the number of students obtaining high percentage marks (between 81% and 90%) has increased from Test 1 to Test 3. This shows that use of concept maps has not only improved the overall performances of the students but has also contributed in improving the quality of the results.

From Table 2, it can be seen that for most students, the marks scored have increased from Test 1 to Test 3. However, for two students (representing 8.3% of the sample), the scores did not change much during the three cycles of the action research, while for five students (representing 20.8% of the sample), the scores have even decreased. This may be attributed to the fact that these students do not have high visual spatial-intelligence, and hence the use of concept maps has not suited their learning style. According to Gardner's theory [15], the different kinds of intelligences can be developed in an individual through proper coaching. Hence, frequent exposure to concept maps in the teaching process may in the future help these students to develop their spatial/visual intelligence.

3.3 *Data Gathered from Students' Questionnaire*

After implementing the three cycles of the action research, a questionnaire was administered to all students involved in the study to gather feedback on their attitudes and opinions on the use of concept maps in classroom teaching. The major findings gathered from the questionnaire are presented below under relevant headings.

- Exposure to concept maps

When asked whether they have been exposed to concept maps prior to this study, it was interesting to note that 19 students, representing 79.2% of the sample, claimed that they had earlier come across concept maps in classroom situation. It is evident from this finding, that teachers are nowadays more concerned with the use of active teaching strategies. Indeed, in pedagogical courses and teacher-training sessions in Mauritius, teachers are exposed to various strategies, which help to make learning more effective and meaningful.

- Students' understanding of the general structure of a concept map

In response to a question about the general structure of a concept map and the basic rules that should be followed when constructing concept maps, all students who asserted they had used concept maps before, knew that the latter show the relationships between pieces of information by linking them with arrows. However, no student was familiar with the terms nodes or cells prior to our study.

- Students' views on how concept maps have helped them in the learning process

Students were asked about the different ways in which concept maps have helped them to learn 'Chemical Periodicity'; they were given six statements and were required to tick whether they strongly agree, agree, disagree or strongly disagree. Findings to this question have revealed that most students (83.4%) agreed or strongly agreed that concept maps represent a good method for summarizing notes; a similar finding was earlier reported by Learning Skills Program [28]. Data from this question have also shown that the majority of the students (75.0%) agreed or strongly agreed that concept maps have helped them to understand relationships between the concepts and have made the lessons more interesting. These results lend support to earlier literature, which stated that concept maps can help to develop an understanding of a body of knowledge by exploring the relationships between the concepts [22] and can make lessons more interesting [7, 23, 26]. It was also interesting to note that 79.2% of the students agreed or strongly agreed that concept maps could be used efficiently to organize large amount of information as earlier observed and reported by Plotnic [18] and Patry [29]. Furthermore, a high percentage of the students (66.7%) has agreed or strongly agreed that concept maps aid memorization and retention of facts; this result is in line with findings from Novak and Wandersee [27]. Nevertheless, eight students (representing 33.3% of the sample) did not agree with the fact that concept maps aid memorization and retention of facts; as stated earlier; this might be attributed to the fact that these students do not possess high visual intelligence. Last but not least, most students (66.7% of the sample) agreed or strongly agreed that concept maps have helped them in retrieval and recall of facts during assessments.

- Students' preference for use of concept maps over other strategies

Our findings have revealed that 75% of the sample students preferred filling blank spaces in concept maps during testing of pre-requisites to questioning, while 66.7%

of the students preferred the use of concept maps for summarizing lessons. In addition, 70.8% of the students stated that they prefer the use of concept maps for presentation of findings following a group activity.

- Rating the different ways of using concept maps as interesting or boring

When asked to rate the different ways of using concept maps as boring or interesting, our findings have shown that 50% of the students stated that following the construction of concept maps by the teacher on the board is interesting, while the rest found it to be either boring (33.3%) or very boring (16.7%). However, it is also clear from our findings that all students found filling of blank spaces in concept maps to be either interesting (66.7%) or very interesting (33.3%). It is also encouraging to note that all students found it interesting (83.3%) or very interesting (16.7%) to construct their own concept maps during the learning process. This last finding points out to the fact that students found the lessons motivating when they are actively involved in the construction of concept maps.

- Students' views on the impact of the use of concept maps on learning

It is interesting to note that most students reported that they were able to retain the facts better when they construct their own concept maps. Indeed, Novak and Wandersee [27] reported that the building up of a structure of knowledge using small units of interacting concepts serves as a scaffold in the organization of knowledge; this enhances meaningful learning and retention of facts. Furthermore, many students preferred construction of concept maps coupled with group work, as the method challenged them to complete the task assigned to them. According to Novak and Cañas [19], constructing concept maps through group work can be challenging and motivating since critical comments are directed at the concept maps, not at the person(s) building the maps; a factor which removes fear of embarrassment and personal insecurities.

- Students' preference for hierarchy or spider concept maps

When asked about the type of concept maps they prefer, it was found that most students (62.5%) preferred hierarchy concept maps to spider concept maps, since the former arrange the concepts more orderly than spider concept maps. Only 12.5% students preferred spider concept maps, while 25% students reported that they appreciated both types of concept maps.

- Students' opinion about use of concept maps in other Chemistry topics

In response to a question about whether concept maps should be used to teach other topics of Chemistry, it was interesting to note that a great majority of the students (62.5%) responded positively as illustrated in Fig. 19.

- Using concept maps for revision purposes

More than 50% (in fact 58.3%) of the students stated that they would use concept maps for revision purposes in the forthcoming examinations, while 41.7% responded negatively as shown in Fig. 20. Learning Skills Program [28] has indeed

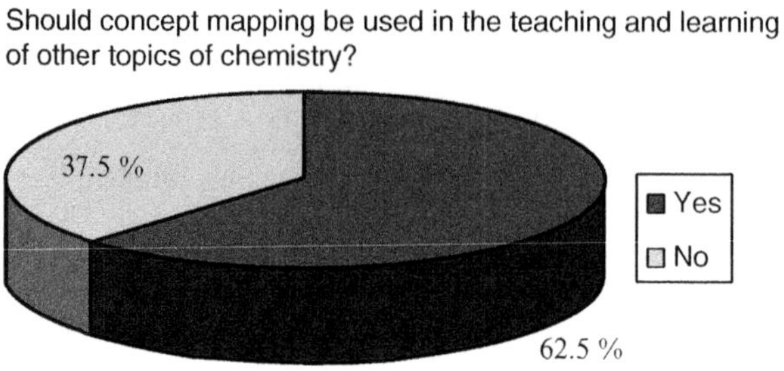

Fig. 19 Students' views about use of concept maps to teach other Chemistry topics

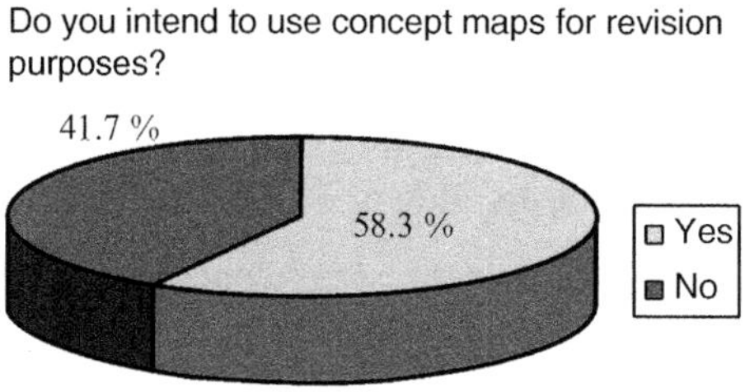

Fig. 20 Students' responses to whether they would use concept maps for revision purposes

reported that concept maps offer an excellent method for revision since they facilitate memorization of facts and recall in the examinations.

- Students' own comments on the use of concept maps in the teaching of 'Chemical Periodicity'

The last question of the questionnaire was an open-ended question that aimed at encouraging the respondents to comment on the use of concept maps in the teaching of 'Chemical Periodicity'. Some students reported that the strategy rendered the lessons less boring, since they were challenged to go through the various concept maps to check if they were able to understand the ideas. Moreover, some reported that since they were required to participate actively by filling and constructing concept maps, the lessons were livelier. However, a few students stated that they preferred linear notes to concept maps since they are more used to access information in linear formats; thus they found it confusing to read the information presented in concept maps. This attitude may be attributed to the fact that these students may have developed their verbal learning style better than the visual learning style.

3.4 Data Gathered from Interview

A group interview was carried out on a sample of five students selected by random sampling method. Findings from the interview have revealed that the students perceived 'Chemical Periodicity' as being a topic loaded with facts, details and descriptions, but they also acknowledged that the concepts involved can be linked to each other and to other topics like atomic structure and chemical bonding. When the interviewed students were asked about the notes provided in the first cycle of the research, they stated that when notes were dictated, they just jotted down whatever they heard without reflecting and understanding these notes; this finding further supports the data gathered from the questionnaire. Nevertheless, all students interviewed claimed that they preferred filling of cloze tests over dictation of notes or projection of notes on overhead projector.

When asked to give their views about the use of concept maps during the lessons, the interviewed students reported that at the start of the study, they had a preference for information given in linear formats, claiming that they have always been exposed to these, and that they were unable to read from concept maps. Moreover all the students advocated that drawing concept maps proved to be quite difficult at the beginning. However, three students acknowledged that with practice, their concept mapping skills had improved and they were able to draw neat concept maps. Furthermore, they all asserted that they now prefer to use concept maps instead of taking bulky linear notes, which they found boring. In addition, they agreed that concept maps decreased the amount of text they need to read, since only the important words are shown and all information are shown on one page.

It is also interesting to note that all students interviewed asserted that concept maps were useful in learning about 'Chemical Periodicity' where a lot of facts need to be memorized. They asserted that concept maps have helped them to organize the concepts in a systematic way and to understand how the different concepts are linked to each other. One student even stated that the concept maps were easily impregnated in his memory so that he could easily retrieve the information during the assessments. Moreover, as revealed in the questionnaire and in the achievement test scripts, the students agreed that they did not have many facts to learn by rote, since they could understand the concepts better and apply them where required when they were taught through use of concept maps. In addition, four students advocated that construction of concept maps has also helped them to retain the facts better.

Findings from the interview have also revealed that all students interviewed advocated that they prefer hierarchy concept maps over spider concept maps, since the latter appear sometimes too complicated and the ideas are 'intertwined'. Hierarchy concept maps, on the hand, were described as being well structured showing the ideas in an orderly sequence. All students interviewed also reported that they prefer to have the pre-structured concept maps on which they can fill the blank spaces during the lessons. They argued that when they drew free-hand concept maps, the latter tended to be untidy and messy, whereas pre-constructed ones were better structured and more attractive. On the contrary, one student stated that he preferred to construct his own concept maps freely without using the pre-structured ones, as the latter restricts his freedom to mould the concept maps according to his will. As revealed in the questionnaire, the interviewed students all acknowledged that they found it easier to construct concept maps when working in groups, since they had the opportunity to discuss the ideas with their friends so that it was easier to find the linking words and concept words.

The interviewed students also claimed that they would be using concept maps to highlight the main points during revisions. It was also interesting to note that two students asserted that constructing concept maps allows them to put down all their ideas on paper and to group similar ideas together. Hence they have started to use concept maps to elicit ideas to write essays in other subjects including General Paper and French.

4 Conclusions

In the present study, concept maps were used in the teaching of 'Chemical Periodicity'. We need to highlight that for certain parts of the lessons, expository teaching was unavoidable to explain the basic concepts upon which students can build up new ones.

In conclusion, our findings have revealed that students had adopted a positive attitude towards the use of concept maps in the teaching process. However, we have also found that for successful implementation of concept mapping as a teaching strategy, it is necessary to ensure that the technique is integrated smoothly in the teaching-learning process; also complex concept maps should be avoided in the initial stages to avoid overwhelming of students.

Moreover, results from our present study have also shown that the use of concept maps in the teaching of 'Chemical Periodicity' has enhanced students' conceptual understanding, leading to an improvement in their performances throughout the action research cycles. Indeed, concept maps promote meaningful learning by allowing learners to build connections between concepts, and understand how the concepts relate to one another. Our findings have also indicated that students' understanding was further enhanced when they were actively involved in constructing their own concept maps. According to Novak and Wandersee [27], concept maps serve as scaffolds, which help to organize and structure knowledge, thus facilitating the construction of new knowledge and promoting meaningful learning. It needs to be highlighted that all students in a classroom do not have the same learning style; it was found that although most students involved in the present study have benefited cognitively from the use of concept maps, a handful of students could not adapt to this technique. These students may have a preference for other learning styles over the visual style. According to Gardner's theory [15], visual intelligence like any other kind of intelligence can be taught. Therefore, if these students are exposed to visual learning often, they can develop their visual intelligence.

Besides, our findings have also revealed that students were more careful when they write answers to questions in terms of relevance, precision and conciseness after they had been exposed to the use of concept maps in the teaching process; this indicates that use of concept maps has promoted the development of higher-order and critical-thinking skills. Our results have also shown that concept maps have helped students to structure their answers in a logical manner and their language had considerably improved.

Moreover, findings from the present research have also shown that during the use of concept maps, students were actively engaged, and the lessons were livelier, a view shared by most students in the questionnaire. Besides, students were very cooperative and there was a lot of sharing among them; thus, use of concept maps has helped them to develop social skills. Mintzes et al. [30] have indeed reported that since concept maps are highly explicit, they serve as ideal vehicles for exchange of ideas, thereby favoring the collaborative construction of new knowledge. Our findings have also shown that during the lessons involving use of concept maps, teacher-student relationship had improved. Even the shy students were more confident to speak up during the lessons, asking for help either from peers or from the teacher. Prnka [21] has indeed reported that use of concept mapping as a teaching strategy improves the communication between teacher and students, since it encourages students to express confidence and ownership in their learning.

In conclusion, it can be noted that concept maps have been an effective tool in motivating students to engage themselves actively during the teaching and learning of 'Chemical Periodicity'. The concept maps have also promoted the development of critical, creative and higher-order thinking as well as metacognitive skills thereby enhancing acquisition of concepts, and improving students' performance. Moreover, the use of concept maps as a strategy has been a stepping-stone to promote students' enthusiasm, interest and self-confidence during the lessons; the students were more confident and expressed ownership in their learning resulting in improved teacher-student as well as student-student relationships.

References

1. Ausubel DP (1963) The psychology of meaningful verbal learning: An introduction to school learning. Grune & Stratton, New York
2. Chinese University of Hong Kong (2005) The use of Concept maps in the teaching learning process. Faculty of Education, the Chinese University of Hong Kong. http://www.fed.cuhk.edu.hk/~johnson/misconceptions/concept_map/cmapguid.html. Accessed 3 October, 2006
3. Earl BL (2007) Concept Maps for General Chemistry. J Chem Educ, 84(11):1788–1789
4. GRAPHIC.ORG. The Graphic Organizer: Concept map. http://www.graphic.org/concept.html. Accessed 3 October, 2006
5. Hyerle D (2000) (cited in Costa, 2001) Developing Minds. Association for Supervision and Curriculum Development, USA
6. University of Cambridge Local Examinations Syndicate, Examiner's Reports (UCLES) (1995, 2001, 2002, 2003). Cambridge, United Kingdom
7. Clark IF, James PR (2004) Using concept maps to plan an introductory structural geology course. J Geosci Educ 52(3): 224–230
8. Hinck S (2004) The Effectiveness of Concept Mapping to Measure Student Learning in a Community-Based Mental-Health Nursing Course. Report of a 2003–04 Project in the SMSU Teaching Fellowship Program. http://adc.missouristate.edu/Hinck%20TFP%20Final%20Report%20for%20FY2004.pdf. Accessed 25 November, 2005

9. Cohen L, Manion L, Morrison K (2000) Research Methods in Education, 5th edn. Routledge Falmer, London
10. Bell J (1999) Doing your Research Project: A guide for first-time researchers in education and social science. Open University Press, Buckingham, Philadelphia
11. Blaxter L, Hughes C, Tigh M (1996) How to Research. Open University Press, Buckingham
12. Cohen L, Manion L (1994) Research Methods in Education, 4th edn. Routledge, London
13. Petty G (2001) Teaching Today. Nelson Thornes Ltd., Cheltenham
14. Fisher R (1995) Teaching Children How to Think. Stanley Thornes Publishers Ltd., London
15. Gardner H (1991) The Unschooled Mind: How Children Think and How Schools Should Teach. Basic Books, New York
16. Fleming ND, Mills C (1992) VARK a guide to learning styles. http://www.vark-learn.com/English/index.asp. Accessed 1 September, 2006
17. Rye JA (1995) Concept maps and Concept mapping. College of Education, The Pennsylvania State University http://www.ed.psu.edu/ci/Papers/STS/gac-6/amapgd.htm. Accessed 25 November, 2005
18. Plotnic E (1997) Concept mapping: A Graphical System for Understanding the Relationship between Concepts. ERIC Digest. ERIC Identifier: ED407938. http://www.ericdigests.org/eric-digests.html. Accessed 30 June, 2006
19. Novak JD, Cañas AJ (2006) The Theory Underlying Concept Maps and How to Construct Them. Florida Institute for Human and Machine Cognition. http://www.ihmc.us/ Accessed 18 November, 2006
20. Striebel JF (2003) Improving Student Metacognition and the Assessment of Conceptual Understanding through Pre and Post Unit Concept Mapping. www.montana.edu/msse/capstone2003.htm Accessed 18 November, 2006
21. Prnka R (2003) The Use of Multiple Assessment Strategies to Enhance Student Achievement in Ninth Grade Physical Science. www.montana.edu/msse/capstone2003.htm. Accessed 18 November, 2006
22. Zimmaro DM, Cawley JM (1998) Concept map module. Schreyer Institute for Innovation in Learning. The Pennsylvania State University. http://www.inov8.psu.edu/faculty/cmap.htm. Accessed 3 October, 2006
23. Trowbridge JE, Wandersee JH (1994) Identifying critical junctures in learning a college course on evolution. J Res Sci Teach 31:459–473
24. Snead D, Young B (2003) Using concept mapping to aid African American students' understanding in middle grade science. Journal of Negro Education http://www.findarticles.com/p/articles/mi_qa3626/is_200307. Accessed 25 November, 2005
25. Ujoodha M (2003) Use of Concept Maps in practical work in Chemistry at Form III Level. A Case Study. PGCE Dissertation, Mauritius Institute of Education, Réduit, Mauritius
26. Scandett JF (2005) The efficacy of Concept Mapping in Aural Skills Training. The School of Education, University of Pittsburg. http://etd.library.pitt.edu/ETD/available/etd-04102005-204914/unrestricted/scandrett_etd05.pdf. Accessed 25 November, 2005
27. Novak JD, Wandersee J (1991) (cited in Novak JD, Cañas AJ, 2006) The Theory Underlying Concept Maps and How to Construct Them. Florida Institute for Human and Machine Cognition. http://www.ihmc.us/. Accessed 18 November, 2006
28. Learning Skills Program (2003). Concept Mapping. University of Victoria. http://ww.counc.uvic.ca.Accessed 3 October, 2006 Lewin K (1946) (cited in McNiff J, 1988) Action Research, Principles and Practice. Macmillan Education Ltd., London
29. Patry J (2004) Effects of short term training in concept-mapping on the development of metacognition http://cmc.ihmc.us/papers/cmc2004-051.pdf. Accessed 18 November, 2006
30. Mintzes JJ, Wandersee JJ, Novak JD (2000) Assessing Science Understanding: a Human Constructivist. Academic Press, New York

Non-Linear POGIL for Developing Cumulative Skills and Multidisciplinary Chemical Concepts for Non-Science and Chemistry Majors

R. T. Criasia, A. B. Lees, M. T. Mongelli, Y.-G. K. Shin, H. Stokes-Huby, and D. E. Vitale

Abstract Engaging non-science majors in chemically related global and civic issues using rigorous chemical principles rather than descriptive discussion has been achieved using POGIL (Process-Oriented Guided-Inquiry Learning) group classroom activities. These activities provide an excellent foundation for developing students' skills, confidence and comfort with scientific content to pursue independent literature projects on global and civic concerns. Based upon researched data, student project reports include a brief summary of an issue and one or more original calculations using conversion factors or stoichiometry to illustrate its magnitude, consequence or resolution. Being able to answer a chemical question of their own making transforms the students' perceptions of their own abilities and of the accessibility of science in their lives [1].

1 Introduction

The learning cycle model [2–4] used for POGIL activities is highly effective in engaging students in thoughtful learning. However, the cumulative nature of chemical concepts is a major challenge for non-science majors. This paper focuses on our beginning efforts to restructure the conceptual content of POGIL activities to support more assiduously the challenge of conceptually cumulative knowledge. A structured multi-concept approach has proven to be highly successful in limited trials. Rather than having a sequence of activities focus upon individual chemical skills, multi-skill and multi-disciplinary classroom activities are being developed to model and reinforce concept mastery, cumulative reasoning and multi-disciplinary significance at the same time. By de-compartmentalizing chemical concepts and treating them as components of a larger problem, students are weaned from their expectation of linear logic and the force-fed pattern-recognition imposed by traditional texts.

R.T. Criasia, A.B. Lees (✉), M.T. Mongelli, Y.-G.K. Shin, H. Stokes-Huby, and D.E. Vitale
Department of Chemistry-Physics, Kean University, Union, New Jersey,
e-mail: blees@kean.edu

M. Gupta-Bhowon et al. (eds.), *Chemistry Education in the ICT Age*,
DOI 10.1007/978-1-4020-9732-4_19, © Springer Science+Business Media, B.V. 2009

Since the content are repeated, multi-concept activities early in the course can foreshadow more of the later concepts and students can be better prepared to master those concepts more rapidly and with greater confidence.

When applied to science majors, multi-skill, multi-disciplinary guided-inquiry activities may provide an ideal mechanism for preparing students for their entire undergraduate curriculum, as an integrated whole, and not as a piecemeal assemblage of sub-disciplinary fragments.

2 Background: Successful Introduction of Guided-Inquiry with Non-science Majors

Comprehensive General Education innovations in 1999 at Kean University resulted from a planned and coordinated 3-year effort to optimize the development and delivery of both curriculum and academic support. It was in this context of General Education curriculum reform that Chemistry 1200, 'Chemistry in Your World', a one-semester, non-laboratory course for non-science majors was approved with the intent of addressing science and math literacy in a framework of civic and global discussion. The composition of the class includes all undergraduate levels from freshmen through graduating seniors, for many of whom science and math phobias have postponed their completing a General Education science elective. Since College Algebra is a course prerequisite, the only student group not well represented is first semester freshmen who are fulfilling this requirement. Class size is capped at 24 students and one POGIL section has been offered each semester.

Many textbooks for this scientific genre are oriented first to teach chemical principles in a traditional format and then apply those principles to chemical issues. With this approach, students spend most of their time in a lecture class where much of the transmitted chemical information is retained no more fully than it may have been in a previous high-school course. In fact, in such an environment, students often rationalize that their poor performance is a result of their inability to 'remember the answers' from high school rather than a result of their inability to apply reasoning and analytical skills.

Even when using one of the few textbooks that incorporate quantitative skills in applied chemical topics, it is still difficult for the students to progress through more than a few topics within the time constraints of a single academic semester. But, the greatest impediment to the educational growth of students, is that no matter how well written, or how integrated the topics, most of the information transmitted to students through most textbooks is ***expository***. As a result, the ability of each student to assimilate new chemical information and transform it into knowledge that they can fully appreciate and apply depends upon the previously developed cognitive skills that each brings with them to the class. Student outcomes then tend to preserve the math and science phobias and the level of performance characteristic of each student's previous efforts, no matter how enthusiastic the students may be about a chemical topic they study.

After teaching this one-semester course, 'Chemistry in Your World', in an interactive, but traditional, lecture format, one begins to recognize and to welcome the following pedagogical challenges:

1. Establishing the mathematical and analytical skills needed for students to understand rigorous chemical principles rather than relying upon more qualitative discussion.
2. Empowering **all** students through active reading, interpreting and solving chemical problems rather than reaching only the most motivated students.
3. Including stoichiometric calculations within the purview of student skills as the heart of chemical knowledge rather than resorting to only arithmetic conversions of simpler quantities.
4. Replacing purported disinterest in science with chemical skills that amaze students with their own abilities.

3 Methodology: POGIL Activities Following a Traditional Conceptual Sequence

To address these challenges, a series of 14 guided-inquiry classroom activities were developed to prepare students with the chemical skills they would need for their independent research projects [5]. Each of these activities was designed to be completed within two class periods of 1 h and 20 min. In order to maintain this schedule, some of the problems from these activities have been assigned as homework for students in those class sections that progressed less rapidly than others.

The class is totally non-lecture-oriented, resorting to mini-lectures when numerous students have difficulty assimilating new concepts in the applications presented. The POGIL activities were constructed to follow the learning cycle model of cognitive development which entails four distinct stages: (i) exploration of information or data; (ii) concept invention; (iii) application of concepts through exercises, problems and research; and (iv) metacognition whereby students identify and verbalize interrelating concepts, principles and procedures and specify their importance [6]. These class activities were conducted in groups of three or four students chosen randomly for each activity. Since a POGIL section has an enrollment of 24 students, there are between six and eight groups for the instructor to monitor during each class. This size preserves personal and constructive academic support without the need for clickers or other reporting mechanisms for group work.

Student chemical skills are measured via two problem-oriented exams during the semester and a cumulative final exam. Student application skills are measured via an independent, research-based project that presents an original calculation on the magnitude or resolution of a documented chemical issue.

3.1 Student Outcomes with POGIL

Only one course section of Chemistry 1200 has been offered each semester in the guided-inquiry format described herein. Because lecture sections have customarily been taught by a rotating staff of adjunct faculty, a consistent comparison of student performance has not been formally established between sections. However, after teaching this course using the combined POGIL and project-based learning formats for six successive semesters, the effectiveness of POGIL activities in developing chemical skills is consistently compelling. On exams that parallel the critical thinking process in classroom activities, most students perform well, with a class average 15–20% points higher than they did on more concrete exams in an interactive, but traditional, lecture class format with the same instructor. In addition, student process skills have been measured using the POGIL format. At least half of the students also do well on previously unseen material. Including the first page of the next scheduled POGIL activity on an exam has been frequently used as a measure of how well each student has advanced in reading and reasoning with new scientific material. Passing rates for the POGIL course are high (usually above 90%), student drop rates after the first week of classes are nearly non-existent, and student performance on project calculations are beyond what the instructor would have previously conceived. Without question, POGIL classroom activities have enabled more students to master basic chemical skills effectively, and, more importantly, enabled them ***to surpass their own expectations***.

A culminating event in this course is the presentation of student project reports based upon researched data. Student project reports include a brief summary of an issue and one or more original calculations using conversion factors or stoichiometry to illustrate its magnitude, consequence or resolution. Being able to answer a chemical question of their own making transforms students' perceptions of their own abilities and of the accessibility of science in their lives. The outstanding strength of sample student projects has been presented previously in the literature [1].

3.2 Unaddressed Pedagogical Challenges

Even though most students adapt readily to a guided-inquiry course format, there is a significant number whose limited skill-level and classroom pre-conditioning greatly hinder their initial progress. POGIL activities try to guide student learning in incremental steps that progress from simple to more encompassing concepts and conclusions. However, the ability to draw conclusions, itself, is for many students a challenging requirement. Drawing conclusions forces students to generate their own mental images that must be clarified and conjoined. No matter how incrementally their thoughts have been guided, those students who have not successfully translated the original concepts into their own mental images are those who are easily 'lost' and most resistant to POGIL. Contrary to our own expectations as faculty, the proportion of seniors struggling with this process is no less than the proportion of freshman. Compounding the problem in the chemistry curriculum is the demand for formulating cumulative conclusions.

The cumulative nature of chemical concepts is a major, and often insurmountable, challenge for many chemistry students and amplified by the often fragmented presentation of chemical topics. The issue of acquiring cumulative knowledge in any academic discipline cannot be overestimated. A professional development conference for faculty was held at Kean University in April 2008 entitled, 'Pedagogy Across Disciplines: Imagining and Delivering the Possibilities'. The keynote speaker, Dr. John Mayher, Distinguished Professor at NYU's Steinhart School of Education, explained how secondary education has actually been de-emphasizing the importance of cumulative knowledge in the methodologies adopted in the classroom [8]. He described how increasing numbers of students never actually read the primary material of study, whether it is a novel, poetry, history or science. Instead, they read summaries, review notes and other learning 'aids' that focus the topic and provide ready-made conclusions for students to regurgitate rather than to formulate.

4 Expanded Methodology to Facilitate Cumulative Learning: An Improved Model of Repeated, Multi-concept, Guided-Inquiry Classroom Activities

The learning cycle model [2–4, 6] used for POGIL activities has been shown to be highly effective in engaging students in thoughtful learning not only at Kean [1], but also at universities throughout the USA. POGIL texts have been developed to address many areas such as General, Organic, Analytical, Physical and Biochemistry courses [8–10]. A more complete list of POGIL resources including problems and activities can be found at the POGIL web site [11].

While POGIL activities for General and Introductory Chemistry directly involve students with their own learning, the conceptual sequence used in POGIL activities has often preserved the topical sequence that has been established in traditional texts. As a result, students still saw the development of chemical concepts from a rather narrow, linear perspective that progresses, not as a result of overriding and interdisciplinary principles, but according to long-established conventions. Consequently, many topics still appear to students as disjoint and not conceptually interrelated.

For a non-science majors course, the traditional textbook sequencing of topics was simply too prolonged to prepare students within one semester to pursue independent research and answer quantitative questions on chemical issues. Building guided-inquiry activities for this course had been defined by the need to reach specific conceptual levels such as redox reactions and stoichiometry as early as possible. To accomplish these objectives chemical topics were reordered, simplified and conceptually integrated. Furthermore, to address the challenge of conceptually cumulative knowledge, the learning cycle model was repeatedly applied to the same chemical concept, so that students had the opportunity to re-invent or re-discover the same chemical principle in more than one context. Reinforcement of the discovery process has proven to be invaluable in supporting students' cumulative understanding of chemical principles. Examples of this expanded methodology will be discussed in the next section.

5 Reorganization of Chemical Concepts and Reinforcement of Conceptual Discovery

As mentioned above, in order to meet the needs of a rigorous, one-semester General Education course for non-science majors, the organization and structuring of chemical concepts that are followed in a traditional textbook had to be reordered, combined, simplified, yet conceptually accelerated in order to prepare students to undertake a quantitative research calculation on the magnitude of a global or civic chemical issue. Consequently, POGIL activities were designed to address conceptually related topics as concisely as possible.

For example, along with the concept invention associated with the structure of atoms, the structure of ions and free radicals were also included in the activity. Once students understood the structure of atoms and ions, ion formation through the gain and loss of electrons could immediately lead to a simple discussion of oxidation and reduction. Without confounding details, students become chemically literate about many everyday processes and can recognize the ***transfer or sharing of electrons as the stabilizing force for molecular formation and the driving force for electrical energy***. Their understanding is more than a piecemeal, pattern recognition that is characteristic of much chemical pedagogy.

Another important insight was learned by developing and incorporating a multi-concept POGIL activity as the beginning activity in the non-science major course. Instead of addressing a single chemical concept, the opening class activity was divided into three parts: (i) reading about the chemical ingredients in a popular confection, the Twinkie, (ii) recalling simple math skills to determine how long a line of trucks would be needed to deliver a year's supply of 500 million Twinkies, and (iii) examining the proportions of ingredients in the Twinkie recipe. Reading about the extracted and synthetic ingredients used in making Twinkies in the first component not only introduced students to scientific information in a comfortable manner, but also established an expectation of the course. The first question they were asked was to ***define*** what is meant by extraction and synthesis from their context in the reading passage. Hence, unlike in most scientific texts, students were not given definitions and expected to apply them; instead, they were asked to construct a definition for themselves. Later they were asked to re-examine their definitions in terms of physical and chemical processes, but the responsibility for conceptualization was placed directly in their lap. The math component was used to prepare students for multiple conversion factors in the subsequent activity, and the recipe component was used as a precursor for understanding mole ratios and limiting reagents in chemical reactions.

While it is common to introduce the concept of limiting reagent by using the number of eggs in a recipe or the number of marshmallows in S'mores, the analogy is usually given immediately before a more extensive discussion of the topic. However, the multi-component 'Twinkie' activity provided an immediate introduction to both physical and chemical processes, stoichiometry and mathematical procedures. The early introduction appeared to develop more interest and more enthusiasm than when the same topics were developed separately, even in the POGIL format. And, without question, subsequent activities on reinventing and applying multiple conversion factors and limiting reagents were the most successful and exhibited the most class-wide student mastery the instructor had ever seen.

All of the activities for our non-science majors course have now been rewritten to include repeating the learning cycle model to several fundamental chemical concepts. In addition to stoichiometry as discussed above, another such topic is oxidation-reduction and its application in electrochemical cells. In an activity entitled, 'The Molar Mass Concept for Atoms, Ions, and Formula Units', the concept of the mole is developed historically using the data from the silver voltameter presented at the 1908 International Electrical Conference of London [12]. Since students 'invented' the concept of atoms gaining and losing electrons from ***atoms*** in an earlier activity, they were ready to 're-invent' the concept of gaining and losing electrons between two silver ***electrodes*** in a silver voltameter. In a multi-component activity that also incorporates Millikan's estimate of the number of electrons in 1 mole [13], students quickly begin to apply the mole concept to mass relationships of elements, compounds and compounds in solution. In a later activity that explores the behaviour of electrochemical cells more fully, students are prepared to 're-invent' the concept of oxidation-reduction once again with even more ease and confidence. Specific activities may be obtained by e-mailing the author designated for correspondence below the title of this paper.

6 Multi-skill, Multi-disciplinary Guided-Inquiry for Chemistry Majors

Based on the insights we have learned with non-science majors, it would appear that a major impediment to chemical learning is imposed by the linear logic assumed by many traditional textbooks and often followed in POGIL activities as well. ***There is a need to transcend traditional chemical pedagogy, to identify unifying chemical concepts, and to develop POGIL activities that reinforce the re-discovery of chemical concepts throughout the course.***

At Kean University, activities developed for the non-science major's class have been tested to a limited extent with General Chemistry students as well. New guided-inquiry activities to a lesser extent have also been written and introduced in

General Chemistry lecture, Organic Chemistry lecture and laboratory section [14] and Physical Chemistry lecture. Though students exhibited enthusiasm and improved understanding on the single topic covered by each activity, their dependence on traditional lecture made it difficult for some to adapt to problem-based learning and many continued to show difficulty with increasingly complex topics that bring together the components they had previously studied separately.

Our success with non-science majors provides the motivation and direction for extending guided inquiry to both the lecture and laboratory components of General Chemistry classes for science majors, courses in which student performance has been poor. For the past several years, at least 40% of the students enrolled in sections of General Chemistry I and General Chemistry II are repeating the course. That means that they have previously either failed or dropped the course before completing it. Passing rates of General Chemistry courses using the POGIL format have been reported to be much higher and consistent with those of the non-science majors in the POGIL section at Kean [15–18].

When applied to science majors, multi-skill guided-inquiry activities can provide an ideal mechanism for preparing students for their entire undergraduate curriculum, as an integrated whole, and not as a piecemeal assemblage of sub-disciplinary fragments. If students make conceptual connections among sub-disciplines during their first year of chemistry classes, they will not only acquire a long-lasting, cumulative knowledge that provides seamless transition into upper level chemistry courses, but also envision chemistry as a central science that is fundamentally connected to other scientific disciplines such as biology, physics and mathematics as shown in Fig. 1.

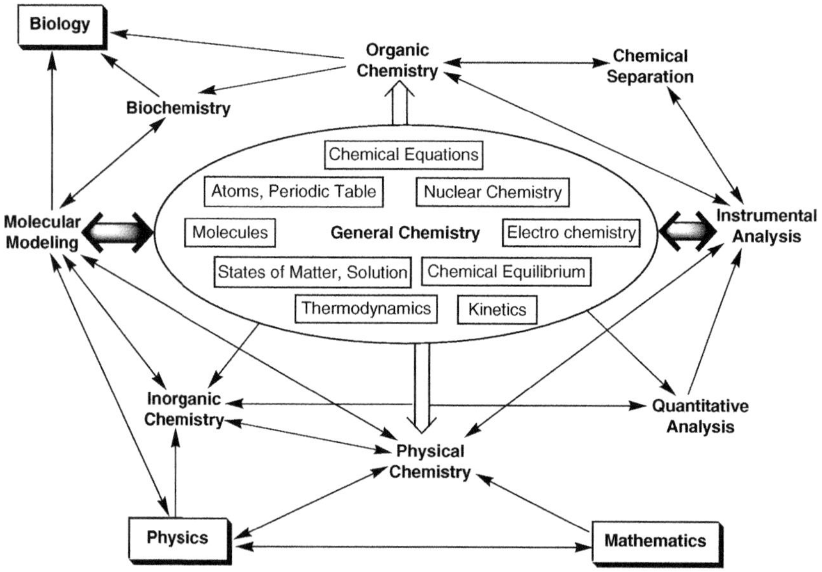

Fig. 1 Data from computational molecular modeling and instrumental analysis are used for group exploration and concept invention in POGIL classroom activities

Our main focus in current and future work is to base guided-inquiry activities for both lecture and laboratory portions of General Chemistry on data that can be obtained from modern chemical instrumentation or computational molecular modeling as illustrated by the shaded bold arrows in Fig. 1. This approach tries to instill an interactive view of chemical sub-disciplines in students' learning.

For example, atomic structure as presented in Chemistry 1200 is expanded to include modern day quantum mechanics where electrons occupy distinct energy levels in x-ray photoelectron spectra. Data from colorimetry and atomic absorption spectroscopy are used for exploring concentration measurements in solutions; data from gas chromatographs are used for comparing the strengths of intermolecular forces; computational data from modeling vibration are used to explore infrared spectra. Our intent is not to overwhelm students or to replace upper-level course content, but to introduce chemical topics in a research context and to better prepare lower-level students for cumulative integration of their chemical knowledge.

6.1 Assessing Student Achievement

Accurate placement of entering students into the appropriate course level has been a fundamental component of Kean's General Education and Learning Assistance Programs [19]. As a result, students entering Kean with SAT scores below a specified level take ACCUPLACER tests in reading, writing and math. Data on placement test scores exist from past students and will be available for future students. With guided-inquiry classroom format, analytical and reading skills applied to the exploration of data, concept invention and problem solving become critically important. Understanding how student preparation impacts their higher-level university performance is essential for our learning how to improve future guided-inquiry activities to maximize cumulative understanding. We are exploring the possible correlations between the placement test skill levels with the grades earned in General Chemistry courses and selected POGIL assessments which include frequent in-class or on-line applications.

The Chemistry-Physics Department has administered a common, cumulative final examination to all sections of General Chemistry I and II for over 20 years. Questions on those examinations have been classified in four levels of chemical skills: (i) recognition of a chemical concept, (ii) direct qualitative application of a chemical concept, (iii) direct quantitative application of a chemical concept, and (iv) multi-step reasoning, either qualitative or quantitative, in applying a chemical concept. The results have consistently shown a much poorer performance in category (iv) than in the other categories, a result that relates to the level of cumulative skill development.

In POGIL sections of non-science majors, student performance on category (iv), multi-step reasoning problems has shown dramatic improvement from performance in non-POGIL sections of the same instructor.

Once POGIL activities are extended to most General Chemistry sections, the same final exam will continue to be administered to measure whether the performance of students on category (iv) questions increases significantly. In addition, a common midterm exam and a supplementary final exam based upon guided-inquiry reasoning will also be used. This exam component will also be used in sections based on traditional lecture techniques to give a comparison to student reasoning when POGIL activities are employed.

7 Conclusions

The student-centered model, as opposed to the current teacher focused pedagogy, yields a college student with better reasoning skills, improved team participation and leadership, more effective oral and written communication skills, and one who assumes more responsibility for their own learning. These results were clearly demonstrated in improved student confidence in developing, writing, revising, and presenting their term projects.

Success in improving the cumulative comprehension of scientific concepts is a major breakthrough in reversing the path to failure followed by so many students in science. It is not a lack of student ability that causes this result, but the alien nature of conceptually cumulative knowledge prerequisite for higher level reasoning skills in any discipline and for applying chemical principles in particular. By mastering the learning cycle to improve personal skills in reading, exploring data and concept invention, students are able to apply their knowledge to increasingly complex and challenging problems.

A multi-concept, guided-inquiry approach has proven to be highly successful in limited trials with non-science majors at Kean University. Reordering the traditional presentation of chemical concepts under unifying principles facilitates student learning. Rather than having a sequence of activities focus upon individual chemical skills, multi-skill classroom activities that 're-invent' chemical concepts reinforce both concept mastery, cumulative reasoning and multi-disciplinary significance at the same time. In addition, multi-concept activities early in the course sequence better prepare students to master more advanced concepts more rapidly and with greater confidence.

As an added feature, we are enjoying improved collaboration of faculty members in the department. To enhance the multi-disciplinary POGIL model of chemical education successfully, the expertise of all chemical specialists are needed – organic, physical, inorganic, analytical and bio-organic. As a team we are learning to focus on a coherent and multi-disciplinary integration of chemical concepts in our chemical curriculum.

Acknowledgements We wish to acknowledge the guidance of numerous POGIL project leaders at various workshops. They include J. Spencer, R. Moog, F. Creegan, J. Lewis, J. Wakefield, K. Martin and K. Butler. We also wish to acknowledge the guidance of F. Glazer in the Project-Based Learning Initiative at Kean University.

References

1. Lees, A. B. (2008) Making science accessible in the lives of non-science majors using POGIL and project-based learning. In: Moog, R. S., Spencer, J. N. (eds.) Process Oriented Guided Inquiry Learning (POGIL), 173–185. American Chemical Society, Washington, DC
2. Karplus, R., Thier, H. D. (1967) A New Look At Elementary School Science. Rand McNally, Chicago, IL
3. Karplus, R. (1980) Teaching for the development of learning. Res. Sci. Edu. 10, 1–10
4. Abraham, M. R., Renner, J. W. (1986) The sequence of learning cycle activities. J. Res. Sci. Teaching 23, 121–143
5. Lees, A. B. (2008) Chemistry in Your World: A Guide to Global and Civic Engagement. unpublished
6. Hanson, D. M. (2006) Foundation of Chemistry: Applying POGIL Principles, 3rd Ed. Pacific Crest, Lisle, IL
7. Mayher, J. S. (1989) Uncommon Sense: Theoretical Practice in Language Education. 328. Boynton/Cook Publishing, Upper Montclair, NJ
8. Moog, R. S., Farrell, J. J. (2006) Chemistry: A Guided Inquiry, 3rd Ed. Wiley, New York
9. Straumanis, A. R. (2008) Organic Chemistry: A Guided Inquiry. Houghton Mifflin, Boston, MA
10. Moog, R. S., Spencer, J. N., Farrell, J. J. (2004) Physical Chemistry: A Guided Inquiry: Atoms, Molecules, and Spectroscopy. Houghton Mifflin, Boston, MA
11. POGIL website. http://www.POGIL.org.
12. Rosa, E. B., Vinal, G. W. (1917) The silver voltameter as an international standard for the measurement of electric current. Proc. Natl Acad. Sci. USA. 3, 59–64
13. Millikan, R. A. (1913) On the elementary electrical charge and the Avogadro constant. Phys. Rev. 2, 109–142
14. Stokes-Huby, H., Vitale, D. E. (2007) Coupling molecular modeling to the traditional "IR-ID" exercise in the introductory organic chemistry laboratory. J. Chem. Educ. 84, 1486–1487
15. Farrell, J. J., Moog, R. S., Spencer, J. N. (1999) A guided inquiry general chemistry course, J. Chem. Educ. 76, 570–574
16. Hanson, D., Wolfskill, T. (2000) Process workshops – a new model for instruction. J. Chem. Educ. 77, 120–130
17. Hinde, R. J., Kovac, J. (2001) Student active learning in physical chemistry. J. Chem. Educ. 78, 93–99
18. Lewis, J. E., Lewis, S. E. (2005) Departing from lectures: an evaluation of a peer-led guided inquiry alternative. J. Chem. Educ. 82, 135–139
19. Best, L., Lees, A. B. (2000) A vision for skill development: the new general education program at Kean University. Research and Teaching in Developmental Education 16, 119–122

Teaching Physical Chemistry in Disadvantaged Contexts: Challenges, Strategies and Responses

L. Mammino

Abstract Physical chemistry courses are often considered among the most difficult ones – if not the most difficult – by chemistry students. The reasons may be traced to their specific requirements, from the extensive familiarity with mathematics and its descriptive role in physical sciences to the expansion of logical thinking into abstract thinking (as an essential tool for the construction and understanding of models) and to the conceptual demands of models and procedures. The difficulties experienced by students increase considerably in contexts that are disadvantaged because of the aftermaths of historical reasons. In these contexts, the combined and mutually enhancing impacts of factors like inadequate background preparation, inadequate or poor mastering of the language that is the medium of instruction and, often, also inadequate visualisation abilities, make students' progress through physical chemistry courses comparatively more arduous. These contexts pose greater challenges for the design of viable teaching strategies. On the other hand, the usefulness of the designed strategies extends beyond the contexts for which they were generated, because their specific objective of facilitating students' approach to concepts and techniques, when such approach is particularly arduous, makes them generally viable.

1 Introduction

Physical chemistry is considered particularly difficult by many chemistry students, in many contexts. The reasons behind such perceptions are multiform: the extensive presence of mathematics, the dominance of conceptual components and the consequent need of expanding logical thinking into abstract thinking, the essential role of conceptual understanding for problem-solving. These features are closely related to the nature of physical chemistry as an interface/overlap area, in which the investigation and interpretation of the behaviour of chemical systems largely utilises concepts and tools pertaining to physics; they are also the features that chemistry

L. Mammino
Department of Chemistry, University of Venda, South Africa
e-mail: liliana@univen.ac.za

M. Gupta-Bhowon et al. (eds.), *Chemistry Education in the ICT Age*,
DOI 10.1007/978-1-4020-9732-4_20, © Springer Science+Business Media B.V. 2009

students perceive as more difficult. The current work aims at offering a comprehensive picture of what could be termed 'the students' side': the difficulties encountered by students, the perceptions they develop about the course material and about their own difficulties, the responses to options and interventions designed on the basis of the diagnosed difficulties or inadequacies and attempting to address them. All the information presented here was collected in a context that – for a variety of historical and socio-economic reasons – can be considered particularly disadvantaged. This provides more detailed and complete diagnostic information on students' difficulties, because a broader range of difficulties is bound to surface in the presence of factors complicating students' approach to learning and diminishing the effectiveness of their learning efforts. It also poses greater challenges for the design of teaching strategies attempting to minimise their impact and being tuned to students' needs. On the other hand, the usefulness of the designed strategies extends beyond the contexts for which they were generated, because their specific objective of facilitating students' approach to concepts and techniques, when such approach is particularly arduous, makes them generally viable.

2 Characterising Features of Physical Chemistry and Students' Perceptions

The nature of the physical chemistry discourse, and the approaches pertaining to it, are perceived as considerably different from the other chemistry courses. The discussion of concepts and models is a dominant component. Approaching models and theories requires considerable abilities to abstract reasoning and to generalisation. Problem-solving requires more than a limited number of algorithms that may become familiar either through practice or through memorisation: it is necessary to derive algorithms from the theory and, therefore, it is necessary to attain adequate understanding of the theory and of its mathematical expression. The laboratory reports require extensive incorporation of theoretical/conceptual components to support the interpretation of the experimental information. The extensive presence of mathematics exceeds the anticipations of many students, who expected a limited presence of mathematics in chemistry courses, also on the basis of the image of chemistry they developed through secondary school; so, they are neither emotionally nor technically (i.e. in terms of acquired abilities and skills) prepared to the way mathematics permeates the physical chemistry discourse throughout, supporting and guiding the development of the conceptual content.

While students often manage to somehow 'simplify' their approach to the content of other courses, narrowing the range of attention they devote to the theoretical discourse and still managing to attain acceptable levels in problem-solving and practical skills, they soon realise that this is not possible for physical chemistry courses. Such realisation enhances the perception that physical chemistry courses are different and more difficult. The ensuing anxiety decreases students' self-confidence in approaching the course, thus complicating the approach itself. Though encountered in several contexts, these aspects have a greater impact in disadvantaged contexts, because they add to the difficulties resulting from the very factors that make the context disadvantaged.

3 The Context

The observations, interpretations and designs of approaches reported and discussed in the current work refer to a context that can be considered particularly disadvantaged - the University of Venda (UNIVEN) in South Africa – and are the outcome of direct engagement in teaching all the physical chemistry courses during the last 10 years. The nature of the disadvantage of this specific context, its main features and their origins and implications, have been comprehensively described in a previous work [1]. They comprise socio-economic disadvantages, with the features that are mostly common to poor rural set-ups; historical disadvantages, related to the fact that UNIVEN was a for-blacks-only university during the *apartheid* regime; and the disadvantages inherent to the use of a second language (a language different from the mother tongue) as medium of instruction. On combining, these factors mutually enhance their undesirable impacts, with heavy consequences on students' preparation and attitudes. The underpreparedness of incoming students concerns both contents and skills: serious inadequacies in the knowledge of the basic contents of the various disciplines sum up with inadequate mastering of instruments that are fundamental for the acquisition of knowledge. The poor mastering of the language through which they are expected to acquire and express knowledge decreases the communication effectiveness of verbal explanations – thus limiting the depth and completeness of understanding of the course material – and generates insecurity, for which students are scarcely available to participate in classroom interactions. Inadequate visual literacy decreases the effectiveness of explanations utilising visualisation and hampers the development of mental images (an essential tool in the process of acquisition of scientific knowledge [2]) and the representation of their own conceptions through images [1]. The overall outcome is a widespread passive attitude equating learning to memorisation, as alternative to unattained understanding.

Teaching physical chemistry in a context of this type poses a number of challenges. It is not easy to find balanced and viable answers to questions like: 'How to select the course content? How to find a balance between the need of not «losing» too many students since the first lectures and the need of not reducing the course content (or lowering its standard) below limits that would be incompatible with the nature of the course itself? How to address the negative impacts of major factors like inadequate familiarity with mathematics, inadequate mastering of the language that is the medium of instruction and inadequate visual literacy? How to do it within the time allocated for the course?'. There are no univocal or already-ready answers. The option selected during these years was that of continuously adapting teaching approaches on the basis of the information provided by interactions with students – diagnoses of difficulties and problems, diagnoses of misconceptions, gradual diagnoses of how student perceive certain interventions and respond to them. This search is still in progress, also because new groups of incoming students (e.g. students who went through some recent pedagogical experimentations in secondary instruction) require new adaptations and new responses. Thus, the current work presents the current progress of a still on-going search – the search for enhancing scientific communication within physical chemistry courses, taking into account as many details as possible of students' difficulties and attitudes, and attempting to address them.

The physical chemistry courses taught at UNIVEN comprise: chemical thermodynamics at second-year level; electrochemistry, chemical kinetics and surface chemistry at third-year level; quantum chemistry in the first semester of the first postgraduate level (called 'B.Sc. Honours'); spectroscopy and statistical thermodynamics in the second semester of the B.Sc. Honours level (an optional course only rarely selected by students).

4 Diagnoses and Documentation

4.1 Sources of Information and Presentation Criteria

Students' works in their course-activities (tests, laboratory reports, etc.) constitute the main sources utilised for the identification of problems and difficulties, because they can be viewed as the most genuine and informative source, corresponding to the outcomes of students' maximum efforts to produce something for which they hope to attain positive assessment. The information from these sources is complemented by interactions with students, both during normal classroom activities and through personal interviews.

Extensive documentation of diagnosed problems and difficulties is considered essential within the current work, to portrait a sufficiently informed picture of the situation, to substantiate analysis and inferences through concrete evidence, and to serve as reference for the presentation and discussion of the designed approaches, of the corresponding students' responses and of the limitations that might be inherent in some of the approaches. For clarity sake, individual sections are devoted to broad categories of identified difficulties. Though the impacts of different types of difficulties are inextricably intertwined and overlapping, their separation under individual sections responds to dominance identifications, that is, to the identification of types of difficulties whose impact is dominant (though never, or nearly never, sole) in a series of instances. Language-related difficulties are considered first because the other types of difficulties often include a considerable language-related component. The examples selected for documentation role are chosen among the most representative ones, so that they document types of errors or confusions that appear frequently. They are numbered progressively, to facilitate references throughout the discussion. When expedient, incorrect words are in italics, to facilitate immediate identification. The very way in which most sentences are worded or constructed brings further documentation of the pervasiveness of the impacts of language-related difficulties. The themes to which the sentences refer identify the course concerned.

4.2 *Impacts of Language-Related Difficulties*

The difficulties inherent in the use of a second language as medium of instruction are the objects of ample investigation and debate, as the awareness of their impact on the acquisition and further utilisation of knowledge increases [3–14]. The impact is actually so heavy as to be considered a major factor hampering development in the sub-Saharan sub-continent [15–20]. In a disadvantaged context, the impact of language-related difficulties may become so extensive as to set severe limits to the depth and completeness with which students can understand the course material. This, in turn, poses so-far-unanswered questions at assessment level, both because of the impossibility of untangling language-related components and concept-related components in the students' errors (impossibility of ascribing individual weights to the two components [21]) and because of the ensuing ethical question as to how 'right' it is to give a poor evaluation for a chemistry course when the student is unable to attain adequate understanding of the course content because of being unable to identify all the information transmitted by a sentence or a text [14].

Two parallel investigations are currently in progress at UNIVEN, to provide systematic documentation of language-related difficulties with specific focus on science courses. They differ by the perspective utilised in the analysis of the material collected: mainly language-related and method-related. The documentation is collected from chemistry students' works and an ample proportion of it is derived from physical chemistry courses.

4.2.1 The Usage of Words and the Building of Sentences

The analysis from a mainly language-related perspective focuses on aspects like the selection and combination of words, or the building of clauses and sentences, respectively relating to the knowledge of individual words and of the grammar and syntax rules [22]. The motivations for this specific type of analysis, the approaches utilised and the expected validity-extent of the resulting inferences are extensively discussed in refs. [22, 23]. The results are progressively organised in a series of articles devoted to individual key-aspects in a sort of increasing-sophistication order, from the simplest to the more complex.

The simplest and most basic aspect is the sound-concept correspondence [23]. Confusions in this regard are never present in the mother tongue, because of the deep internalisation of the correspondence; but they may appear, and also become frequent, within a second language, resulting in incorrect utilisation of words whose sounds can be perceived as similar by non-native speakers, and they may be responsible for conceptual uncertainties or serious conceptual misunderstandings.

They often involve the use of a more familiar term in place of the less-familiar correct one:

1. In ideal gases there are intermolecular *reactions*.
2. The change in entropy is given by the *reaction* $\Delta S = \int dq/T$.
3. In a chemical reaction some substances are *transferred* into other substances.

The use of the *reaction* term in place of other terms – like *interaction* (1) or *relationship* (2) – is very frequent [23, 24] and is often associated to inadequate understanding of the concept expressed by the other term. The confusion (3) between the verbs *to transfer* and *to transform* [23] generates difficulties in the understanding of the very nature of chemical reactions (a *transformation*) or some of their aspects (e.g. electron *transfer* in redox reactions). At assessment level, it is difficult to guess whether the perceptions or mental images that students have formed correspond to the meaning concerned, or remain blurred because of inadequate distinction between the meanings of two different terms.

The next aspect (in terms of increasing complexity) is the selection and combination of individual words having a key role in individual clauses, like the subject–verb, subject–verb–object and adjective–noun combinations [25]. Errors in this regard imply incorrect expression of aspects associated both with the immediate content and with the scientific method: confusion [25, 26] between systems and processes (4), use of verbs typical of events in relation to objects or to physical quantities, like *to occur* in relation to *rate* (5), use of verbs expressing something that the given subject cannot do (6–8), or omission of key words in pairs that need to be coupled to convey a meaning, e.g. the omission of the noun in «adjective + noun» pairs (*state* and *spin* are omitted in (9) and (10) respectively):

4. *Electrolytic cell is a process* where a non-spontaneous redox reaction is driven by electric current.
5. The higher the temperature, the more the reaction *rate occurs* faster.
6. At lower temperature the *reactants* took more time to *become complete.*
7. Observations: the *copper* rod *turned to silver*.
8. The zinc electrode is corroded and the zinc ions are *deposited* into the solution.
9. The heat capacity depends on the *physical* of the system.
10. Fermions are particles having *half-integral.*

The logic of individual clauses relies on the accurate use of the connectives identifying the different functions. Confusions in the use of prepositions affect the understandings or the expression of important chemistry concepts [27]. A typical example is the frequent confusion between the prepositions *in* and *between*, resulting in confusions between intramolecular and intermolecular bonds on reading or answering, and discussed extensively in ref. [14]. Other errors in the use of prepositions pose doubts as to whether the student has developed a correct image of what he/she is describing, like in examples (11) and (12), referring respectively to the phenomena taking place at the electrodes and to the meaning of considering derivatives from a starting molecular structure:

11. The copper ions move to the cathode and are deposited *into* the copper rod.
12. The molecule of interest is 2,2,4-trimethylhexane, and *in* this molecule 19 derivatives were made.

A number of grammar rules are neglected nearly routinely, probably because of inadequate awareness of the implications of their neglect on the information that the sentence conveys. Most of these rules would not be neglected within the mother tongue, because the internalisation of the correspondence between mode of expression and meaning conveyed would have a dominant role. An example is the very frequent neglect of the distinction between the singular and plural of nouns [22] that may lead to sentences whose literal meaning is seriously incorrect from a chemical point of view. For example, sentence (13) transmits the meaning that particles in a crystal vibrate around one single centre and sentence (14) shows inadequate awareness of the fact that a mixture consists of at least two components and that partial pressures refer to each component and the total pressure to the mixture.

13. Particles vibrate around the *centre* of the crystal lattice called the *node*.
14. Dalton's law of partial *pressure* is the *total partial* pressure of a *mixture of the gas* is equal to the sum of the partial *pressure* in a *mixture of that gas*.

The neglect of the grammar rules relative to the expression of comparisons affects the clarity of the comparison mental-operation or the recognition of its relevance in specific contexts, and the vagueness in its general perception expands also to cases that would not require the change in the ending of an adjective or an adverb. Sentence (15) contains simply a grammatical error; sentence (16) neglects the comparison – fundamental in thermodynamics – between the initial state and the final state (the solid and the liquid in the specific case); sentence (17) neglects the fact that comparisons can only refer to entities of the same nature; sentence (18), by omitting *relative* before *tendency*, neglects the fact that electrochemical phenomena are related to how the tendencies of different elements to be in an oxidised state compare (although, in this case, the experiment itself showed that this comparison is the key factor):

15. The Born-Oppenheimer approximation is justified by the fact that nuclei move much *slowly* than electrons.
16. Entropy increases on melting because there is *high* disorder in liquid.
17. The final *volume* will be below the initial *pressure*.
18. The objective of the experiment is to determine the *tendency* of an element to be oxidised or reduced.

4.2.2 The Expression of Fundamental Method-Related Aspects

The second type of documentation and analysis currently in progress at UNIVEN singles out key issues that can be considered mainly method-related, that are fundamental for the clarity and correctness of content understanding, and whose identification and expression requires some sophistication in the level of language mastering. The impact of inadequate awareness of the presence, nature and implications of these issues is particularly heavy within physical chemistry courses, because the conceptual demands of their contents require consistency with key method–related aspects, for the discourse to be meaningful. These are issues like the various aspects concerning numbers and values [28], the distinction between physical quantities and their changes [29], the distinction between individual cases and statements with general validity [30], the cause–effect relationship [31], and also the distinction and interplay between microscopic and macroscopic descriptions [32, 33] that is fundamental throughout any chemistry discourse [32–34]. Ascribing values to a system (19) highlights inadequate awareness of the nature of the measure concept and of the fact that we can ascribe values only to physical quantities [28]:

19. State functions take the *values of the system*, not the *values of the path* followed by the system.

The neglect of the distinction between physical quantities and their changes – widely spread among students, probably from the combined effects of the tendency to words omission (in this case, the *change* word) and of inadequate attention to comparisons – hampers clear understanding of individual concepts whose definitions refer to changes of specified quantities (e.g. the heat capacity (20) is defined in relation to a temperature change); it also hampers full understanding of fundamental aspects like the role of references selected by definition (the zero enthalpy value assigned to elements in standard conditions or the 0 V reduction potential assigned to the standard hydrogen electrode) and can even affect the processing and interpretation of experimental results, e.g. (21) the recognition of a mass decrease (while a mass cannot be negative, a mass difference can):

20. The heat capacity of a substance is a proportionality constant of the heat supplied to that substance and the *temperature* of the substance.
21. When calculating the mass difference between the beginning and the end of electrolysis, the mass differences for the anode had negative values, which is impossible, because a mass cannot be negative.

Inadequate awareness or internalisation of the distinct functions of statements with general validity and of the consideration of individual cases leads to a variety of confusions: generalisation of properties specific to individual entities, like the diffuse habit of referring to 0°C whenever discussing melting or freezing, as if it were the melting/freezing point of all substances (22, 23); the provision of solely a specific example in place of general-type explanations (like (24) and (25), respectively answering questions asking to explain what we mean by 'spontaneous process' and what is an eigenfunction), or in place of mathematical-type derivations of certain conclusions (in which case, students invent or recall numerical values); the provision of general-type statements or descriptions when a question or a context would require attention to individual examples (e.g., several answers to a question asking for the dimensions of the rate constant in a second order reaction consisted of the reproduction of the entire procedure for the integration of a second order rate law, memorised from the lecture handouts; and more than half of the answers to a question, asking students to describe how they would proceed to gold-plate a ring, consisted solely of a regurgitated memorised description of electroplating without any reference to features of the specific case, like where the ring would be placed in the electrolytic cell, or which ions should be present in the solution):

22. When a substance is melting, temperature remains constant at *0°C.*
23. A substance is liquid *above 0°C.*
24. The system cools down to the temperature of the surroundings.
25. Eigenfunction is a wavefunction and is given by $\psi = A\ e^{ikx} + B\ e^{-ikx}$.

The cause–effect relationship is fundamental in the scientific approach. Its identification is essential both for conceptual understanding and for the interpretation of experimental information. Errors in the expression of cause–effect relationships are symptoms of serious misconceptions, as illustrated by the examples below: the presence of intermolecular interactions is the cause (not the result) of the different behaviour of real gases (26); our descriptions, like equations (27), or the fact that we determine certain values experimentally (28), cannot be the cause of the behaviour of a system; work is done in the reaction between iron and HCl (29) because there is the formation of a gas and, therefore, a volume change:

26. The presence of intermolecular interactions is the main *result* that real gases behave differently.
27. Real gases behave differently from ideal gases *because they have several state equations*, while ideal gases have only one.
28. Real gases behave differently from ideal gases *because they are determined experimentally.*
29. There is work done in the reaction between iron and HCl, *because they form iron chloride.*

The clarity attained with regard to the distinction and interplay between the microscopic and the macroscopic levels are highlighted both by the selection of individual words and by the overall meaning conveyed in students' answers. Difficulties surface in all the courses. The examples selected illustrate some of the most common problems: difficulties at keeping the wording consistent with the nature of the two levels, even in simple statements informing about what gases (macroscopic concept) do or what molecules do (30, 31); difficulties at understanding and expressing the nature of the microscopic-level randomness inherent in the entropy concept (32); incorrect assignation of macroscopic-character thermodynamic quantities to microscopic entities (31, 33); difficulties at distinguishing the two levels even when describing direct observations in laboratory reports, resulting in frequent statements literally implying the possibilities of observing microscopic entities with naked eyes (34, 35) or highlighting confusions between performed operations and the theoretical interpretation of phenomena (36, referred to an experiment in which students measured the time needed, at various temperatures, for a potassium permanganate solution to become colourless, so that the reference to a determination based on the average kinetic energy of the molecules is totally not consistent with the operations performed):

30. Ideal *gases* move in straight lines and collide with each other.
31. *Molecules* in real gases are having high pressure and low temperature.
32. In a solution there is a higher entropy than in a pure solvent because in the solution there is the presence of intermolecular interactions.
33. In the cell, *each tin ion has its own activity*.
34. Observations: *copper ions* were sticking to the cathode.
35. It was *observed that zinc atoms* were releasing electrons.
36. The objective of this experiment is to determine the effect of temperature changes on the rate of reaction *based on the average kinetic energy of the molecules*.

As the complexity of the relationships between pieces of information increases, the difficulties in their expression increase sharply and their impact becomes particularly serious for those logical relationships whose expression requires compound sentences. Probably the most important among these is the hypothesis–inference relationship in its three forms – reality, possibility and unreality. When language-mastering inadequacies prevent their identification – e.g., prevent the realisation that a certain thing would have happened if a certain other thing were true, but it does not actually happen because the second thing is not true – the meaning of models and interpretations may remain largely unclear [35]. At the expression level, syntactical oversimplifications result in deeply altered meanings. Answer (37) shows non-identification of the hypothesis–inference relationship in a statement like 'if it were possible to run a Carnot cycle having 0 K as cooler temperature, then the efficiency of that cycle would be one'. In the case of models, syntactical oversimplifications often point to perceived coexistence of mutually excluding models (38–40) and generally fail to capture how and why models change during the historical evolution of science – e.g. fail to capture the difficulties posed by the realization that some phenomena could not be described by classical physics, at the dawn of the study of the atom. It is the case, for example, of the oversimplification of a statement like 'If Thomson model of the atom had been true, all the α particles should have behaved in the same way in Rutherford's experiment. But this did not happen, showing that the atom is not homogeneous' into statement (38), or the oversimplification of a statement like 'Rutherford's model was not consistent with classical physics because, according to classical physics, the electron should have fallen on the nucleus. The problem was overcome by Bohr's model of the atom' into statement (39) – which appears to assume that Rutherford's model actually included the falling of the electron of the nucleus:

37. The absolute temperature is zero if, when is utilised to cooler temperature, it enables the efficiency of the Carnot cycle.
38. In Thomson model all the α particles behave in the same way. In Rutherford's model they behave in different ways.
39. In Rutherford's model the electron falls on the nucleus. In Bohr's model the electron does not fall on the nucleus.
40. In classical physics, the electrons approach the nucleus and eject the energy in a continuous manner. In quantum physics, the electron remains in the same orbit.

In this way, an essential feature of the scientific approach – the fact that our models are just approximations in the attempts to understand reality and have to adapt whenever new experimental information appears – remains somehow not-conveyed to students and does not integrate into their vision of science, even when they reach advanced courses like those of the post–graduate level.

4.3 *Impacts of Passive Memorisation and Passive Attitudes*

Passive memorisation is the most common – practically generalised – response by students to the difficulties inherent in not attaining adequate level of understanding of the course material and in not being sufficiently able to express things when answering questions or writing other types of texts, like laboratory reports. It is also a habit developed from prior instruction and so deeply internalised that memorising becomes synonym of learning and students do not conceal that, for them, the most important component of classroom activities is what the lecturer writes on the board, because they can copy and memorise it. Everything becomes an object of memorisation: the course handouts, the answers to previous years tests, examples discussed during the lectures. That the memorisation is passive is proved by the way it is utilised – through mechanical regurgitation, and not as a sort of 'storage' from which to critically select relevant information. This concerns all the components of the material, and the effects on students' performance are disastrous.

Passive memorisation of the theoretical component does not provide tools for answering proposed questions. Some answers remain limited to a single short sentence reproducing the beginning of the treatment of the given topic in the study material, or to some parts of the introductory information, without real focus on the given question. For example, answer (41) – to a question asking to explain the meaning of wave–particle duality – reproduces only the first sentence of the treatment in the course handout, and answer (42), besides adding a comment before reproducing that sentence, proceeds with the list of the differences between particles and waves from a classical point of view; but only a small proportion of answers extended beyond the recalling of this introductory part, to actually discuss the wave–particle duality of elementary particles. Common answers (like 26–28 and 43) to the question 'Explain why ideal gases behave all in the same way with respect to changes of pressure, volume or temperature, whereas real gases behave in different ways that depend on their chemical nature', imply that aspects of our modelling activities (like state equations) can be responsible for what happens in nature; (longer answers to the same question consisted of word-by-word reproductions of the descriptions of the ideal gas model and the real gas model; no answer really focused on the question given). Answer (44), to the second part of a question asking to select the reactions apt for a galvanic cell, from a proposed set, and then to justify the selection, consists of a single memorised statement, while answer (45) combines different memorised pieces, unrelated to the focus of the question. In more extreme cases, answers consist of assemblages of memorised fragments, mostly not consistent among themselves (pertaining to different descriptions) and not always relevant to the given question (46, 47):

41. Waves and particles are completely different.
42. Wave particle duality – this is the difference between a particle and a wave. According to classical physics, waves and particles are two different entities.
43. Ideal gases have only one state equation, while real gases have many state equations.
44. A galvanic cell utilises a spontaneous redox reaction.
45. A galvanic cell utilises reaction to occur. This device utilises a spontaneous redox reaction to produce electric current. In this reactions the e.m.f. of a galvanic cell, we can measure the potential differences between the two electrodes, not the contribution of each electrodes.
46. Entropy increases on vaporization because the compound form perfect crystalline which makes an entropy to be a positive value.
47. The melting point of a solution is lower than the melting point of the pure solvent because the total volume occupied by the gas is negligible in comparison with the volume of the gas.

Similar phenomena are observed with regard to descriptions, because of the difficulties students encounter in organising a text that needs to develop through several sentences and to maintain a logical thread throughout them. Therefore, they try to avoid composing a whole description, and resort to the usual assembling of memorised parts when writing a test, or to straightforward copying from some source when writing a laboratory report. This often results in failure to discuss the specific issue considered, as the memorised parts are reproduced as they appear in the initial source, without any adaptation to the specific issue. In laboratory reports, it may even result in neglect of the experimental information that the student has collected. For example, the procedure of an electrolysis experiment with copper electrodes dipped in a copper sulphate solution, without addition of acids to the solution, required also to measure the pH of the solution at the beginning and at the end of the experiment. In the observations section of the lab-report, all students reported the pH values and noted that the pH had not changed. Yet, in the treatment-of-results section of the report, more than two thirds of the students copied the example discussed in the course handout (referred to the electrolysis of a copper sulphate solution using inert electrodes) and passively reproduced not only the electrode half-reactions, but also the description of the outcomes, including the information that 'the concentration of H^+ ions in solution increases as the process proceeds, and the solution gradually turns into a sulphuric acid solution'; nobody realised the contradiction between this statement and having observed no pH change.

Passive memorisation of solved problems may lead to extreme cases in which the answers to problems proposed in a test, and being similar to problems discussed in a class, but with different values of the variables involved, were word-by-word reproductions of the answer built in the classroom, including the numerical values, as if those students had not read or noticed the values in the test question. The damages become more evident in the case of thermodynamics problems requiring the formal splitting of a given process into two or more steps, because identifying the correct series of steps for such splitting requires careful analysis of the process considered in the given question, and the reference to memorised answers to analogous – but not identical – questions often results in the inclusion of steps that cannot pertain to the process considered, or the neglect of steps that pertain to it. For example, an example analysed in the class concerns the calculation of the entropy change of a process in which a certain amount of ice is melted at 0°C, then the liquid water is heated to 100°C, boiled to complete vaporisation at 100°C and finally the vapour is expanded isothermally to a certain volume. When students are given similar questions, with different sets of steps (e.g. not including the expansion of the vapour after vaporisation, or starting at a temperature lower than 0°C, so that the heating of the ice becomes the initial step, or considering only the melting of the ice and the heating of liquid water up to a temperature below its boiling point), several students include all the steps of the classroom-example into their answers, or neglect the heating of ice if the starting temperature is below 0°C.

The inadequate development of individual reflection abilities induced by passive attitudes fosters a dominant compartmentalisation preventing the identification, and even the very search, of relationships between information encountered under different 'titles' – whether different topics in the same course, or different courses. For example, the 'practical' component of the quantum chemistry course involves the study of the geometry and charge distributions of selected molecules using inexpensive calculation methods [36], and students are invited to identify all the possible connections between the data obtained from the calculations and what they can know about those molecules on the basis of the information learnt within organic chemistry courses. However, the passive attitude prevents fruitful reflections, and most students simply report the computed data, without identifying or discussing even simple features like the bond-length differences between single and double C–C bonds, or the differences in bond angles associated with the different hybridizations of the C atoms, or the reactivity of sites with high charge-density.

In summary, passive memorisation prevents the ability to relate different pieces of information and to identify connections between them. It prevents the recognition of connections between pieces of information encountered within the same course, as the course progresses. It prevents connections between information encountered in different courses – what is particularly negative for physical chemistry courses, as the courses that provide the theoretical foundation to the overall chemical discourse and for which, therefore, such connections are important to highlight application paths. It even prevents connections between pieces of experimental information pertaining to the same experiment.

4.4 *Impacts of Inadequate Visual Literacy*

Visualisation plays fundamental roles in the teaching/learning activity [37]; within chemistry courses, it is also an essential tool for building basic familiarity with the microscopic world of atoms and molecules [38, 39]. The visual literacy of students entering UNIVEN is often seriously inadequate. The background roots of the inadequacies, their impacts on the efficiency and quality of learning at tertiary level, and the way they decrease the effectiveness of explanations utilising visualisation and practically prevent the resort to its potentialities as an alternative communication tool that could be valuable for addressing some language-related difficulties, are discussed in detail in [1]. The heaviest impacts within physical chemistry courses stem from the combination of inadequate visual literacy and inadequate mathematical literacy, hampering the drawing and interpretation of diagrams (Section 4.5), and from the scarce utilisation of images in general.

The absence of the habit to utilise visualisation on tackling problems, for which such resort would be informative, becomes particularly negative within the quantum chemistry course. Students do not draw models of systems, when asked to write their Schrödinger equation, and this unavoidably results in the writing of incorrect equations. For example, in a test in which no student drew a model of the hydrogen molecule, all the answers missed the presence of two nuclei, resulting in equations apt for the H^- ion or in totally meaningless equations. Since the discussion of examples at classroom level always starts with the drawing of a model and proceeds with continuous references to it, it appears reasonable to infer that the non-utilisation of models by students relates to the combined effect of two factors: the habit to memorisation, for which they rely on reproducing memorised equations, and a practical discrimination of the components of the approaches utilised at classroom level, based on perceptions viewing drawing as not sufficiently sophisticated to be utilised at tertiary level [1]. In this way, they fail to realise how arduous it is to memorise Schrödinger equations without focusing adequate attention on the characteristics of the system considered, and renounce the main tool that can enable a systematic analysis leading to correct equations.

4.5 *Impacts of Inadequate Familiarity with Mathematics*

Since mathematics is an integral component permeating the physical chemistry discourse throughout, the impacts of inadequate familiarity with mathematics are heavier than for other chemistry courses, and are often perceived as the dominant ones. Inadequacies in what could be called the required mathematical literacy generate difficulties with regard to several important aspects, like: performing certain calculations (above all when they involve exponentials or logarithms); identifying or perceiving the correlations between the mathematical discourse and the physical discourse; and deriving equations – an operation that, in the case of physical chemistry, is largely controlled by the physical meaning of each step. The combination of these difficulties with language-related difficulties and with inadequate visual literacy poses serious obstacles both to the understanding of physical chemistry contents and to its expression or utilisation at problem-solving level or for the interpretation of experimental information.

The handling of values is the simplest mathematics-related aspect. Since values are related to measurements of physical quantities, an essential aspect is their reasonability.

However, when performing calculations, students often handle numbers without reference to their physical meaning, thus accepting results like a negative pressure, a negative absolute temperature, a negative rate constant, a concentration of a reactant at a certain time after the beginning of the reaction higher than the initial concentration, etc. – while the realisation of the inconsistency with their physical meaning would have been clear indication that either the procedure or the calculations contained errors.

The handling of physical quantities in general, and the understanding of their roles in relation to definitions and procedures, or to the characterisation of a system, is affected by insufficient familiarity at relating mathematical and physical aspects. This results in a variety of confusions: confusion in the identification of the dependent and independent variables, even in the discussion of experiments like monitoring how the concentration of a reactant changes during a chemical reaction (48) or how the time, needed for a reaction to complete, changes at different temperatures (49); incorrect identification of the quantity that remains constant in a given situation (e.g. considering constancy of the heat capacity, while the constancy of the volume is responsible for no work being done (50)); confusions regarding the distinction between a variable and a specific value (e.g. between the concentration [A] of a reactant and the value $[A]_0$, it takes at time zero (51)):

48. Time increases as the concentration of the reactant decreases.
49. The diagram of the time taken by the reaction versus the absolute temperature of the reactants indicates that the absolute temperature decreases as time increases.
50. If the heat capacity is constant, no work is done.
51. I draw the diagram of $[A]_0$ versus time.

The impact of incorrect definitions provided at pre-university level remains throughout the courses, confirming that the internalisation of the first definitions encountered reaches deeper levels than subsequent attempts to correct them. The most important ones concern the proportionality concepts [40]. Students interpret as 'direct proportionality' any case in which the dependent variable increases as the independent variable increases (52), and as 'inverse proportionality' any case in which the dependent variable decreases as the independent variable increases (53), regardless of the type of dependence, even when they refer explicitly to the equation expressing the dependence (54); and even the absence of dependence may be viewed as direct proportionality (55):

52. Increasing temperature in a reaction lead to increase in the rate of reaction. *Temperature and rate of reaction are directly proportional to each other.*
53. The graph of the time taken by the reaction versus absolute temperature shows that the time decreases with the temperature increase and therefore *the time is inversely proportional* to temperature.
54. In a first order reaction, $[A]_t = [A]_0 e^{-kt}$ and, therefore, the concentration of *A* is inversely proportional to time.
55. Since the equation of the half-time of a first order reaction does not depend on the initial concentration, we can conclude that the half-life is *directly proportional* to the concentration of reactant *A*.

The last examples highlight also another major problem: the absence of the habit of 'reading' equations, whether from a mathematical or from a physical point of view. Equations are often memorised without sufficient attention to their physical meaning, and questions related to the latter become difficult. For example, many answers to the question 'A certain ideal gas occupies a volume V_1 at a pressure P_1. If the pressure is doubled, what can you say about the final volume?' remain vague (56, 57), are incorrect (58) or consist of a memorised fragment without complete meaning (59), and only a small proportion of students answer that the final volume will be half the initial volume. Difficulties increase further when the equation/s concerned have been learnt recently, and the question requires the combination of physical reasoning and mathematical reasoning and implies a comparison with a situation considered previously. Thus, the answers to the question (given subsequently to the previous one): 'A certain real gas occupies a volume V_1 at a pressure P_1. If the pressure is doubled, what can you say about the final volume?' show attempts to provide answers differing from those provided for the case of ideal gases, but with scarce clues as to how to identify the differences (60–63), and only a small proportion of students say that the volume will decrease, but the final volume will not be half the initial volume, and its value must be calculated with a real gas equation, or determined experimentally:

56. If the pressure is doubled, the final volume will be low.
57. The final volume will decrease.
58. The final volume will be equal to the first volume.
59. Assume that the temperature is constant and the volume and pressure is proportional.
60. If the pressure is doubled, the final volume will remain constant, or say will be zero.
61. The final volume would remain the same for real gas if the pressure is doubled it won't affect the volume.
62. The final volume will increase.
63. The final volume of the real gas will be double the initial volume.

The absence of the habit of reading equations in terms of their physical meaning, or simply in language terms, generates also frequent errors with regard to the interpretation of the symbols appearing in the equation. This may result in total misinterpretation of the meaning of an equation, above all if coupled with the effect of the diffuse lack of attention to the distinction between capital and small letters, like reading the v appearing in Graham's law as volume (64), or in combinations of the meanings of a symbol that may denote different entities, like the S symbol, which can denote both for the overlap integral and the total spin of a system (65):

64. For a given mass of ideal gas at a constant pressure, the volumes is directly proportional to the square of the masses, $V_1/V_2 = \sqrt{(M_2/M_1)}$.
65. S is the overlap integral spin.

Difficulties at reading equations appear to combine language-related difficulties and mathematics-related difficulties, as illustrated, for example, by attempts to 'read' the equation expressing Graham's law (66–71). Because of these difficulties, when asked for the statement of something that can be expressed through an equation, students often reproduce the sole equation, without any explanation through words (even at advanced level, for example, as answers to questions asking to state De Broglie hypothesis or the uncertainty principle):

66. Graham's law states that the diffusion velocities of two gas molecules are inversely proportional to the square root of ratio of their molecular masses.
67. The *proportion* between the velocity of the gases is equal to the square root of the inverse *proportionality* of their molecular masses.
68. The diffusion of the velocities of the gas are directly proportional to the square root of their molecular masses.
69. The ratio of the diffusion velocities of two gases is equal to the square root inverse of the ratio of the molecular masses.
70. Graham's law states that the ratio of the diffusion velocities of two gases equal to the *square* of the inverse of their *molar* masses.
71. Graham's law is the proportional ratio of volumes multiplied by the square root of the ratio of number of molecules.

The understanding of equations and of the corresponding physical meanings is affected also by other difficulties identified as language-related in Section 3.1. The difficulty in identifying, handling and expressing comparisons extends to physico-mathematical-character comparisons, as illustrated by the answers to a question asking students to compare the state equation of ideal gases and the state equations of real gases (72–76), with answers (75) and (76) highlighting also the additional effects of inadequate familiarity with method-related aspects, as they miss the distinction between physical reality and our mathematical description (a gas cannot *consist* of equations or *contain* parameters):

72. The state equation of an ideal gas is the only one state $PV = nRT$. They cannot have parameters because they did not depend on the nature of the gas. While the state equation in real gases are the van der Waals equation and the virial coefficients virial equation. They contain parameters because their physical states depend on the nature of the gases.
73. In ideal gases only one state equation is valid. The equation is $PV = nRT$, it does not contain terms that are related to the chemical constant of the gas. In case of real gas, the state equation contains related terms of parameters of the nature gases and those values are determined experimentally.
74. The van der Waals equation consists of a and b which are parameters depending on the nature of gases.
75. Real gases consist of different equation, e.g. van der Waals equation and virial equation. There are parameters that are involved in a real gas which differ according to the different gases used which means that different gases have different parameters.
76. The real gases contain the parameters that related to the nature of gas, therefore they are different in different gases.

Difficulties with the comparison concept (or comparison mental operation) are probably responsible also for difficulties in identifying which quantities can be equal in a given situation, above all when the comparison involves ratios. For example, when solving problems on the classical experiment of entering NH_3 and HCl at the two ends of a glass tube, and calculating the distances travelled by the two gases when they meet and form the NH_4Cl ring on the tube wall, students find it difficult to specify that, since the two gases are entered simultaneously into the tube, the time taken to travel the respective distances is the same and, therefore, the ratio of their velocities and the ratio of the two distances travelled are equal; they often equate quantities that are not equal (like the two distances, 77) or that cannot be equated (like velocity and distance, 78). The same difficulties hamper the interpretation of experimental results. For example, after determining the reaction rate at several temperatures, increasing by 10°C in the 10–70°C interval, students are invited to consider the ratios of the velocities for consecutive temperature values (20°C and 10°C, 30°C and 20°C, etc.) and derive inferences; however, they often confuse the comparison of the velocities with the comparison of the ratios, missing the fact that the ratios tell us by how much the velocity has increased between the two temperature values; e.g. with his ratio for 20°C and 10°C being 2.19 and that for 30°C and 20°C being 2.21 (thus showing that the reaction rate doubled for each 10°C temperature increase) a student writes comment (79):

77. The two gases enter in the tube at the same time, the distance travelled by HCl to cover d_{HCl} is equal to the distance travelled by NH_3 to cover d_{NH3}.
78. Since the two gases were entered at the same time inside the tube, their speed will be equal to their distance travelled.
79. The reaction rate for these cases, the change is very small, we can even consider them as having the same rate because the difference between them is negligible.

The combination of inadequate visual literacy and mathematics-related difficulties hampers the handling and interpretation of diagrams. Diagrams are memorised as individual items, without connecting them with a mathematical or a physical meaning. Inadequate internalisation of aspects like the correspondence between functions and graphs results in incorrect inferences about types of dependence. An experimental diagram that is not a straight line is more frequently considered hyperbolic, even when it corresponds to a known law (like the exponential dependence of the concentration of a reactant on time for a first order reaction). The answers to questions asking for diagrams to be drawn on the basis of chemical reasoning, like a general-type diagram of the concentration of a product versus time as a reaction proceeds, highlight total dichotomy with chemical information, as most students draw continuously increasing diagrams, varying from linear to parabolic to exponential. References to diagrams within a text (e.g. an answer or a report) often do not specify the nature of the quantities reported on the axes. On handling experimental information, a considerable number of students select only the values through which a straight line can be fitted and discard all the others, thus thoroughly altering the results and their interpretation; answer (80) refers to a situation in which the values appearing on the diagram had been manipulated with respect to those reported in the data section:

80. The results obtained are in agreement with Arrhenius law because the graph predicted by *Arrhenius is a straight line* and the graph obtained of reaction rate versus absolute temperature is also a straight line.

5 Strategies and Responses

5.1 General Criteria in the Design of Strategies

Any attempt to address identified difficulties needs the active participation of the learners to have possibilities of success. Therefore, promoting students' active engagement is considered an essential component on designing strategies, and all the strategies are designed for implementation within interactive teaching/learning options. They are also designed to be integrated into the normal course activities (lectures, laboratory practicals), so that they can be viewed as normal components of these activities. This responds to the need that students may, at all times, feel engaged with the demands of the course and, therefore, with the course content - a feature that is particularly important in view of the dominance of the marks-accumulation perspective in students' attitude. For instance, making separate time to intervene on the language-related aspects would not be successful, as students would feel that this is not part of what will be the direct object of tests or exams questions; and the awareness that these interventions would be important to enhance the level of understanding would be comparatively scarce, in view of the internalised conception that memorisation is what enables to pass tests and exams and in view of the acquired/internalised habit of not devoting attention to language and to the mode of expression. On the other hand, when the options utilised to address language-related problems are an integral part of the normal explanations or other standard activities, they will reach students in the same way as the other components of the classroom activities.

The next sections present the main features of the strategies explored in the attempts to address the major identified categories of difficulties. Their implementation is realised in an integrated way, building what can be considered a holistic approach.

5.2 Promoting Students' Active Participation

Interactive options are utilised extensively in all the courses, as the options that can maximise the efforts to engage students actively. The advantages of classroom interactions are well known [41, 42]: they stimulate and maintain students' attention towards the material that is the object of consideration at a given moment and enable students to catch details that otherwise would remain unnoticed. Simultaneously, they have a valuable diagnostic role, by offering real-time feedback to the teacher as the work on a given theme progresses, thus enabling immediate responses both in terms of specific clarifications and in terms of more refined 'tuning' of the overall explanation to the students' needs.

Since the theoretical component of physical chemistry courses often involves contents that are totally new to students and cannot be linked to any aspect of their expected prior knowledge, the integrated combination of presentation of new information by the lecturer and derivation of inferences on an interactive basis – guiding students to 'discover' the implications of the new information – has proved the most viable strategy. An extant example in this regard is offered by the quantum chemistry course, whose content does not relate to expected prior knowledge, and for which even the derivation of inferences or the identification of implications follows patterns that are not familiar from any other previous experience; therefore, interactions require extensive guidance by the lecturer, to facilitate the contact with an approach to the interpretation of physical reality that is totally new both from a conceptual and from a mathematical point of view [43].

Because of the characteristics of the context, interactions need to be somehow 'gently forced' on students [1], by requiring their contributions as a necessary component of the classroom activities. The extensive use of questions is a major tool, apt both for diagnosing the types and details of pre-existing conceptions and for guiding the search for inferences and implications of newly provided information [44]. Since, in most cases, students do not volunteer verbal answers, it has become a nearly common practice, within physical chemistry courses, to request written answers from all the students; the answers are then quickly surveyed to single out the features that need clarifications and further analysis; these, in turn, are also realised through guiding questions (e.g. along error analysis patterns). Though a higher extent of verbal interactions would undoubtedly have a number of benefits, the provision of written answers has the advantage of requiring deeper reflection from the student, what can by itself be a tool for clarification or for identifying aspects that require clarification, thus becoming a tool that favours learning [45–47].

Attempting to break the loop of passive 'memorisation plus regurgitation' requires also interventions on the setting of the test-questions or on the guidelines for the laboratory reports. Specifically designed multiple-choice questions on conceptual aspects, and questions asking for provisions of examples of the student's own choice, are particularly suitable as questions that require active reflection and cannot be answered on the basis of passive memorisation. Similarly, the guidelines for the laboratory reports are organised in such a way as to require that students reflect on all the aspects of the experiment.

5.3 *Attempting to Address Language-Related Difficulties*

The impact of the use of a second language that students do not master adequately extensively affects classroom activities. The same difficulties that students encounter on reading books (failing to understand the meaning of sentences that are not extremely simple in their syntactical structure – besides often failing to understand the meaning of individual words that play key-roles in conveying a message [48]) apply to classroom explanations: students manage to follow only very simple sentences, and often need to see them written on the board. A number of interventions concerning the language level are therefore integrated into the explanation activity, incorporating them into the way of presenting the material [49]. The conviction that rigour in the mode of expression is by itself a valuable pedagogical tool [50–53] fosters the use of rigorous wording throughout any explanation. This is basically an indirect way of addressing the problem that might have more effects through the sentences written on the board (sentences that students copy and memorise). It is however also expected that hearing only modes of expressions that respond to the requirements of the language-of-science [54] may contribute – though to an extent that is practically impossible to estimate – to build some habits, at least in terms of combination of words and expression of basic concepts. To increase the efficacy of the approach, explanations of the meaning of individual words are added whenever it may be doubtful that the meaning is familiar to all students, and explanations of the reasons to select a certain term, and not another, are added whenever previous experience shows that students have somehow become accustomed to using a non-rigorous term.

The analysis of errors is utilised as frequently as time constraints allow, because it is a precious explanation tool, enabling simultaneous analysis of concept-related and language-related components of errors, and being particularly apt for interactive options [21, 55]. In order to keep it as active as possible, it is mostly carried out through questions. An incorrect statement or diagram is reported on the board by the teacher and students are asked to detect the errors; if the task appears too difficult for them, the incorrect parts are underlined or other hints are provided; a direct disclosure by the teacher remains only a last resort, and is followed by the request that students find a correct statement. For instance, the frequently encountered statement reported in example (30) is written on the board and proposed for discussion. When students fail to identify the error (and they usually fail in this specific case), two gas cylinders are drawn on the board [1], with arrows showing that they are moving towards each other, and the drawing is complemented by the information that a gas is a macroscopic system and, therefore, the literal meaning of the sentence is the one illustrated by the drawing. At this point, some students reach the conclusion that the subject should be 'gas molecules' and not 'gases'. In this way, the combined use of visualization and some humour tries to convey the importance of the distinction between macroscopic systems and microscopic objects, and the relevance of associating subject and verb in a way that remains consistent with the distinction.

The difficulties towards understanding logically complex statements or logical structures need addressing options specifically tailored for each individual case. Complex statements are analysed nearly word by word, explaining the role of each bit of information and the logical relationship between them. Flow charts prove particularly useful to illustrate the logical trends and the relationships between individual pieces of information of sets of reasoning that lead to certain conclusions, or of complex operational procedures. A flow chart illustrating the set of experimental observations and related inferences that led to the conclusion that electrolytic solutions contain ions is reported in [1]. Examples of logically complex procedures are offered by procedures typical of quantum chemistry, like the Hartree procedure for the study of multi-electron atoms or the procedure for the determination of the coefficients of the expansion of molecular orbitals in the LCAO approach. Language-related difficulties prevent most students from understanding the logic of the procedure (the very meaning of the sentences describing it, and the connections between subsequent logical steps) on reading the textbook. Flow charts with boxes containing very simple statements have proved the best option to highlight the overall logic [56], and a collaborative building of the flow chart, formulating and entering each new statement in close association with the reading of the text describing the procedure in the textbook, maximise students' active participation [1] and simultaneously contribute to training in the reading of science texts.

Other difficulties stemming from inadequate language mastering appear more complex to address. These most often concern the communication of fundamental aspects of the scientific approach. Students' difficulties in following and understanding statements with a minimum internal complexity pose severe limitations to the efficacy of any form of communication. They pose severe limitations to the efficacy of oral communication, by decreasing its efficiency not only as a direct explanation tool, but also within attempts to build skills with regard to other tools – e.g. to build visual literacy [1]. They also pose severe limitations to the possibility of exploring additional pathways, of enriching the course with reflections and discussions. This also highlights the limitations inherent in the nature of the addressing approaches designed so far and described in the previous paragraphs. As long as wording is restricted to syntactically very simple sentences expressing simple statements, important logical relationships remain excluded. This hampers the communication of important aspects of the generation of theories and models, of the building of conceptual and interpretation frameworks. Even during interactions, expansions to the consideration of method-related aspects, or of fields open to explorations and investigations, remain nearly impossible, because most students would not follow the literal meaning of the discourses (the very consideration of examples like (38–40) highlights how essential components of the discourse are lost in what students perceive of it). Thus, the addressing approaches designed so far can be functional to the building of basic physical chemistry literacy, but they appear inadequate for the communication of the richness of the subject and the variety of its investigation pathways.

5.4 Attempting to Address Mathematics-Related Difficulties

Addressing inadequacies deriving from prior instruction is not easy, whether it refers to replacing incorrect definitions with correct ones, or to the improvement of calculation skills. In the former case, the deep internalisation of the concepts and mental images formed on the first encounter with a specific topic prevents an effective and lasting replacement, even when the use of visualisation – e.g. through the consideration of diagrams showing different types of dependence, all responding to the incorrect definition of direct or inverse proportionality students have acquired – appears to be convincing during classroom discussions. As for the improvement of calculation skills, it is not easy to include this activity into the time devoted to the course, and students are not easily available to accept the need of some remedial classes. Thus, explanations on standard calculation procedures are incorporated into some examples that offer suitable opportunities, and target those skills for which inadequacies have been diagnosed extensively, like the handling of exponential forms or of some simple integrals, the graphs of some functions, or the utilisation of diagrams for interpolation.

The physical chemistry courses are the chemistry courses that make more extensive use of the role of mathematics as a fundamental description tool [57], and this is the single major aspect for which students perceive physical chemistry as particularly difficult. Pursuing the objective of making mathematics as student-friendly as possible develops through a variety of interventions [58, 59], mostly responding to the criterion of refraining from making mathematics the dominant vehicle of communication of information. Equations are always accompanied by detailed explanations of their physical meaning, as well as by their formal 'reading', to try and ensure that their literal meaning is clearly understood. Whenever possible, diagrams are added to complement the reading through the alternative instrument offered by visualisation. In this way, the approach tries to integrate the three major communication vehicles: verbal expression, mathematical formalism and visual representation. Complete derivations of equations are developed for selected significant cases and each step is discussed thoroughly, thus attempting to balance the need to expose students to as many important aspects of physical chemistry as possible, and the need to maintain the course accessible to most students (removing derivations completely would suppress exposure to a fundamental component of the physical chemistry approach; including many derivations would complicate the progress of the course, because the time for a detailed discussion of each of them would not be available).

The difficulties in searching for viable options to address mathematics-related inadequacies increase considerably for the quantum chemistry course. The inadequacy of students' familiarity with the mathematics required by that course is a rather common situation for quantum chemistry courses, also in other contexts. The course contents usually make provision for this, by including the development of familiarity with the needed mathematics (operators etc.) into the course. However, the characteristics of the UNIVEN context drastically reduce the viability of such option, because of the gap between students' attained familiarity with mathematics, and what would be needed to cope with the mathematics for quantum chemistry. It is therefore opted to maximise the focus on the conceptual aspects and on the description of systems and behaviours, while only few mathematical procedures (e.g. the solution of the Schrödinger equation for the hydrogen atom) are presented, to provide at least some exposure to the ways of proceeding of quantum chemistry.

5.5 *Fostering the Ability to Relate Theory and Observations*

The need of adequate conceptual understanding for the interpretation of experimental information is a typical feature of physical chemistry. Difficulties affecting the level and clarity of understanding unavoidably affect the handling and interpretation of experimental information. Other aspects highlighted by the laboratory reports refer to inadequate awareness of the roles of the various components of experimental work: observations, operations immediately following the collection of observations (calculations of the values of other quantities from those measured directly, drawing of diagrams, and all the aspects that can be included within the 'treatment of results' concept), identification of the theoretical aspects involved and of how they relate to the observations; identification of sources of experimental errors or of accuracy-decrease. Additional frequent errors relate to the habit of copying from often-randomly-selected sources (including secondary school books) and to the resurfacing of incorrect information provided in previous instruction (e.g. the fact that several students consider that there must be an 'evaporation' part in the diagram of the heating curve of water, located after the horizontal segment corresponding to boiling and the subsequent temperature rise corresponding to the heating of the vapour, and they include that part in the diagram, adjusting the interpretation of their data to this purpose [60, 61]).

Interventions to attempt to foster basic familiarity with relating observations and theory, and also with the very description of observations, include several options. Detailed guidelines to the laboratory report are provided for each experiment, to ensure that the report maintains internal logic and to stimulate students to reflect on all the significant aspects of the given experiment. The briefing before each experiments focuses specifically on the recall of all what was mentioned in the classroom and relates to the experiment. Major errors in the laboratory reports are discussed at classroom level.

6 Discussion and Conclusions

The difficulties students encounter in physical chemistry courses are enhanced by the inadequacies of a disadvantaged contextual situation. Eradication of the difficulties would require major interventions at pre-university instruction, to build those fundamental skills (language mastering, visual literacy, habit to reflections and enquiries) that are essential for further acquisition of knowledge. When these skills have not been developed adequately, tertiary-level instruction can try to decrease their impacts through a variety of integrated *ad hoc* interventions that, however, can only mitigate the problems. For the physical chemistry courses, these interventions can enable the attainment of basic physical chemistry literacy for a considerable proportion of students, while other desirable effects, like the development of an inquiring mind, remain restricted to a small number of students. In particular, the development of real interest in the subject is hampered by the difficulties students experience and struggle to overcome, with the outcome that only exceptionally gifted students reach a level of understanding and competencies that can stimulate them to undertake research in physical chemistry for their professional careers.

The latter problem is not restricted to heavily disadvantaged contexts, but concerns several contexts in sub-Saharan Africa. The awareness of the risk that the very presence of physical chemistry might be endangered in some universities by the lack of specialists had prompted a formal expression of concern at the *Fifth International Chemistry Conference in Africa* (Gaborone, July 1992). The continued scarcity of physical chemists shows that the problem remains. On the other hand, the demand for some types of expertise is increasing – e.g. the demand for expertise in computational chemistry, in relation to the possibilities it offers for complementing other types of research with information about the molecules involved. This underlines the urgency of the preparation of specialists in the main areas of physical chemistry, to meet the demands of chemistry research, chemical education and the chemical industry. It is therefore necessary to find ways for stimulating and maintaining students' interest towards physical chemistry – what, in turn, requires the design of options to equip students with the needed tools (language mastering, basic familiarity with mathematics, etc.), without which their approach to physical chemistry is bound to be complicated by objective difficulties that prevent the development of interest.

References

1. Mammino L. (2008). Teaching chemistry with and without external representations in professional environments with limited resources. In Gilbert J.K., Reiner M. and Nakhlekh M. (Eds.), *Visualization: Theory and Practice in Science Education*. Dordrecht: Springer, 155–185.
2. Harre R. (1970). *The Principles of Scientific Thinking*. University of Chicago Press, Chicago.
3. Case S.M. (1968). The language barrier in science teaching. *Teacher Education in New Countries*, 9(1), 15–26.
4. Zepp R.A. (1981). Relationships between mathematics achievements and various English language proficiencies. *Educational Studies in Mathematics*, 12, 59–70.
5. Zepp R.A. (1982). Bilinguals' understanding of logical connectives in English and Sesotho. *Educational Studies in Mathematics*, 13, 205–221.
6. Brodie K. (1989). Learning mathematics in a second language. *Educational Review* 41 (1), 39–53.
7. Chipere N. (1993). Dependence and the transition from high school to university in Zimbabwe. *Society for Research in High Education International News*, 19, 4–7.
8. Nangu N.C. (1994). How can we be expected to learn science in English? *Southern African Association for Research in Mathematics and Science Education Proceedings*.
9. Nyapfeme K. and Letseka M. (1995). Problems of learning among first year students in South African Universities. *South African Journal of High Education*, 9(1), 159–167.
10. Mammino L. (1998). Science students and the language problem: suggestions for a systematic approach. *Zimbabwe Journal of Educational Research*, 10(3), 189–209.
11. Seepe S. (2000). A pedagogical justification for mother tongue instruction. In Seepe S. and Dowling D. (Eds.), *The Language of Science*. Johannesburg: Vyvlia Publishers, 40–51.
12. Mammino L. (2000). Studying the Details of the Transition from the Mother Tongue to the Second Language, In Seepe S. and Dowling D. (Eds.), *The Language of Science*. Johannesburg: Vyvlia Publishers, 94–101.
13. Rubanza Y.I. (2002). Competition through English. The Failure of Tanzania's Language Policy. In Prah K.K. (Ed.), *Rehabilitating African Languages*, CASAS Book series No 18, 39–51.

14. Mammino L. (2006). *Terminology in Science and Technology – An Overview Through History and Options. Thohoyandou*: Ditlou Publishers.
15. Prah K.K. (1993). *Mother Tongue for Scientific and Technological Development in Africa.* German Foundation for International Development. Education, Science and Documentation Centre, Bonn.
16. Prah K.K. (1995). *African Languages for the Mass Education of Africans.* German Foundation for International Development. Education, Science and Documentation Centre, Bonn.
17. Prah K.K. (1998). The missing link in African education and development. In Prah K.K. (Ed.), *Between Distinction and Extinction.* Johannesburg: Witswatersrand University Press, 1–16.
18. Prah K.K. (2002). The Rehabilitation of African Languages. In Prah K.K. (Ed.), *Rehabilitating African Languages*, CASAS Book series No 18, 1–6.
19. Mammino L. (2004). L'impatto, sull'apprendimento della chimica, dell'uso di una lingua diversa da quella materna. *Il pensiero Chimico e la Formazione Scientifica nelle Riforme della Scuola*, Assisi (Italy), December 2004.
20. Prah K.K. (2008). The Language of Instruction Conundrum in Africa. *Meeting on The Implications of Language for Peace and Development (IMPLAN). University of Oslo, May 2008.*
21. Love A., Mammino L. (1997). Using the Analysis of Errors to Improve Students' Expression in the Sciences, *Zimbabwe Journal of Educational Research*, 9(1), 1–17.
22. Mammino L. (2005). Language-related difficulties in science learning. I. Motivations and approaches for a systematic study. *Journal of Educational Studies*, 4(1), 36–41.
23. Mammino L. (2006). Language-related difficulties in science learning. II. The sound-concept correspondence in a second language. *Journal of Educational Studies*, 5(2), 189–213.
24. Mammino L. (2004). Dominancia del concepto de reacción, y de los terminos asociados, en las percepciones de alumnos de quimica. *Anuario Latinoamericano de Educación Quiacute;mica* XVIII, 46–52.
25. Mammino L. (2007). Language-related difficulties in science learning. III. Selection and combination of individual words. *Journal of Educational Studies*, 6(2), 199–214.
26. Mammino L. (2002). La percepción de la distinción entre sistemas y procesos por parte de los alumnos de quimica. *Anuario Latinoamericano de Educación Quimica*, XV, 125–129.
27. Mammino L. (Submitted manuscript). Language-related difficulties in science learning. IV. The use of prepositions and the expression of related functions.
28. Mammino L. (2002). Actitud hacia números y valores en cursos de quimica general. *Anuario Latinoamericano de Educación Quimica*, XVI, 5–10.
29. Mammino L. (2001). Physical quantities and their changes. Difficulties and perceptions by chemistry students. *SAARMSTE (Southern African Association for Research in Mathematics, Science and Technology Education) Journal*, 5, 29–40.
30. Mammino L. (2000). General y particular en química. *Anuario Latinoamericano de Educación Química*, XIV, 25–28.
31. Mammino L. (2005). Cause and effect – a relationship whose nature involves conceptual understanding, scientific method, logic and language. *Anuario Latinoamericano de Educación Química*, XXI, 33–37.
32. Mammino L. (1997). Estudio de la forma en que los alumnos perciben la distinción y las mutuas relaciones entre el mundo micróscopico y el mundo macróscopico. *Anuario Latinoamericano de Educación Química*, 9(2), 178–189.
33. Mammino L., Cardellini L. (2005). Studying students' understanding of the interplay between the microscopic and the macroscopic descriptions in chemistry. *Baltic Journal of Science Education* 1(7), 51–62.
34. Lijnse P.L., Licht P., Vos W. & Waarlo A.J. (1990). *Relating Macroscopic Phenomena to Microscopic Particles. A Central Problem in Secondary Science Education.* Utrecht: CD-β Press.
35. Mammino L. (2005). Method-related aspects in an introductory theoretical chemistry course. *Journal of Molecular Structure (Theochem)* 729, 39–45.
36. Mammino L., Kabanda M.M. (2006). Semiempirical methods as introduction to computational chemistry. *18th International Conference on Chemical Education.* Seoul (Korea), August 2006.

37. Gilbert J.K. (2008). Visualization: An emergent field of practice and enquiry in science education. In Gilbert J.K., Reiner M., Nakhleh M. (Eds.), *Visualization: Theory and Practice in Science Education*. Dordrecht: Springer, 3–24.
38. Gabel D.L., Samuel K.V., Hunn D. (1987). Understanding the particulate nature of matter. *Journal of Chemical Education* 64(8), 695–697.
39. Gabel D.L. (1993). Use of the particle nature of matter in developing conceptual understanding. *Journal of Chemical Education* 70(3), 193–194.
40. Mammino L. (1999). Semplificazione nella presentazione dei concetti: necessità ed esigenze. *Edichem'99: Chemistry in the Perspective of the New Century*. Bari, 341–356.
41. Brewer I.M. (1985). *Learning More and Teaching Less*. Guildford, Surrey: SRHE & NFER-Nelson.
42. Forman E.A., Cazden C.B. (1985). Exploring Vygotskian frameworks in education: the cognitive value of peer interaction. In Wertsch J.V. (Ed.), *Culture, Communication and Cognition: Vygotskian perspectives*. Cambridge: Cambridge University Press, 323–347.
43. Mammino L. (1998). Enseñanza interactiva en cursos de Quimica teórica. *Anuario Latinoamericano de Educación Química*, XI, 275–278.
44. Mammino L. (2005). The use of questions in the chemistry classroom: an interaction instrument with maieutic nature. *Anuario Latinoamericano de Educación Química*, XXI, 241–245.
45. Cooper M.M. (1993). Writing/an approach for large enrolment chemistry courses. *Journal of Chemical Education*, 70(6), 476–477.
46. Beall H., Trimbur J. (1993). Writing as a tool for teaching chemistry. *Journal of Chemical Education*, 68(1), 148–149.
47. Castro E.A. (1995). Escribir sobre quimica para aprender quimica. *Panamerican Newsletters*, 8(4), 8–9.
48. Mammino L. (1995). Teaching/learning theoretical chemistry at undergraduate level. *Southern Africa Journal of Mathematics and Science Education*, 2(1&2), 69–88.
49. Mammino L. (2002). Integrating chemistry teaching and language teaching. *17th International Conference on Chemical Education*. Beijing, August 2002.
50. Bradley J.D., Brand M., Gerrans G.C. (1987). Excellence and the accurate use of language, symbols and representations in chemistry. In *Widening the Scope of Chemistry*, IUPAC Blackwell Scientific Publications.
51. Mammino L. (1995). Algunas reflexiones sobre el rigor conceptual y formal en la enseñanza de la química. *Anuario Latinoamericano de Educación Química*, 8(2), 375–384.
52. Mammino L. (1998). Precision in wording: a tool to facilitate the understanding of Chemistry. *CHEMEDA, the Australian Journal of Chemical Education*, 48 & 49 & 50, 30–38.
53. Mammino L. (2000). Rigour as a pedagogical tool. In Seepe S. and Dowling D. (Eds.), *The Language of Science*. Johannesburg: Vyvlia Publishers, 52–71.
54. Mammino L. (1995). *Il linguaggio e la scienza*. Torino: Societá Editrice Internazionale.
55. Mammino L. (2000). Educating towards the Language of Science. In Seepe S. and Dowling D. (Eds.), *The Language of Science*. Johannesburg: Vyvlia Publishers, 72–93.
56. Mammino L. (1998). Evidenciación de estructuras lógicas en la enseñanza de la química teórica. *XXIV International Conference of Theoretical Chemists of Latin Expression*, Puebla (Mexico), September 1998.
57. Mammino L. (1997). La matematica come strumento di descrizione. *Nuova Secondaria*, 10, 94–95.
58. Mammino L. (2000). Making mathematics student-friendly in theoretical chemistry courses. *16th International Conference on Chemical Education*. Budapest, August 2000.
59. Mammino L. (2002). Mathematics in Chemistry: a Challenge for Tertiary Level Chemical Education. *17th International Conference on Chemical Education*. Beijing August 2002.
60. Mammino L. (2000). Alternative Conceptions and Misconceptions about the Phases of Matter. *SAARMSE Conference 2000*. Port Elizabeth (South Africa), January 2000.
61. Mammino L. (2007). Chemistry Students' perceptions of the Factors Influencing Phase Transitions of Pure Substances, as Expressed in Laboratory Reports. *Anuario Latinoamericano de Educación Química* XXIII, 41–46.

Reduction in Chemical Oxygen Demand and Color Intensity of Dye-Contaminated Wastewaster Using Visible Irradiation and ZnO-Assisted Advanced Oxidation Process: A Green Laboratory Experiment for Wastewater Treatment

B. Pare and P. Singh

Abstract This experiment depicts a very simple and benign chemistry involving photodegradation of model dye pollutant in wastewater. In the past 2 decades, photocatalytic degradation has become a hot topic for environmental engineers combating with aquatic pollution. In this paper, we have designed a very simple and safe laboratory experiment for undergraduate students in order to have a better understanding of decolorization and degradation processes. The decolorization and degradation of *Lissamine Fast Yellow (model dye contaminant)* has been confirmed by reduction in color intensity (Lambert Law or Beer Law) and chemical oxygen demand (Closed Reflux method). The dye solution completely decolorizes in 180 min of irradiation time while complete degradation has been found in 8 h of irradiation time. Students develop in-depth understanding of the degradation process by analyzing effects of various process variables on the pseudo-first-order rate constant for degradation of dye. Complete decolorization of dye takes place in 3 h while reduction in chemical oxygen demand (COD) approaches zero in 8 h of irradiation time in the presence of ZnO. This experiment can be performed in two lab sessions of 4 h. This paper explains the contribution of chemistry in solving environmental problems relating to water contamination.

B. Pare (✉) and P. Singh
Laboratory of Photocatalysis, Govt. Madhav Science College
Vikram University, Ujjain, (Madhya Pradesh) 456010 India
e-mail: brijesh_pare@hotmail.com or pardeepchem@gmail.com

M. Gupta-Bhowon et al. (eds.), *Chemistry Education in the ICT Age*,
DOI 10.1007/978-1-4020-9732-4_21,

1 Introduction

Since the beginning of industrialization, the variety and quantity of pollutants emitted into the environment have steadily increased. While the rates of development and waste production are not likely to diminish, efforts to control and dispose of wastes appropriately are rising [1]. One of the key concerns is the elimination of dyes and other colorants from wastewater effluents of industries and domestic sewage. Recently, advanced oxidation processes (AOPs) have shown alternatives to conventional physico-chemical methods of wastewater treatment. Among these processes, photocatalytic degradation involving semiconductor particles under UV or visible light illumination has become a promising technology for the wastewater treatment leading to complete mineralization of organic pollutants present in aquatic environment.

Upon irradiating aqueous suspension of ZnO, valance band electrons are excited to conduction band leaving a hole behind. These electron-hole pairs interact separately with other molecules. Valence band holes (h_{VB}^{+}) may react with surface-bound H_2O or OH^- to produce hydroxyl radical ($OH^{\bullet}$) and conduction band electron (e_{CB}^{-}) combines with molecular oxygen to generate superoxide radicals, as shown in Fig. 1. It has been suggested that the hydroxyl radicals ($OH^{\bullet}$) and superoxide anions ($O_2^{-\bullet}$) are the primary oxidising species in the photocatalytic degradation processes. These oxidising species would lead to decolorization of dye. Secondly, sensitization of dye molecule by visible light leads to excitation of dye molecule in singlet or triplet state, subsequently followed by electron injection from excited dye molecule onto conduction band of semiconductor particles [2]. This process results in enhancement of production of hydroxyl radicals and superoxide radicals resulting in complete mineralization of organic molecules.

The kinetics of photocatalytic degradation of the dye on the surface of the catalyst follows *pseudo* first-order kinetics. It is rationalized in terms of the Langmuir-Hinshelwood model, which is modified to accommodate reactions occurring at a solid-liquid interface, as shown by Eq. 1 [3]:

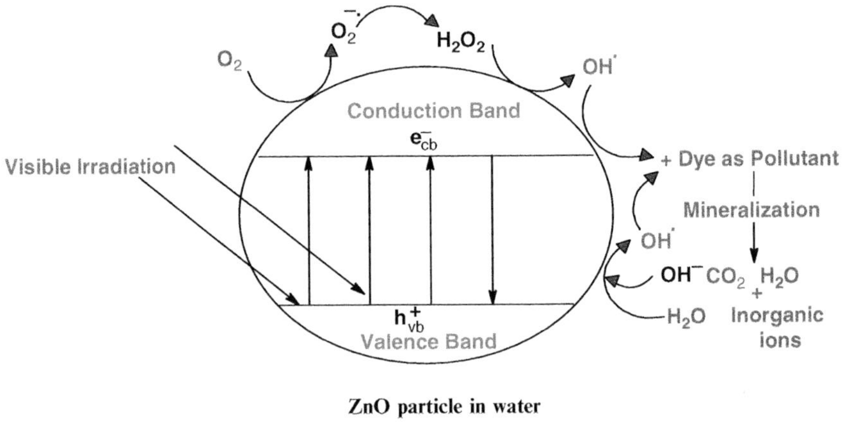

Fig. 1 General mechanism of photocatalytic degradation

Where r_0 represents the reaction rate for initial dye concentration (C_0). K represents the equilibrium adsorption constant of the dye on ZnO particles and K_r represents the limiting rate of the reaction.

$$r_0 = \frac{k_r K C_0}{1 + K C_0} \tag{1}$$

at maximum coverage under the experimental conditions. The integrated form is reported by Eq. 2

$$t = \frac{1}{Kk_r}(\ln C_0 / C) + \frac{(C_0 - C)}{k_r} \tag{2}$$

where t is the time required for the initial concentration of dye C_0 to become C. At low concentration of the dye, the second term in Eq. 2 is negligible compared to the first. On neglecting the second term, the final form of the equation reduces to

$$\ln(C_0 / C) = \mathrm{Kt}\, k_r = kt \tag{3}$$

where k is the apparent rate constant of the photocatalytic degradation.

The plot of ln C_0/C versus t is a straight line and its slope is k.

A good photocatalyst should be photoactive, able to use visible or near UV light, biologically and chemically inert, photostable, and non-toxic. Over the years, many semiconductors with photocatalytic properties have been investigated of which TiO_2 and ZnO are prominent. TiO_2, a UV active photocatalyst, can use only 3.5% of the visible spectrum while artificial UV sources are expensive and difficult to handle [4]. The main advantage of using ZnO photocatalyst is its photoactivity in visible region and non-toxicity.

Photocatalytic treatment of dye pollutants involves mainly three processes.

I. Decolorization process due to removal of chromophoric group or change in their position in organic ring leading to conversion of dyes to colorless leuco dyes
II. Degradation of colorless leuco dyes involving the conversion of organic dye skeleton into lower aliphatic compounds.
III. Formation of stable inorganic ions leading to complete mineralization of dye molecules.

Decolorization of dye can be ascertained by decrease in dye absorbance using visible spectrophotometer. The degradation of dye can be confirmed by reduction in COD and increase in CO_2 of dye solution.

In the present experiment, undergraduate students understand and investigate the nature of photocatalytic degradation process in two lab sessions of 4 h. Through the proposed experiment the undergraduates will be able to understand decolorization and the degradation process during photocatalytic degradation of dye. Inclusion of visible light-induced photocatalysis in undergraduate chemistry and environmental study is of great importance for better understanding of the context of green chemistry. To understand the chemistry involved in both studying and solving environmental problems has an immediate relevance, and it provides the students with significant motivation to understand the underlying chemical principles and the interdisciplinary aspects of the issue [5].

2 Methodology

2.1 Reagents

Lissamine fast yellow (LFY) dye was purchased from Sigma-Aldrich Company (India) and was used as received. All chemicals used were of AR grade. The photocatalyst ZnO used in the study obtained from Merck was about 99% pure. All solutions were prepared in double-distilled water. The chemical structure of LFY is given in Fig. 2.

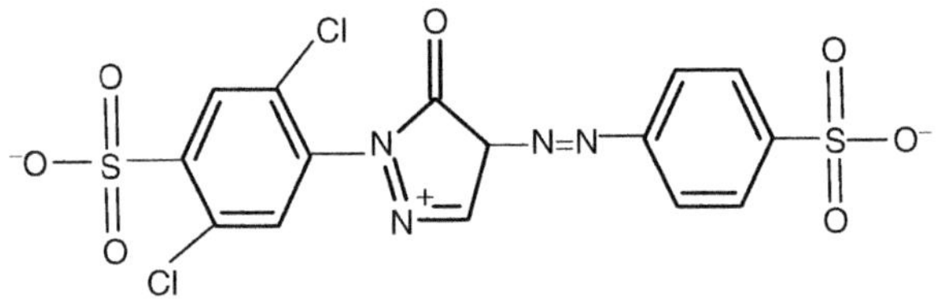

Fig. 2 Lissamine fast yellow

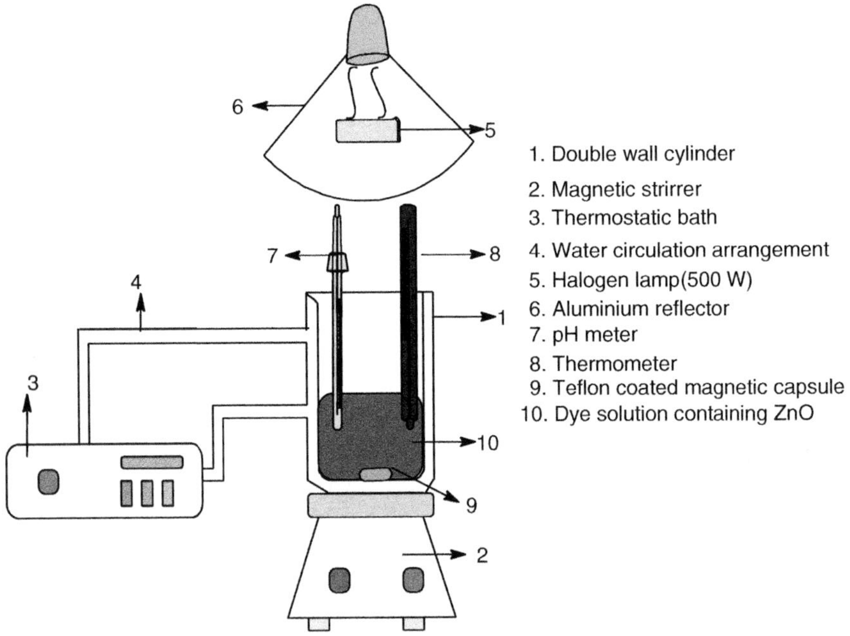

Fig. 3 Schematic representation of photoreactor

2.2 *Photoreactor and Radiation Source*

The photocatalytic experiments were carried out in a slurry-type batch reactor having a Pyrex vessel with dimensions of 7.5 × 6 cm (height × diameter). The Pyrex vessel was surrounded by thermostatic water circulation arrangement to keep the temperature in the range of 30 ± 0.3 °C (Fig. 3). Irradiation was carried out using 500 W halogen lamp surrounded with aluminium reflectors to avoid loss of radiation.

2.3 *Procedure and Analysis*

During the photocatalytic experiment, after stirring for 10 min, the solution composed of dye solution and catalyst was placed in the dark in order to establish equilibrium between adsorption and desorption. After this, the solution was transferred to a Pyrex reactor, and the lamp was switched on to initiate the reaction. During irradiation, the Pyrex reactor was mounted on a magnetic stirrer to keep the suspension homogeneous. At specific time intervals, aliquot (2 mL) was withdrawn and centrifuged for 2 min at the rate of 3,500 rpm to remove the ZnO particles in order to assess the extent of decolorization photometrically. After spectrophotometric analysis, both the centrifuged dye solution and ZnO particles were put back in the reaction solution. The intensity of visible radiation was measured by a digital lux-meter (Lutron LX –101). The detection was realized at 405 nm. Chemical oxygen demand (COD) was measured by the closed reflux method employing potassium dichromate as the oxidant under acidic condition. The unreacted oxidant was determined by titrating with ferrous ammonium sulphate using ferroin indicator [6]. The amount of CO_2 was estimated by using NaOH as titrant with phenolphthalein as indicator [7]. The percentage of decolorization was calculated as follows:

$$(\%)\ \text{decolorization} = (C_0 - C) / C_0 \times 100 \qquad (4)$$

where C_0 and C are the initial and final absorbance of dye solution.

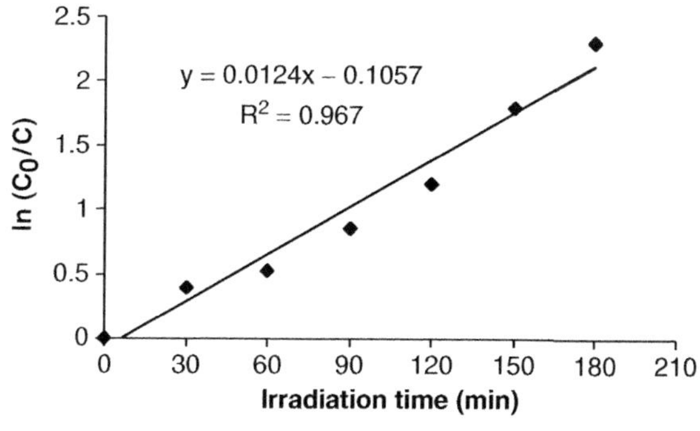

Fig. 4 Pseudo first-order kinetics for dye decolorization [LFY] = 5.0×10^{-5} mol dm^{-3}; ZnO = 300 mg/100 mL; pH = 8.1; Irradiation intensity = 134×10^2 lux

3 Results and Discussion

3.1 Decolorization of Lissamine Fast Yellow

A group of three students can perform this experiment under the supervision of the instructor. They obtain a plot of ln (C_0/C) *versus* time (where C is the absorbance of dye at time t and C_0 is the initial absorbance). The straight line with slope equal to k was obtained (Fig. 4). In the present study, decolorization of dye takes place in 180 min. The obtained data shown in Table 1 give the rate constant value of 3.87×10^{-4} s^{-1}. Students should be encouraged to observe the changes taking place during the process and to probe into detail the kinetics of degradation of dye.

3.2 Effect of ZnO

In order to understand the nature of heterogeneous photocatalysis, students observe the effect of ZnO loading in aqueous solution of LFY. In the present experimental study, the rate constant increases from 2.14×10^{-4} s^{-1} to 3.87×10^{-4} s^{-1} with increase in catalyst loading from 100 mg/100 mL to 300 mg/100 mL. However, further increase in the amount of catalyst results in decrease in the value of the rate constant (Table 2). The increased rate constant of LFY with higher catalyst loading is attributed to increment in the active sites available on the surface of the catalyst for the reaction, which enhances the rate of radical formation of hydroxyl radicals and superoxide radicals. Above optimal concentration, turbidity of ZnO suspension impedes further penetration of light in reaction. Increase in catalyst loading beyond 300 mg/100 mL also results in the deactivation of activated molecules due to collision with ground state molecule as shown below [8, 9]:

$$ZnO^{*} + ZnO \rightarrow ZnO^{\#} + ZnO \quad (5)$$

Table 1 Effect of irradiation time

[LFY] = 5.0×10^{-5} mol dm^{-3}; ZnO = 300 mg/100 mL; pH = 8.1; Irradiation intensity = 34×10^{2} lux

Irradiation time (min)	Absorbance	% Decolorization
0	0.902	0
30	0.501	44
60	0.266	70
90	0.150	83
120	0.100	88
150	0.040	95
180	0.001	100

Table 2 Effect of ZnO
[LFY] = 5.0×10^{-5} mol dm^{-3}; pH = 8.1;
Irradiation intensity = $134 \times 10^2 \pm 10$ lux

ZnO mg/100 mL	$k \times 10^{-4}\ s^{-1}$
100	2.14
200	2.72
300	3.87
400	2.68
500	2.34
600	2.00
700	1.90

Table 3 Effect of dye concentration
ZnO = 300 mg/100 mL; pH = 8.1;
Irradiation intensity = $134 \times 10^2 \pm 10$ lux

[Dye] $\times 10^{-5}$ mol dm^{-3}	$k \times 10^{-4}\ s^{-1}$
3	2.2
4	2.6
5	3.8
6	2.7
7	2.1

ZnO^*: ZnO with active species adsorbed on its surface.
$ZnO^{\#}$: deactivated form of ZnO^* shielding by ZnO.

3.3 Effect of Dye Concentration

While performing photocatalytic degradation, the variation of dye concentration affects the observed degradation rate constant. In the present study, the initial concentration of dye varies from 3×10^{-5} mol dm^{-3} to 7×10^{-5} mol dm^{-3}. The results are presented in Table 3. The rate of photodegradation ($9.07 \times 10^{-4}\ s^{-1}$) increases with increase in dye concentration up to 5×10^{-5} mol dm^{-3}. This is due to the fact that more dye molecules are available in photoactive volume for the photodegradation process. The rate of photodegradation decreases with further increase in concentration of dye, that is, above their optimum concentration. The decrease is attributed to the fact that the dye itself will start acting as filter for incident irradiation, thereby reducing the photoactive volume. Excessive adsorption of dye molecule on the catalyst surface hinders the competitive adsorption of OH^- ions and lowers the formation rate of hydroxyl radicals [10, 11]. Consequently, above optimal concentration, the rate of degradation decreases with increase in dye concentration.

3.4 Estimation of Chemical Oxygen Demand and CO_2 During Dye Degradation

Chemical Oxygen Demand test determines the oxygen required for the chemical oxidation of most organic matter and oxidizable in inorganic substance with the help of oxidant. Free carbon dioxide accumulates in this solution during the degradation of dye. CO_2 is determined by titrating the sample with strong alkali to pH 8.3.

The students investigate the COD and CO_2 of dye solution at different time intervals during photocatalytic treatment of LFY. In the present case, LFY (1×10^{-4} mol dm^{-3}) has been illuminated in the presence of ZnO (300 mg/ 100 mL). The results are presented in Table 4. The value of COD decreases from 66 mg/100 mL to 0 in 8 h of irradiation time. Maximal of 46 mg/100 mL of CO_2 has been estimated after 8 h of irradiation time. A slight decrease in pH and gradual increase in conductivity has been observed during 8 h of photocatalytic process. The reduction of COD confirms the degradation of LFY dye into CO_2 and inorganic ions under chosen experimental conditions.

3.5 Logistics

A group of undergraduate students can perform this experiment in two lab sessions of 4 h with two photoreactors, one spectrophotometer and one COD reflux assembly. Students can run two photoreactor assemblies simultaneously. They put the dye solution and ZnO in the first photoreactor and estimate COD of different time intervals of 0, 2, 4, 6, 8 h of irradiation time. Dye variation, effect of ZnO and decolorization process can be studied by students by using a second photoreactor in the time between two consecutive time intervals for COD determination. The standard concentration of dye solution, amount of ZnO, intensity of light and pH are given in Fig. 4 and in Tables 1–3. In the case of COD, in the first session, COD and CO_2 may be estimated for 0, 2, and 4 h of irradiation. After the first lab session, the reaction can be stopped by switching off the halogen lamp and putting the solution in the dark. The next day, the dye solution is magnetically stirred to make a homogeneous suspension of ZnO,

Table 4 Mineralization of dye
[LFY] = 1.0×10^{-4} mol dm^{-3}; ZnO = 300 mg/100 mL: pH = 4.75; Irradiation intensity = $134 \times 10^2 + 15$ lux

Irradiation time (h)	COD (mg/100 mL)	CO_2 (mg/100 mL)	pH	Conductivity (mS/cm)
0	66	–	4.7	0.256
2	45	22	4.7	0.374
4	32	33	4.7	0.457
6	12	44	4.1	0.672
8	0	46	3.9	0.873

then COD and CO_2 are estimated for 6 and 8 h. The instructor should consider the level of instruction in detail for the standard experiment given to students. All the standard solutions should be prepared 1 day before the experiment. The experiment should be conducted in the presence of an instructor in order to save time. Information regarding (i) wavelength at which to monitor dye absorbance, (ii) the time interval needed to study degradation process, and (iii) instructions for confirming Beer Law should be provided prior to conducting the experiment.

3.6 Hazards

This experiment is safe and simple. Artificial visible irradiation can be handled easily and ZnO is a non-toxic novel photocatalyst. However, Material Safety Data Sheets for dyes as well as other chemicals used should be consulted. Due to its corrosive nature, concentrated H_2SO_4 should be handled carefully during reflux of dye sample for COD estimation.

4 Conclusions

The primary goal of this paper is to enhance an understanding of photocatalytic degradation process among undergraduate students. This experiment describes the environmental application of heterogeneous photocatalysis. In the present experiment, decolorization of dye takes place in 3 h. COD reduces to half of its value in 4 h. COD reduces from 66 mg/100 mL to 0 in 8 h of irradiation in two lab sessions of 4 h. The proposed experimental exercise is safe and simple and can be performed by undergraduate students to better understand the kinetics of photodegradation of pollutants present in aquatic environment.

Acknowledgements The authors acknowledge the support and laboratory facilities provided by Dr. U. C. Maiwal, Head of Chemistry Department, and Dr. U. Srivastava, Principal, Govt. Madhav Science College, Ujjain (Madhya Pradesh) India.

References

1. Kahn A (2002), TiO_2 Coated Activated Carbon for Water Recovery. *J Undergrad Res* 3 No. 10 (http://www.clas.ufl.edu/jur/200206/papers/paper_khan.html)
2. Konstantinou I K, Albanis T A (2004), TiO_2- assisted photocatalytic degradation of azo dyes in aqueous solution: kinetic and mechanistic investigation. *Appl Catal B: Environ* 49:1–14
3. Poulios I, Micropoulu E, Panou et al. (2003), Photooxidation of eosin Y in the presence of semicoducting oxides. *Appl Catal B: Environ* 41:345–355
4. Anpo M (2000), Use of visible light: second-generation titanium oxide photocatalysts prepared by the application of an advanced metal ion-implantation method. *Pure Appl Chem*, 72(9): 1787–1792

5. Kegley S, Stacy A M, Carroll M K (1996), Environmental chemistry in the general chemistry laboratory, Part I: A context-based approach to teaching chemistry. *Chem Educ* 1(4)
6. American Public Health Association (APHA) (1985), Standard methods for examination of water and wastewater (sixth edition) 535–537
7. Maiti S K (2001), Handbook of methods in environmental studies: water and wastewater analysis (ABD pub Jaipur, India), 1, 50–51
8. Pare B, Jonnalagadda S B, Tomar H S et al. (2007), Advanced oxidation process in general chemistry laboratory: ZnO catalysed degradation of acridine orange dye in visible light. *Aus J Educ Chem* 68:30–33
9. Pare B, Jonnalagadda S B, Singh P et al. (2008), ZnO assisted photocatalytic degradation of acridine orange in aqueous solution using visible irradiation. *Desalin* 232: 80–90
10. Pare B, Singh P, Jonnalagadda S B (2008), Visible light induced heterogeneous advanced oxidation process to degrade pararosanilin in aqeous suspension of ZnO. *Indian J Chem Sec A* 47: 836–842
11. Saien R, Soleymani A R (2007), Degradation and mineralization of Direct Blue 71 in a circulating upflow reactor by UV/TiO_2 process and employing a new method in kinetic study. *J Hazard Mater*, 144(1–2):506–512

Designing Effective E-Learning Environments – Should We Use Still Pictures, Animations or Interactivity?

S. Schmid, A. Yeung, A. V. George, and M. M. King

Abstract There has been much research interest in the area of multimedia learning and using e-learning to assist student learning. A range of tools and teaching materials are used to develop students' understanding in different domains. We have developed an online chemistry module with three different versions, designed to determine the most effective way for enhancing student learning and addressing misconceptions students may have. One version incorporates the use of still pictures, one version uses animations and simulations, and the final version uses animations and simulations together with interactivity. The present study investigated students' perception of the modules and performance associated with using these different versions of the module. While students rated the animated and interactive versions of the module more highly than the one that had still pictures only, this did not translate to increased performance in associated tests.

Abbreviations CHEM1102: Chemistry 1B; CHEM1109: Chemistry 1 Life Sciences B; CHEM1611: Chemistry A (Pharmacy)

1 Introduction

Over recent years, information and communications technology has increasingly been incorporated into teaching activities in higher education where student populations are diverse. A range of tools and teaching materials are used to develop students' understanding in different domains. Consequently, best practices for design of multimedia instructional materials and for establishing effective e-learning

S. Schmid (✉), A. Yeung, and A.V. George
School of Chemistry, The University of Sydney, NSW, 2006, Australia
e-mail: s.schmid@chem.usyd.edu.au

M.M. King
Faculty of Education and Social Work, The University of Sydney, NSW, 2006, Australia

M. Gupta-Bhowon et al. (eds.), *Chemistry Education in the ICT Age*,
DOI 10.1007/978-1-4020-9732-4_22, © Springer Science+Business Media B.V. 2009

environments have been investigated. Previous research has shown that concurrent presentation of text with either animations or pictures can enhance learning [1] and that multimedia materials, which include interactive elements may lead to enhanced student engagement [2, 3]. Interactive multimedia has the potential to provide students with new ways of interacting with various representations of information and allows them to explore ideas in innovative ways [4]. Higher quality learning occurs when students are actively and cognitively involved in the learning process, and engagement promotes such involvement [5, 6]. Thus, user engagement is an integral part of learning [7, 8]. The complexity of interactive multimedia, however, can lead to problems and design of such environments must be based on the needs and interests of the user. The limitations and capabilities of multimedia need to be understood well, before the potential benefits can be realized [2, 9]; however, evidence addressing the effectiveness of different approaches for promoting engagement with multimedia technology remains limited.

In this context we developed online learning materials for use as pre-laboratory instruction or stand-alone learning modules for first-year chemistry students. To inform development at an early stage a study was undertaken to investigate the most effective design of online chemistry modules for enhancing student learning and addressing misconceptions. Initially we have developed three different versions of an online chemistry module:

- Static version: text and still pictures
- Animated version: text and animations/simulations
- Interactive version: text, animations/simulations and interactivity

The outcomes of this study are of significant interest, since students often use online learning as a supplement to lectures. If one particular method of delivery is more effective than another, then such information is likely to have a significant impact on the design of online materials in the future. Not only can this information help enhance student performance, it can also allow the process to be more efficient and cost effective.

The results reported in this paper draw on student surveys and participants' assessment results.

1.1 Online Learning Module

A number of design issues were considered when creating the three versions of the online chemistry module used in this investigation. The topic of acids and bases was chosen since past examination performance [10] and anecdotal evidence suggest that students find this is a difficult topic and common misconceptions are held. The module focused on five main aspects that covered areas often associated with such common student misconceptions:

- The difference between strong and weak acids and bases
- The difference between concentrated and dilute acids and bases

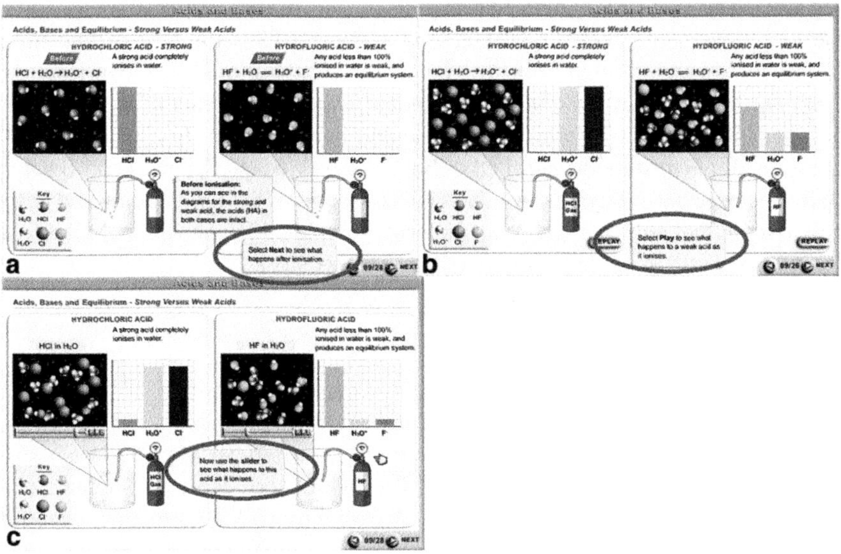

Fig. 1 Examples of screens captured from the (a) static (b) animated and (c) interactive versions of the online chemistry learning module. The static version uses still pictures only. The animated version requires students to select 'Play' (see highlight in (b)) and watch an animation that shows a representation of the ionization of an acid. The interactive version gives students control of the animation that shows the ionization of an acid by moving a slider bar (see highlight in (c)). In this version, students are able to stop the animation as they please to help them with their understanding

- The relative strengths of acids and bases and their conjugates
- Simple calculations of pH involving strong acids and bases
- The relationship between pH, K_a, concentration and strength of acids and bases

Every effort was made to ensure that the content presented was similar in all three module versions, with the only difference being the design of the module, i.e., layout, type of visuals or type of interactivity. Moreover, animations and interactivity were only included if the additional features added to the presentation of content and ideas. Otherwise animations and interactivity would simply distract students' attention away from what they are intended to learn [11]. Figure 1 gives an example of a corresponding screen from each version of the module.

2 Methodology

2.1 Participants

First-year chemistry students participated in this study at the University of Sydney in semester 1 and semester 2, 2007. In semester 1, 2007 all students were enrolled in either CHEM1102 (Chemistry 1B), or CHEM1611 (Pharmacy). CHEM1102 is the second of two linked first-year chemistry courses offered to students undertaking mainstream science qualifications. CHEM1611 is the first of two linked chemistry courses offered only to students undertaking qualifications towards a Bachelor of Pharmacy. In semester 2, students who participated in the study were enrolled in CHEM1109, which is the second of two linked first-year chemistry courses offered to students enrolled in Bachelor of Medical Sciences and Bachelor of Science specialist degrees. All units cover aspects of the topic "acids and bases" and assume that students have satisfactorily completed Higher School Certificate (HSC) chemistry (i.e., university entry level chemistry).

2.2 Study Design

Participation in this study was voluntary. Participants from CHEM1102, CHEM1611, and CHEM1109 were randomly assigned to one of three groups. Students completed a pre-test to determine their knowledge prior to working through the module. Each group completed a different version of the online chemistry module. After completion of the module, students were asked to do a post-test, which used the same questions as the pre-test. Feedback, answers and marks were only given after students had completed the post-test.

Marks were recorded for both the pre-test and the post-test associated with the online module. A total of 65 students in semester 1, 2007 and 168 students in semester 2, 2007 completed all three aspects of the study (i.e., pre-test, module, post-test) and their results were used in the analysis. The data for students in both CHEM1102 and CHEM1611 were analyzed together because all aspects of the module were identical for both units of study.

2.3 Student Survey

Students were asked to complete an online module rating survey after they finished the online modules in an attempt to determine students' views of the module. Specific questions the survey sought to answer were those related to students' motivational level, students' perception of the modules helpfulness, user friendliness, and how students use the modules to assist their learning.

2.4 Pre- and Post-Tests

The pre- and post-tests sought to test students' understanding of the concepts presented in the module, rather than just the memorization of facts. The multiple choice questions that were asked ranged from simple low-road transfer questions to more complex conceptual high-road transfer questions relating to the module. Examples include:

- Which of the following is a strong base?
  (a) HNO_3
  (b) HCOOH
  (c) H_2O
  (d) NH_3
  (e) NaOH

- Using the table below, which of the following solutions is the strongest acid?
  (a) HA
  (b) HB
  (c) HC
  (d) HD
  (e) It is not possible to tell from these data

Acid	Concentration (M)	pH
HA	0.100	3.4
HB	0.010	3.4
HC	0.050	4.6
HD	0.080	3.7

3 Results

3.1 Student Survey

The student survey had a number of questions for students to answer on a 5-point scale and some free format questions.

Figure 2 shows the distribution of survey responses for a few selected questions from semester 1, 2007. A larger proportion of students who completed the animated and interactive versions of the module find the explanations easier to follow, the material more engaging and interesting than students who completed the static version of the module. Moreover, a larger proportion of students who completed the animated version of the module responded by saying that the module was helpful for their learning. Students who completed the static version were the only students who did not find the module engaging or interesting at all. Additionally, a larger proportion of these students found the explanations in the module difficult to follow.

Chi-square analyses were conducted to determine whether there were statistical differences in the distribution of survey responses for students who completed different versions of the module. However, due to the small number of responses (41 in total), the differences between the three groups (see Fig. 2) did not reach statistical significance.

A similar result was found in semester 2, 2007, when 120 students completed the student survey. A more distinct trend emerged where a larger proportion of

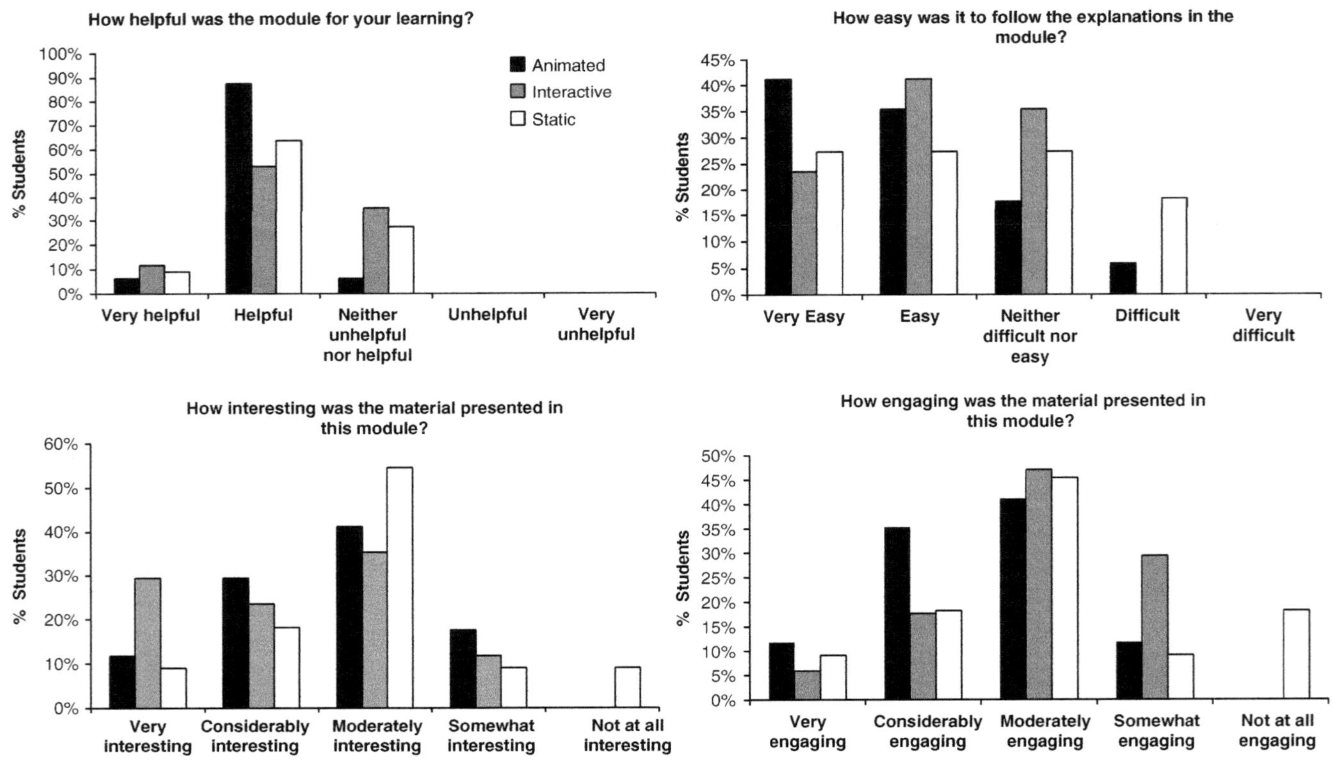

Fig. 2 Student responses to selected questions of the module rating survey for each version of the module for semester 1, 2007 (N_{static} = 11, $N_{animated}$ = 14, $N_{interactive}$ = 16)

students in the animated and interactive groups found the module engaging, interesting, and helpful for their learning. Those in the static group did not have such positive responses about that version of the module. The distributions of responses for selected questions are shown in Fig. 3.

Chi-square analyses were performed, as the sample size was much bigger in semester 2, 2007. Significant differences in students' responses depending on the version of the module they completed were found. A summary of the results is shown in Table 1.

In addition the free format questions were analyzed to see whether differences were evident for responses from students who were assigned to different versions of the module.

In general, students who completed either the animated or interactive versions had *positive* comments about the module, for example:

- "Use of animations and videos help with understanding"
- "I like them! I like the interactivity of online modules, especially animations, which help to shape my understanding of the concerned topics"
- "It's colourful, animated and interesting!" and "Far more interesting and motivating"
- "Allows me to learn at my own pace, allows me to learn in a comfortable environment – in my own comfort zone"
- "I enjoy the online modules as they reinforce lecture material. The animations help me understand reactions which would otherwise be more difficult to visualize"
- "It is more enjoyable than slaving away at a textbook, because it is more interactive"
- "I thought that it was an engaging and perfectly adequate module"

Students who completed the static version had *negative* comments only or did not answer the free response questions about the module. Some of their comments include:

- "Perhaps less text or text more spread out"
- "Its quite strenuous to look at the screen and learn. ... It makes my eyes really tired ... but the online activities should be more interactive"
- "Walls of text are bad"

3.2 Academic Performance

3.2.1 CHEM1611 and CHEM1102

One-way ANOVA was conducted to determine whether there was any difference in performance in the pre-test or post-test for groups of students who completed different versions of the module. Subjects were divided into treatment groups depending on which version of the module they completed: static version, animated version, or interactive version. It was found that there was no statistically significant difference in academic performance between the treatment groups for the pre-test [$F_{2,75}$ = 0.257, $p = 0.774$] (Fig. 4a). Post-hoc comparisons using Tukey HSD confirmed that there were no significant differences between the treatment groups.

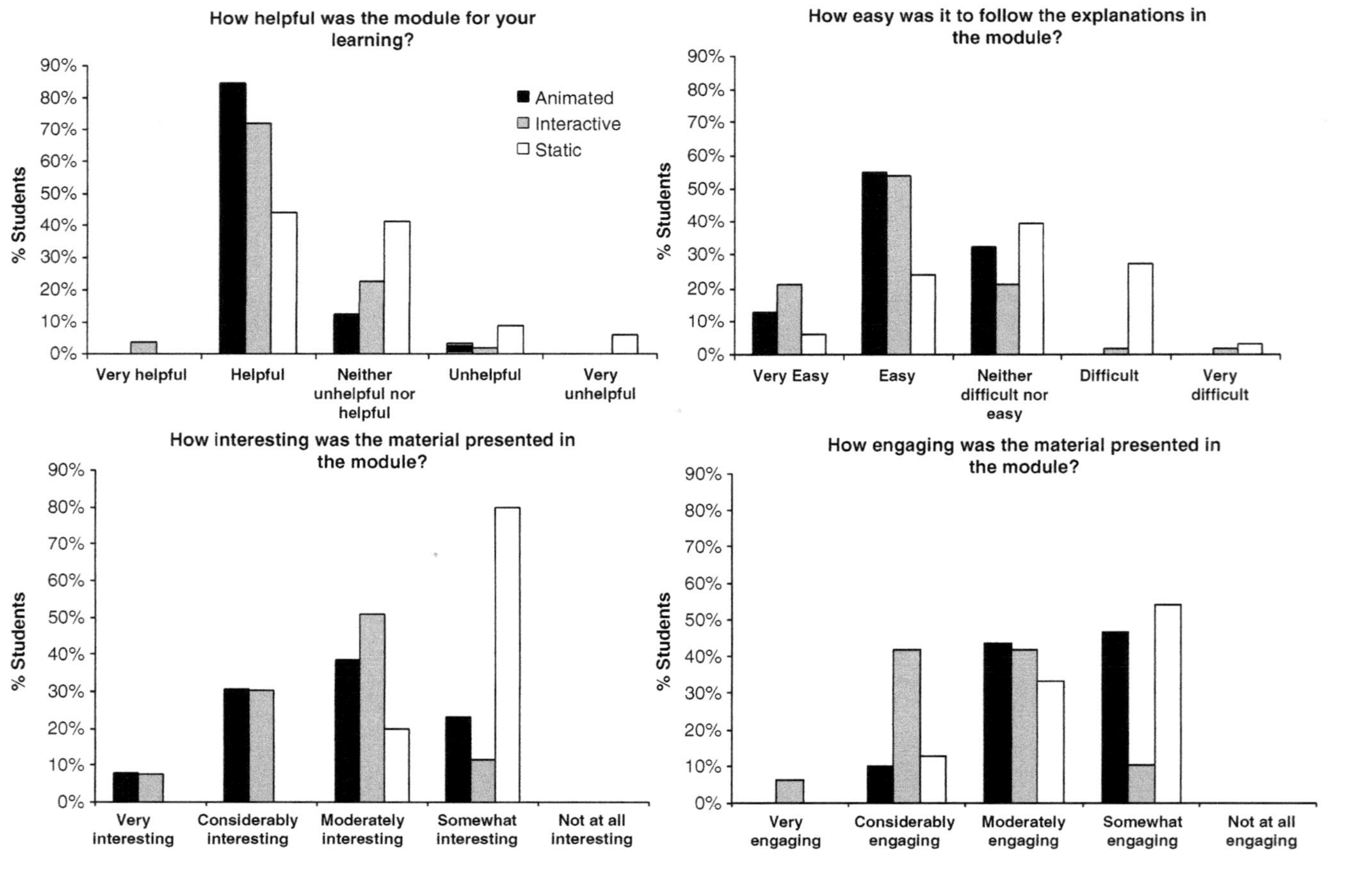

Fig. 3 Student responses to selected questions of the module rating survey for each version of the module in semester 2, 2007 ($N_{static} = 35$, $N_{animated} = 32$, $N_{interactive} = 53$)

Table 1 Summary of results from chi-square analysis of student responses to selected questions from module rating survey distributed in semester 2, 2007

	χ^2	df	*p*-value
Helpfulness of the module	16.5	4	0.00247
Explanations in module	11.7	2	0.00286
Interesting material in module	16.7	2	0.000233
Engagement with module	35.0	4	4.54×10^{-7}

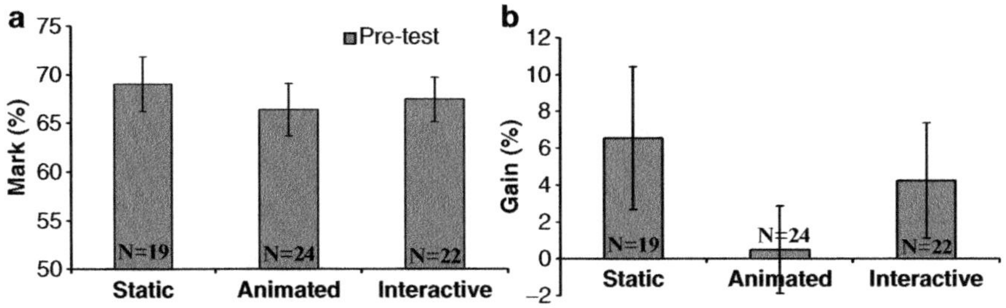

Fig. 4 (a) Mean marks for the pre-test for each version of the module and (b) mean gain for students who completed different versions of the module in CHEM1102 and CHEM1611

To determine the effect of different versions of the module on academic performance, the difference between a student's pre-test and post-test marks was determined. This "gain" gave an indication of the benefit students may have received from the module. Subsequently, one-way ANOVA was conducted to determine whether there was a difference in gain between students who completed the different versions of the module. It was found that there was no statistically significant difference in *gain* for the treatment groups [$F_{2,62} = 0.876$, $p = 0.422$] (Fig. 4b). Post-hoc comparisons using Tukey HSD confirmed that there were no significant differences between treatment groups.

Since there was no significant difference between the three treatment groups in terms of the gain students received, a paired-samples t-test was conducted to determine if there was a difference in students' marks in the pre- and post-tests for each treatment group. No significant differences in marks were found for static ($t_{18} = 2.08$, $p = 0.0510$), animated ($t_{23} = 0.134$, $p = 0.895$), and interactive ($t_{21} = 1.79$, $p = 0.870$) versions.

3.2.2 CHEM1109

One-way ANOVA was conducted to determine whether there was any difference in performance in the pre- and post-test for the module. Again, subjects were divided into treatment groups depending on which version of the module they completed: static version, animated version, or interactive version. It was found that there was no statistically significant difference in academic performance between the treatment groups for the pre-test [$F_{2,168} = 1.96$, $p = 0.144$] (Fig. 5a). Post-hoc comparisons using Tukey HSD confirmed that there were no significant differences between the treatment groups.

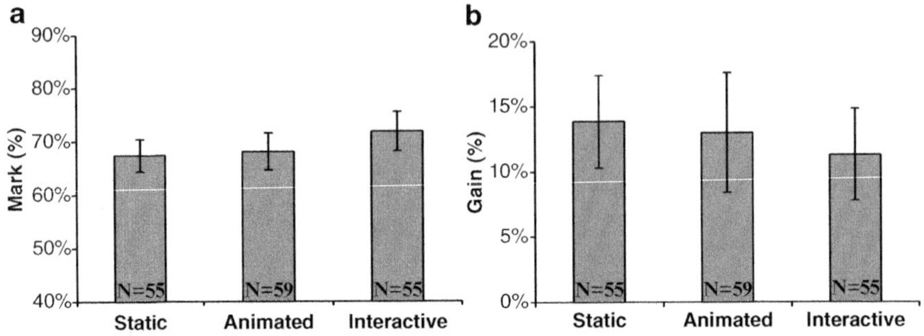

Fig. 5 (a) Mean marks for the pre-test for each version of the module and (b) mean gain for students who completed each version of the module in CHEM1109

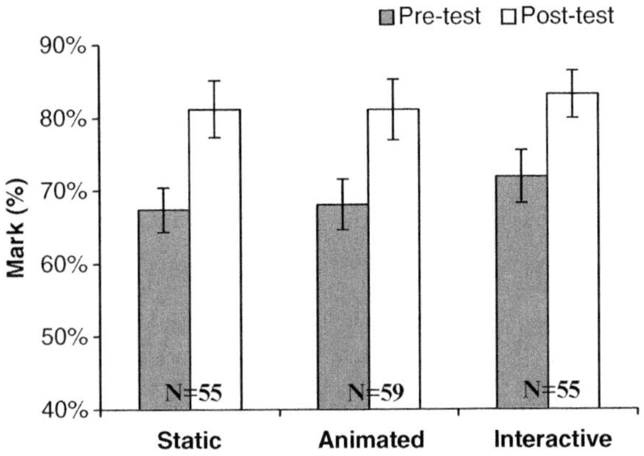

Fig. 6 Mean marks for pre- and post-test for CHEM1109 students in each treatment group

One-way ANOVA was then conducted to determine whether there was a difference in *gain* in academic performance between students who completed the different versions of the module. It was found that there was no statistically significant difference in *gain* for the treatment groups [$F_{2,168} = 0.410$, $p = 0.664$] (Fig. 5b). Post-hoc comparisons using Tukey HSD confirmed that there were no significant differences between treatment groups.

A paired-samples t-test was conducted to determine if there were significant differences in performance between the pre- and post-tests for students who completed each version of the module. It was found that there were significant differences between pre- and post-test marks for students in the static group ($t_{54} = 6.312$, $p = 0.000$), animated group ($t_{58} = 7.674$, $p = 0.000$), and interactive group ($t_{54} = -5.570$, $p = 0.000$) as shown in Fig. 6. This result is different to the semester 1, 2007 findings.

4 Discussion

This study provides strong evidence that there are differences in student rating for different versions of the investigated online learning module. Those who completed the interactive and animated versions overall had much more positive responses in the module rating survey compared to those who completed the static version. Statistical analyses confirmed significant differences were evident for the CHEM1109 cohort which was the largest cohort involved in this study. Student comments in the free-response section of the module rating survey also indicated that students who completed the animated and interactive versions had a very positive experience compared to negative or no comments from students who completed the static version. These findings are similar to those found in a study conducted by Lustria [12] where a significant relationship was found between the level of interactivity and attitudes towards online modules. While it is clear that the static version was not preferred the situation is less clear for the differences between the animated and interactive versions of the module. For these no clear trend could be determined from this study for either semester.

The preference for, and higher engagement with, the animated and interactive versions of the module compared with the static version does not seem to translate into improved academic performance. Surprisingly, when looking at the mean *gain* for students who completed the different versions in semester 1, 2007, it appears that those who completed the static version gained more from doing the module compared with the other two versions. Due to the small number of participants in this study, differences in the academic performance of students who completed the three different versions of the module did not reach statistical significance. The result, however, conflicts with results from the module rating survey, as one would expect that students who are more engaged will perform better [6]. Similarly, Lustria [12] also found that higher interactivity did not lead to higher comprehension, the measure of performance used in that study. Since engagement did not lead to better performance in our study, further study on a larger sample is required to determine whether this result was an artifact of the small sample size or can be confirmed. Otherwise, this may confirm other studies where multimedia applications fail to live up to developers' claims of providing enhanced learning [13, 14]. Previous findings have shown that learners can tend to focus on dynamic elements such as video clips, rather than fully engaging in the content of the material [14].

Another fact that needs to be taken into consideration is that CHEM1611 students generally have a reasonably good chemistry background and are academically very able students. For this reason, the module may not assist them as much in their learning as expected, as these students may not require the extra help. This is demonstrated with the relatively high pre-test marks. Consequently, further research on a larger sample of students with varying levels of prior knowledge would be desirable. In turn, this will then allow a better indication of the effect of each version of the module on student performance and better inform us of the most effective design of online chemistry learning modules.

The larger study conducted in semester 2, 2007 showed similar results in that preference for the animated and interactive versions of the modules did not translate into greater *gains* in performance, i.e., students who completed a particular version of the module did not perform better than those who completed the other versions of the module. However, significant differences were found between post-test mark and pre-test mark. Similar to the CHEM1611 students, CHEM1109 students also have a relatively good chemistry background, which may need further investigation.

5 Conclusion

The aim of this study was to determine the most effective format of an online chemistry module to enhance student learning. This study has shown that students have a preference for, and are more engaged with, the animated and interactive versions of the online chemistry learning module that we have designed, compared with the static version. However, differences in student rating for the interactive and animated versions could not be determined with certainty. For this reason, further research on a larger sample is required to determine whether any differences between the animated and interactive versions exist. In-depth interviews and focus groups should be conducted and may provide more detailed data on students' authentic experiences of the module and better determine the effects of the different versions on student performance.

Surprisingly, students who completed the static version of the module had a greater mean *gain* between the pre and post-tests of the module in semester 1, 2007, although one would expect more engaged students to do better. In semester 2, 2007 no significant difference in gain between those completing the different versions of the module was observed at all. This warrants further research using a much larger sample, as most differences did not reach statistical significance. Furthermore, research into the influence of the different versions of the modules on the academic performance of students with different chemistry backgrounds would also be an important factor to investigate. Students involved in the present study all have a relatively good chemistry background, which may account for the non-significant differences in performance between treatment groups. Ultimately, we want to be better informed about the design of online chemistry modules to effectively enhance student learning and engagement for students of all levels of prior knowledge.

Acknowledgments The authors would like to thank all the students at the University of Sydney who participated in the study. Data collections described in this work were authorized by the University of Sydney Human Research Ethics Committee (Project reference number 04–2007/9910). The WebCT helpdesk at the University of Sydney has also been of tremendous help in this process. The authors would finally like to thank the University of Sydney for providing funds towards the development of the online module

References

1. Mayer, RE and Moreno, R (1998) A Split-Attention Effect in Multimedia Learning: Evidence for Dual Processing Systems in Working Memory. J Educ Psychol 90:312–320.
2. Craincross, S and Manion, M (2001) Interactive Multimedia and Learning: Realising the Benefits. Innov Educ Teach Int 38:156–164.
3. Littlejohn, A, Suckling, C, Campbell, L et al. (2002) The Amazing Patient Tutor: Student's Interactions with an Online Carbohydrate Chemistry Course. Brit J Educ Technol 33:313–321.
4. Masterman, E and Rogers, Y (2002) A Framework for Designing Interactive Multimedia to Scaffold Young Children's Understanding of Historical Chronology. Instr Sci 30:221–241.
5. Kennedy, GE (2004) Promoting Cognition in Multimedia Interactivity Research. J of Interact Learn Res 15:43–61.
6. Scott, B (2001) Conversation Theory: A Constructivist Dialogical Approach to Educational Technology. Cybern Hum Knowing 8:1–25.
7. Giardina, M (1992) Interactivity and Intelligent Advisory Considerations. In: Giardina, M (ed.), Interactive Multimedia Learning Environments, pp. 48–66. Springer-Verlag, Berlin.
8. Laurillard, D (1995) Multimedia and the Changing Experience of the Learner. Brit J Educ Technol 26:179–189.
9. Norman, D (1988) The Psychology of Everyday Things. Basic Books, New York.
10. Read, J, George, AV, Masters, A et al. (2004) Proceedings of the Scholarly Inquiry into Science Teaching and Learning Symposium. In: Merrett, D (ed.) Students' Perceptions of Their Understanding in Chemistry 1 for Veterinary Science, pp. 45–51. UniServe Science, Sydney, NSW.
11. Hutchings, GA, Hall, W and Colbourn, CJ (1993) Patterns of Students' Interactions with a Hypermedia System. Interact Comput 5:295–313.
12. Lustria, MLA (2007) Can Interactivity Make a Difference? Effects of Interactivity on the Comprehension of and Attitudes toward Online Health Content. J Am Soc Inf Sci Tec 58:766–776.
13. Aldrich, F, Rogers, Y and Scaife, M (1998) Getting to Grips With "Interactivity": Helping Teachers Access the Educational Value of Cd-Roms. Brit J Educ Technol 29:321–332.
14. Rogers, Y and Scaife, M (1997) How Can Interactive Multimedia Facilitate Learning? In: Proceedings of the First International Workshop on Intelligence and Multimodalities in Multimedia, pp. 123–142. The Live Oak Press.

Students' Learning Styles and Academic Performance

S. Schmid, A. Yeung, and J. R. Read

Abstract Learning style preference data were collected from 1,252 first-year chemistry students at the University of Sydney in 2004/05. Analysis of the preference score distributions strongly supports the adoption of a trait-based personality model, in line with modern five-factor model research, rather than older dichotomous type-based models. Using this interpretive framework, significant and strong correlations have been found between dimension preference scores and academic performance in end-of-semester examinations. In general, students' average examination performance improves as introvert personality characteristics increase, and a similar trend is observed as thinking (as opposed to feeling) characteristics become dominant. These findings have substantial implications for pedagogical practice, and raise equity issues relating to assessment.

Abbreviations CHEM1001: Fundamentals of Chemistry 1A; CHEM1101: Chemistry 1A; CHEM1901: Chemistry 1A (Advanced); CHEM1002: Fundamentals of Chemistry 1B; CHEM1102: Chemistry 1B; CHEM1902: Chemistry 1B (Advanced); E: Extrovert; EI: Extroversion/Introversion; F: Feeling; FT: Feeling/Thinking; HSC: Higher School Certificate; I: Introvert; J: Judging; JP: Judging/Perceiving; MBTI: Myers Briggs Type Indicator; N: Intuition; P: Perceiving; PLSI: Paragon of Learning Styles Inventory; S: Sensing; SN: Sensation/Intuition; T: Thinking.

S. Schmid (✉) and A. Yeung
School of Chemistry, The University of Sydney, NSW, 2006, Australia
e-mail: s.schmid@chem.usyd.edu.au

J.R. Read
School of Chemistry and Physics, The University of Adelaide, SA 2006, Australia

M. Gupta-Bhowon et al. (eds.), *Chemistry Education in the ICT Age*,
DOI 10.1007/978-1-4020-9732-4_23, © Springer Science+Business Media, B.V. 2009

1 Introduction

It has long been recognised that individual differences such as learning style preferences influence students' interactions with their learning environment [1], and individuals' approaches to assessment tasks also exhibit variation. Students' adoption of either mastery or performance goals [2] will influence study strategy [3, 4] and academic performance [5]. Levels of situational or individual interest [6] influence learner engagement [7], as do teacher choices about learning activities [8, 9]. These differences do not raise equity issues as they either relate to factors (such as learning tasks) that are identical for all students, or they reflect student-made choices. But, what about the influence of non-controllable student characteristics like personality? This paper first considers the appropriate personality model to use as a theoretical basis for learning style research, and then investigates the relationship between learning styles and examination performance, discussing the potential equity issues, which may arise as a consequence.

Students' learning style preferences refer to the ways in which they respond to learning stimuli and to their characteristic ways of acquiring and using information. Learning styles recognise not only that individuals learn in different ways, but also that individual characteristics like personality influence learning [10]. This recognition is in line with constructivist notions of learning [11–13], recognising that factors like learning styles and prior knowledge may (for example) influence judgements of the importance of information presented and also influence the interpretation given to that information [14]. As early as the 1970s, attempts to include learning style considerations in pedagogical practice were being described, with the intention of better satisfying needs, which may differ between students [15, 16]. Learning style differences were originally believed to relate to different modes of thought, or different cognitive styles [17], but research findings do not, in general, provide convincing evidence for the existence of cognitive styles [18]. Nevertheless, knowledge of students' learning styles still has educational utility, provided they are interpreted as reflections of attitudes, actions, preferences and habits connected to personality characteristics, and not seen as being indicative of differences in cognitive architecture.

Alignment between students' learning styles and an instructor's teaching style has been reported to be beneficial for recall, understanding and more positive post-course attitudes [19]; improved student performance [20–22]; and improved student engagement/interest [23]. Such alignment will help to maintain situational interest [24], which in turn leads to the use of deeper learning strategies, increased self-regulation and ultimately to improved academic performance [25]. For this reason it is desirable to match instructional practice to the learning style preferences of students. The individuality of learning style preferences does pose a challenge for classroom instruction, however, as the most effective mode of instruction will differ between students. Fortunately, teaching is most effective when it caters for a range of learning styles, in part because occasionally having to learn in a less preferred style helps to broaden students' range of skills [26].

It is likely that learning style preferences are well established in the context of tertiary education [27], creating a risk that incompatible learning environments may lead to students struggling, adversely impacting on learning and attitudes towards education [26], thereby resulting in lower grades and increased drop-out rates [19]. In short, a balance must be struck that allows student engagement in the subject to be maintained, whilst catering for the range of learning style preferences present within a cohort.

1.1 The Paragon of Learning Styles Inventory

Several instruments exist for investigating students' learning styles, differing in psychological bases or target audience, and it is important to select an appropriate instrument to obtain valid and useful results. The Paragon of Learning Styles Inventory (PLSI), used in this work, has been successfully used with university-level chemistry students [28]. The PLSI design originates from the Myers-Briggs Type Indicator (MBTI), developed in 1962 [29–32] and is based on Jung's theories of personality [33]. Like the MBTI, the PLSI was designed to classify people along the four Jungian psychological personality dimensions, thereby giving a measure of cognitive and perceptual preferences. It is a reliable and validated inventory for assessing personality, developed for use in educational settings for students aged 8 and older [34]; studies with adults have shown excellent instrumental stability and reliability, with reports of split half reliability for each dimension between 0.90 and 0.94 [35].

The Jungian dimensions relate to extroversion/introversion (EI), sensation/intuition (SN), feeling/thinking (FT) and judging/perceiving (JP), and each is examined by the PLSI. Each PLSI item relates to a single dimension and is scored 0 or 1 according to the characteristic favoured, and a dimension score is then obtained by summing item scores for that dimension. A dichotomous type-based interpretation of a dimension score simply classifies it according to the predominant characteristic (i.e. more zeroes or more ones). As such, students fit into one of 16 possible distinct psychological types, covering every possible combination using one style from each dimension. A trait-based interpretation considers dimension scores as measurements of the strength of preferences along a continuum, with lower scores indicating preference for the first characteristic named.

Using type-based interpretations, PLSI results have been used to provide teachers with valuable insights into their classes [29], allowing them to tailor their instruction to best engage their students. For example, a classroom dominated by students who exhibit predominantly thinking and extrovert characteristics would respond well to a classroom debate of issues – but such a discussion risks alienating those with stronger inclinations towards introvert characteristics (who prefer quiet, less active classrooms and solitary activities) and feeling characteristics (who dislike conflict). Similarly, assessments involving oral presentations would be less positively received by a classroom dominated by introverts.

2 Methods

Study participants at the University of Sydney in 2004 were enrolled in one of the semester 1 first-year chemistry units available to students undertaking mainstream science courses. These units were CHEM1001 (Fundamentals of Chemistry 1A), CHEM1101 (Chemistry 1A) and CHEM1901 (Chemistry 1A – Advanced). All three units cover similar general chemistry material, but differ in the level of assumed prior knowledge and the level of presentation. CHEM1001 students have either not completed Higher School Certificate (HSC) chemistry (university entry level) or achieved poor results. CHEM1101 students have satisfactorily completed HSC chemistry, whilst CHEM1901 students have achieved a HSC chemistry mark above 80%. The 48 item PLSI was distributed to 1,143 students during their laboratory sessions and produces dimension scores out of 12.

In 2005, the EI dimension items from an expanded (13 items per dimension) PLSI were distributed to 824 students enrolled in one of the semester two first-year chemistry units offered for mainstream science students. These units were CHEM1002 (Fundamentals of Chemistry 1B), CHEM1102 (Chemistry 1B) and CHEM1902 (Chemistry IB – Advanced). Each of these units assumes successful completion of a corresponding semester one unit, and includes both general and organic chemistry content.

3 Results and Discussion

Response rates are summarised in Table 1, along with comparative performance data relating to end-of-semester examination performance of respondents and non-respondents. In the 2004 cohorts, these results suggest that the CHEM1001 and CHEM1101 respondent groups outperformed non-respondents on the end-of-semester examinations, with the CHEM1901 respondent group being representative of the entire cohort. In 2005, the samples appear representative of the student populations, despite the substantially lower response rates occurring with semester two data collections. The distributions of results in grade bands were also examined. At the University of Sydney, five grade bands (fail to high distinction) are used, with cutoffs at 50%, 65%, 75% and 85% – marks at the cutoff value are included in the higher band. In 2004, there was no statistically significant difference between respondent and non-respondent examination grade band distributions in CHEM1901 ($\chi^2 = 4.17$, df = 3, $p = 0.243$), confirming that the advanced student sample was representative of the CHEM1901 cohort. For the other two units, the results indicate that the samples were representative for all passing students (CHEM1001: $\chi^2 = 2.80$, df = 3, $p = 0.423$; CHEM1101: $\chi^2 = 1.11$, df = 3, $p = 0.774$), but that students in the fail category are systematically underrepresented in the respondent group - this explains the differences in average examination results seen in Table 1. Given the very high response rates in 2004, the results are useful and informative. In 2005, there were no statistically significant differences between

Table 1 Response rates and comparative end-of-semester performance data from the 2004 and 2005 first-year chemistry cohorts

		Mean end-of-semester examination result		
Unit and Year	Response rate	Respondents	Non-respondents	Two-sided t-test
CHEM1001, 2004	292 (78%)	57.6%	50.4%	$t_{374} = -3.98, p < 0.001$
CHEM1101, 2004	497 (81%)	59.2%	54.4%	$t_{615} = -3.33, p = 0.001$
CHEM1901, 2004	122 (85%)	70.5%	70.5%	$t_{141} = 1.48, p = 0.140$
CHEM1002, 2005	60 (32%)	45.8%	51.1%	$t_{186} = -2.03, p = 0.0430$
CHEM1102, 2005	218 (46%)	52.7%	54.9%	$t_{469} = -1.54, p = 0.124$
CHEM1902, 2005	63 (58%)	62.5%	60.1%	$t_{106} = -1.61, p = 0.111$

respondent and non-respondent examination grade band distributions for CHEM1002 ($\chi^2 = 3.01$, df = 3, $p = 0.391$), CHEM1102 ($\chi^2 = 3.85$, df = 4, $p = 0.427$) and CHEM1902 ($\chi^2 = 2.85$, df = 3, $p = 0.416$). When combined with the average performance data in Table 1, it seems clear that these samples are representative of the student cohorts in each of these units.

3.1 Personality Traits or Types?

The PLSI was originally designed to categorise personality characteristics into dichotomous types, with dimension scores considered meaningful only insofar as they allow the type to be identified, because these types are qualitatively distinct and represent mutually exclusive groups of people [36]. Several important consequences follow: The midpoint of each dimension is not an arbitrary cutting point, but rather represents a true zero point – for this reason, a dimension pairs list should be used to clarify a preference for respondents at the midpoint when the instrument has an even number of items for that dimension [37]; the number of respondents at this midpoint should be comparatively low; and, the distribution of preference scores should be bimodal. It is this claim of bimodality, which leads to many of the criticisms of MBTI-based investigations. If the PLSI is truly measuring types, then it is reasonable to suggest that a person's responses will be consistent with their type. Arbitrarily setting the probability of answering to type at 75%, and using a reported [34] population estimate of 60% extroverts, the preference score distribution can be simulated by combining the component (introvert and extrovert) binomial distributions. The resulting bimodal distribution is shown in Fig. 1a,

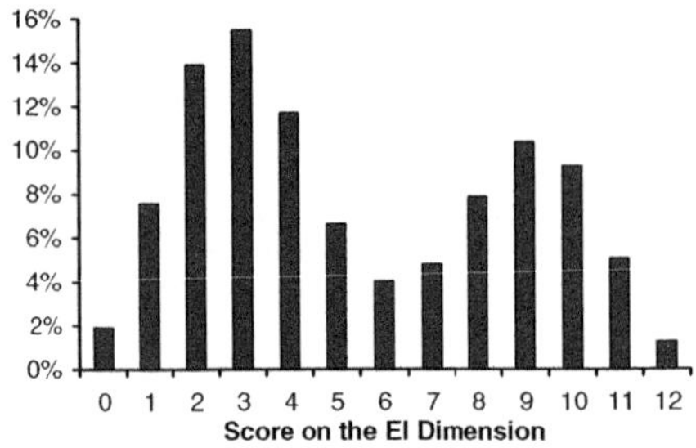

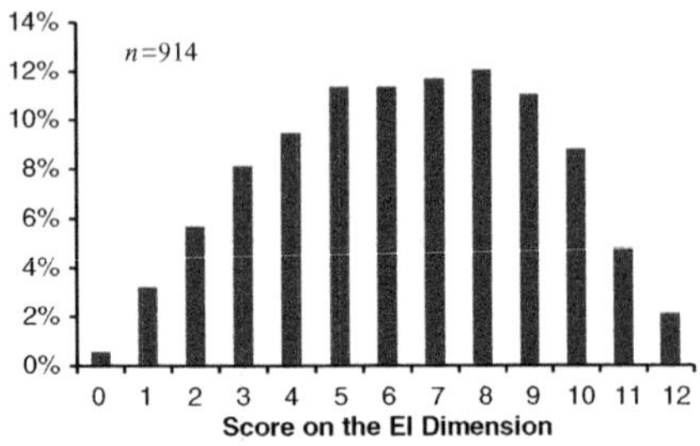

Fig. 1 (a) Expected population distribution along the EI dimension for a type-based personality model, and (b) actual distribution of 2004 study respondent scores along the EI dimension

whilst Fig. 1b shows the actual distribution of EI dimension scores from the 2004 respondents. The observed distribution is clearly centre-weighted and unimodal, and significantly different from that expected for a type-based model ($\chi^2 = 373$, df = 12, $p < 0.0001$). Similar results were found when the extended PLSI EI items were used with the 2005 cohort. Coupled with the fact that the claim of bimodality has been questioned previously [36, 40], similar centre-weighted and unimodal distributions have been reported [41], and previous reports of bimodality having been shown to result from analysis artefacts [42], it is clear that PLSI dimension scores are inconsistent with a type-based personality model.

According to Cronbach and Meehl [43], construct validity concerns whether the empirical relations between test scores match theoretical relations in the nomological network [44]. It follows that the lack of bimodality in dimension scores raises questions about the construct validity of the MBTI [38]. However, this does not mean that either the MBTI or the PLSI is invalid, but it does require that the problem with the theoretical model used for interpreting the results be addressed. Modern personality theories reject the notion of dichotomous types, instead viewing characteristics as traits existing on a continuum of preferences running from one extreme to the other. The widely used five-factor model [39, 40] describes personality by reference to sliding scales of extroversion, agreeableness, conscientiousness, emotional stability (neuroticism) and openness to experience. A high degree of correlation and convergence has been reported [36] between the four MBTI dimensions and four of the five dimensions in the five-factor model (the neuroticism scale has no MBTI equivalent); this has been used to explain the utility of MBTI scales for applications in occupational and educational settings [45]. Similar five-factor model interpretations of MBTI scales have been described [46]. This paper adopts the position that PLSI dimension scores should be interpreted as indicators of the strength of trait preferences, and that type-based analyses of PLSI data should be avoided.

The relationships between PLSI dimension scores and academic performance will now be considered. Since significant relationships were not observed on the SN and JP dimensions, only results from the EI and FT dimensions will be discussed.

3.2 Academic Performance and the Extrovert–Introvert (EI) Dimension

Within each unit in both years, mean end-of-semester examination results, along with their 95% confidence intervals, were calculated at every EI dimension score. Figure 2 shows the correlation between EI score and average examination mark for CHEM1102, the largest of the units examined with the extended PLSI. This linear correlation shows that students with high levels of introversion tend to perform better than those with high levels of extroversion with an average difference of up to ten marks across the dimension. This correlation is strong ($R^2 = 0.815$), and has a very high level of statistical significance ($t_{12} = 7.42$, $p < 0.0001$). All other units showed qualitatively identical results – in each case, there is a strong correlation, which has a high level of statistical significance, and higher examination performance amongst introverted students. These empirical results, reproducible despite differences in cohort and unit content, stand on their own.

Academic achievement requires the capacity to deal intensively with concepts and ideas, which can be expected to favour people with introvert characteristics. On this basis, and using a type-based model, it has been predicted that introverts should outperform extroverts on tests of academic aptitude [29]. This prediction has been independently verified in the domains of engineering [26, 47, 48] and economics [49]. The present study not only extends this finding to the domain of chemistry, but also reports the previously unknown linear correlation between students' examination performance and the strength of their inclination towards introvert characteristics. Given the, at times, abstract and theoretical nature of chemistry [50], this finding is consistent with expectations. Furthermore, it is difficult to maintain substantial levels of social interaction in large chemistry classes when they predominately involve

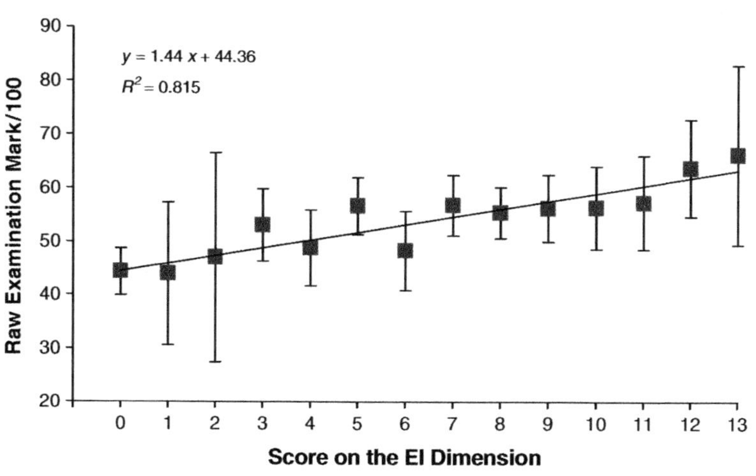

Fig. 2 Mean end-of-semester examination performance data and EI dimension score for responding CHEM1102 students in 2005

lecture-based instruction, which favours those with introvert learning style preferences. By contrast, students exhibiting strong extrovert characteristics prefer to learn theories and concepts that are connected to their own experiences and learn best through group activities and cooperative projects. These types of activities are less common with large university chemistry cohorts, despite the large amount of laboratory work that is involved [51]. It seems clear that assessment programs which include a substantial written examination component risk favouring students with high levels of introversion over those with highly extrovert characteristics. Empirical data from different studies in chemistry, engineering and economics all lead to the same qualitative finding, irrespective of whether a type- or trait-based interpretative framework is used. As such, the data raise important issues in pedagogical and assessment practice, which seem likely to extend across a range of domains.

3.3 Academic Performance and the Feeling–Thinking Dimension

There have been previous reports of a gender difference in learning style preferences for the feeling–thinking (FT) dimension [34], therefore males and females have been separated in this analysis. Indeed, for our cohort, there is a statistically significant difference ($\chi^2 = 24.1$, df = 10, $p = 0.00742$) between the distributions for each gender with females showing a higher degree of 'feeling' characteristics, and males showing a higher degree of 'thinking' characteristics. Importantly, independent sample t-tests confirm that this small shift in the modal position does not result in any significant gender-based difference in mean examination performance ($t_{436} = -1.93$, $p = 0.054$).

The relationship between examination result and FT dimension score for the 2004 cohort was analysed in a similar fashion to that used for the EI dimension. Scores along the FT dimension also show centre-weighted and unimodal distributions, as shown in Fig. 3, again supporting the use of a trait-based interpretive framework. Figure 4 shows the moderately strong ($R^2 = 0.514$) and statistically significant ($t_{11} = 3.27$, $p = 0.00841$) correlation between FT dimension score and average end-of-semester examination mark for the CHEM1101 unit – the largest unit involved in this study. Students with predominant thinking characteristics clearly tend to outperform those with predominant feeling characteristics, with an average difference of up to eight marks between the extreme scores.

In engineering, it has been previously suggested that differences in academic performance associated with the FT dimension might be expected [29]. Since engineering is a domain based on the application of principles from across the scientific disciplines, the same suggestion could reasonably have been made concerning science students. However, prior to this work there have been no research findings to support the existence of such a relationship in either domain – perhaps because of the relative statistical insensitivity inherent in type-based approaches. The present study provides the first unambiguous empirical evidence of a link between academic performance and FT dimension score in chemistry, in particular showing that students for whom thinking characteristics predominate, do outperform those for whom feeling characteristics predominate in examination assessment.

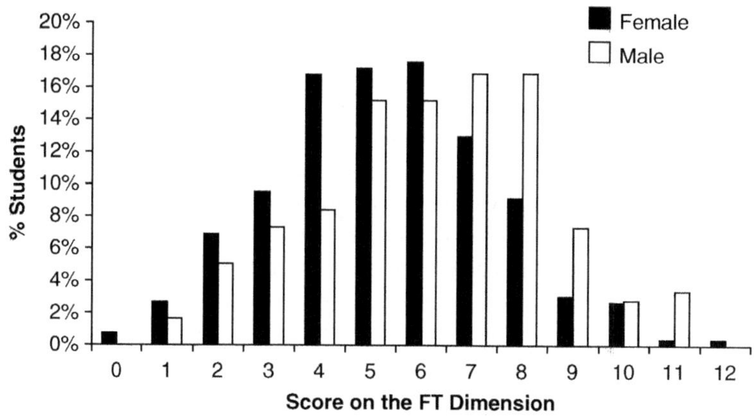

Fig. 3 Distribution of male and female respondents' scores along the FT dimension (2004 study)

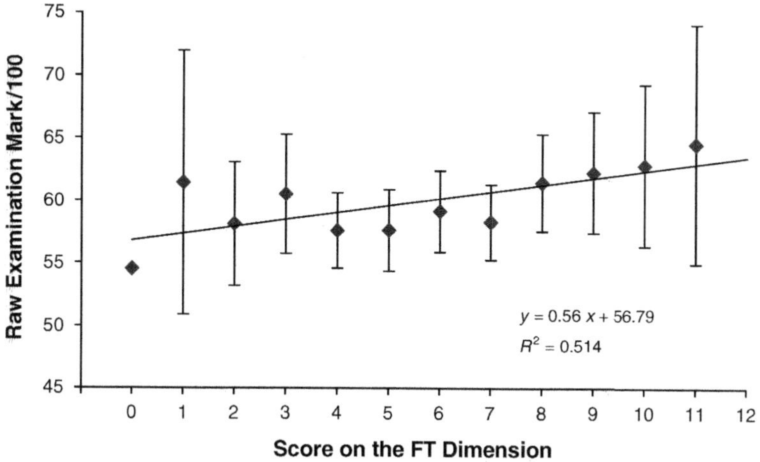

Fig. 4 Mean end-of-semester examination performance and FT dimension score for responding CHEM1101 students in 2004

3.4 Implications for Assessment Practice

There are differences in examination performance associated with students' positions on both the EI and FT dimensions. Therefore, consideration needs to be given to making changes to current teaching and assessment strategies to improve academic performance amongst students who exhibit dominant extrovert and feeling characteristics. Felder [52] has suggested that the use of different teaching strategies would be advantageous, especially when trying to accommodate the learning styles of a student cohort that has a varied distribution.

Universities with large first-year chemistry cohorts typically include substantial lecture-based instruction incorporating relatively few group-based activities, even if laboratory exercises are often being done in pairs. It has been established empirically that students who favour extrovert and feeling characteristics have a high preference for collaborative learning environments [53] and as such, the learning style preferences of more extroverted students would be better accommodated by the introduction of collaborative learning or group-based assignments. Group work may also provide the harmonious social interaction preferred by those with strong feeler characteristics, and has been found to promote motivation in general [8]. Oral examination of group work, for example, allows students with extrovert characteristics to showcase their skills, thereby actively engaging them in the learning process and building their confidence, but this may also disadvantage students with introvert qualities who do not perform as well in such activities [54]. Since appropriate levels of challenge have been shown to motivate students [8], the instructor can try to maintain the interest of more introverted students by presenting the activity as an opportunity to develop important communication skills.

Inclusion of a poster presentation as part of the assessment of group work would also be welcomed by more extroverted students, whilst providing all students with practice in another form of scientific communication. Huddle [55] has reported that the inclusion of a poster session as part of the assessment in organic chemistry enhanced student learning. Wimpfheimer [56] has also shown that poster presentations have helped second-year chemistry students understand and organise information concisely and it follows that such an assessment modification has benefits beyond those associated with assisting the students with extrovert and feeling characteristics. Some of the framework necessary for implementing such a program already exists as approaches to assessing poster sessions have been described [57], and such activities can also include a peer-assessed component.

The introduction of a wider variety of teaching strategies than currently used would not only cater for the students who do not perform as well but also help develop skills of students who are quite comfortable with current teaching practices. As mentioned earlier, students should be taught both in the style they prefer, where they are most comfortable and engaged, and in less preferred modes, allowing them to develop diverse strengths, which will help them to function more effectively in their careers [26]. For example, group presentation assessments that require each member of the group to present a part of the assignment encourages students with high levels of introversion to develop not only oral presentation skills but also team work skills, both of which are important generic attributes [58].

Furthermore, if a variety of teaching strategies are incorporated, then students' learning styles are more likely to be compatible with the instructors teaching style. Consequently, a larger number of students can be expected to not only achieve higher grades but also have a more positive course attitude in chemistry. As a result, students with a variety of learning styles and skills may continue to study chemistry, hopefully increasing the diversity of the science population. A growing trend of students with the potential and capabilities to become good scientists switching to non-scientific fields has been reported [19]; this might be attributable in part to feelings of disinterest, and an inability to perform well, produced by these students' learning style preferences not being accommodated. The changes suggested here have the potential to assist in addressing the problem of declining student numbers and if successful may well have a significant impact.

4 Conclusions

The findings in this paper can be divided into two distinct categories – those relating to the instrument and the choice of a trait-based analysis, and those relating to student learning and assessment in chemistry. Concerning the PLSI, the findings of this study suggest that it can be used to measure traits, despite having been originally developed as a type-based instrument. Indeed, the dimension score distributions produced by this instrument are inconsistent with a dichotomous type interpretation – this finding is important for anyone who may wish to use this instrument in the future. Follow-up work from this research can include conducting a longitudinal study by following a cohort of students through their undergraduate degrees to determine whether their learning styles change as they move through the higher years of chemistry. A cross-sectional study could also be conducted to determine whether the distributions of learning style preferences differ in the different year groups as well as the academics. These findings would be of great interest.

Using a trait-based analysis to compare each of the three 2004 unit cohorts' learning style preferences and end-of-semester examination performance, it is clear that academic performance was significantly correlated with both increasing levels of introversion and increasing levels of thinking (as opposed to feeling) characteristics. Along the EI dimension, the same correlation has also been found for each of the three 2005 cohorts. Related findings have been reported in engineering and economics, albeit following type-based analyses. Collectively, these results suggest a high degree of domain generality, which strongly implies that learning styles do need to be considered when addressing both questions of pedagogical practice and questions of assessment design. These issues raise questions as to equity in assessment practice, and support the adoption of mixed-mode assessment methodologies. A few ways in which knowledge of students' learning style preferences could be used to modify learning environments to more effectively engage and support student learning have been described. Several suggestions concerning how students favouring extrovert and feeling characteristics might be better accommodated have also been made. Finally, some advantages of diversification of teaching strategies have been considered: addressing a wider variety of learning styles would help to minimise mismatches between learning style preferences and modes of instruction, whilst ensuring that the needs of students with less common learning style preferences are accommodated. Such an approach has the potential to not only improve teaching and learning but might also increase student engagement and retention rates.

Acknowledgments The authors would like to thank all University of Sydney students who agreed to participate in this study. Data collections described in this work were authorised by the University of Sydney Human Research Ethics Committee, project reference numbers 04-2004/7353 and 08-2005/4/8461.

References

1. Riding, R (2005) Individual Differences and Educational Performance. Educ Psychol 25: 659–672.
2. Wolters, CA (2004) J Educ Psychol 96:236–250.
3. Lindblom-Yianne, S and Lonka, K (1999) Learn Instr 9:1–18.
4. Nolen, SB (1996) Why Study? How Reasons for Learning Influence Strategy Selection. Ed Psych Rev 8:335–355.
5. Nist, SL, Simpson, ML and Olejnik, S (1991) The Relation between Self-Selected Study Processes and Test Performance. Am Educ Res J 28:849–874.
6. Hidi, S and Renninger, KA (2006) The Four-Phase Model of Interest Development. Educ Psychol 41(2):111–127.
7. Fredricks, JA, Blumenfeld, PS and Paris, AH (2004) School Engagement: Potential of the Concept, State of the Evidence. Rev Educ Res 74:59–109.
8. Paris, SG and Turner, JC (1994) Situated Motivation. In: Pintrich, PR, Brown, DR and Weinstein, CE (eds.) Student Motivation, Cognition and Learning, pp. 213 – 237. Erlbaum, Hillsdale, NJ.
9. Skinner, EA and Belmont, MJ (1993) Motivation in the Classroom: Reciprocal Effects of Teacher Behavior and Student Engagement Across the School Year. J Educ Psychol 85:571–581.
10. De Raad, B and Schouwenburg, HC (1996) Personality in Learning and Education: A Review. Eur J Pers 10:303–336.
11. Bodner, GM (1986) Constructivism: A Theory of Knowledge. J Chem Educ 63:873–878.
12. Phillips, DC (1995) The Good, the Bad and the Ugly: The Many Faces of Constructivism. Educ Res 24:5–12.
13. Tobin, K (1993) The Practice of Constructivism in Science Education. Erlbaum, Hillsdale, NJ.
14. Bailey, PD and Garratt, J (2002) Chemical Education: Theory and Practice. U Chem Ed 6:39–57.
15. Biggs, JB (1979) Individual Differences in Study Processes and the Quality of Learning Outcomes. High Educ 8:381–394.
16. Entwistle, N (1987) A Model for the Teaching-Learning Process. In: Richardson, TE, Eysenck, MW and Piper, DW (eds.) Student Learning: Research in Education and Cognitive Psychology, pp. 13–28. Open University Press, Milton Keynes, England.
17. Miller, A (1987) Cognitive Styles: An Integrated Model. Educ Psychol 7:251–268.
18. Rezaei, AR and Katz, L (2004) Evaluation of the Reliability and Validity of Cognitive Styles Analysis. Pers Indiv Differ 36:1317–1327.
19. Felder, RM (1993) Reaching the Second Tier: Learning and Teaching Styles in College Science Education. J Coll Sci Teach 23:286–290.
20. Ford, N (1985) Learning Styles and Strategies of Postgraduate Students. Brit J Educ Technol 16:65–79.
21. Ford, N and Chen, SY (2001) Matching/Mismatching Revisited: An Empirical Study of Learning and Teaching Styles. Brit J Educ Technol 32:5–22.
22. Riding, R and Douglas, G (1993) The Effect of Cognitive Style and Mode of Presentation on Learning Performance. Brit J Educ Psychol 63:297–307.
23. Larkin, TL (2003) Learning Style in the Classroom: A Research-Guided Approach. In: Proceedings of the International Conference on Engineering and Computer Education, pp 1–8. Sao Paulo, Brazil.

24. Mitchell, M (1993) Situational Interest: Its Multifaceted Structure in the Secondary School Mathematics Classroom. J Educ Psychol 85:424–436.
25. Ryan, RM and Deci, EL (2000) Self-Determination Theory and the Facilitation of Intrinsic Motivation, Social Development, and Well-Being. Am Psychol 55:68–78.
26. Felder, RM, Felder, GN and Dietz, EJ (2002) The Effects of Personality on Engineering Student Performance and Attitudes. J Eng Edu 9:3–17.
27. Robotham, D (1999) The Application of Learning Style Theory in Higher Education Teaching. http://www.chelt.ac.uk/gdn/discuss/kolb2.htm. Accessed 13 July 2004.
28. Tasker, R, Miller, J, Kemmett, C et al. (2003) Analysis of Student Engagement with Online Chemistry Modules Using Tracking Data. In: Crisp, G, Thiele, G, Scholten, I, Barker, S and Baron, J (eds.) Proceedings of the Interact, Integrate, Impact: 20th Annual Conference of the Australasian Society for Computers in Learning in Tertiary Education, pp 505–514. ASCILITE, Adelaide.
29. Briggs-Myers, I and McCaulley, MH (1985) Manual: A Guide to the Development and Use of the Myers-Briggs Type Indicator. Consulting Psychologists Press, Inc, California.
30. Lawrence, GD (1993) People Types and Tiger Stripes (3 ed.). Centre for Applications of Psychological Type, Inc, Gainesville, FL.
31. Quenk, NL (2000) Essentials of the Myers-Briggs Type Indicator. Wiley, Canada.
32. Zywno, MS and Waalen, JK (2002) The Effect of Individual Learning Styles on Student Outcomes in Technology-Enabled Education. Global J Engng Educ 6:35–44.
33. Jung, CG (1923) Psychological Types, or, the Psychology of Individuation. Routledge & Kegan Paul, London.
34. Shindler, J and Yang, H (2002a) Plsi: Four Dimensions. Paragon of Learning Styles Inventory. http://www.oswego.edu/Acad_Dept/s_of_educ/curr/plsi/4dimen.htm. Accessed 25 February 2004.
35. Shindler, J and Yang, H (2002b) Plsi: About Plsi. Paragon Learning Style Inventory. http://www.oswego.edu/Acad_Dept/s_of_educ/curr/plsi/plsinfo.htm. Accessed 25 February 2004.
36. McCrae, RR and Costa, PT (1989) Reinterpreting the Myers-Briggs Type Indicator from the Perspective of the Five-Factor Model of Personality. J Pers 57:17–40.
37. Shindler, J and Yang, H (2003) Plsi: Directions. http://www.oswego.edu/Acad_Dept/s_of_educ/curr/plsi/direction.htm. Accessed 25 February 2004.
38. Pittenger, D (1993) The Utility of the Myers-Briggs Type Indicator. Rev Educ Res 63: 467–488.
39. Digman, JM (1990) Personality Structure: Emergence of the Five-Factor Model. Ann Rev Psychol 41:417–440.
40. McCrae, RR and Costa, PT (1999) A Five-Factor Model of Personality. In: Previn, LA & Johns, OP (eds.) Handbook of Personality: Theory and Research, 2nd edn. pp. 139–153. Guilford Press, New York.
41. Harvey, RJ and Murry, WD (2002) Scoring the Myers-Briggs Type Indicator: Empirical Comparison of Preference Score Versus Latent Trait Methods. J Pers Assess 62:116–129.
42. Bess, TL and Harvey, RJ (2002) Bimodal Score Distributions and the Myers-Briggs Type Indicator: Fact or Artifact. J Pers Assess 78:176–186.
43. Cronbach, LJ and Meehl, PE (1955) Construct Validity in Psychological Tests. Psychol Bull 52:281–302.
44. Trochim, WMK (2006) The Nomological Network. http://www.socialresearchmethods.net/kb/nonmonet.php. Accessed 18 December 2006.
45. Briggs-Myers, I (1993) Introduction to Type. (5th ed.) Consulting Psychologists Press Inc., California.
46. Harvey, RJ, Murry, WD and Markham, SE (1995) A "Big Five" Scoring System for the Myers-Briggs Type Indicator. In: Proceedings of the Annual Conference of the Society for Industrial and Organisational Psychology, pp 1–9. Orlando.
47. O'Brien, TP, Bernold, LE and Akroyd, D (1998) Myers-Briggs Type Indicator and Academic Achievement in Engineering Education. Int J Engng Ed 14:311–315.
48. Rosati, P (1998) The Learning Preferences of Engineering Student from Two Perspectives. http://fie.engrng.pitt.edu/fie98/papers/1016.pdf. Accessed 17 February 2004.

49. Borg, MO and Shapiro, SL (1996) Personality Type and Student Performance in Principles of Economics. J Econ Educ 27:3–25.
50. Marais, P and Jordaan, FJ (2000) J Chem Educ 77:1355–1357.
51. Buntine, MA, Read, JR, Barrie, SC et al. (2007) Advancing Chemistry by Enhancing Learning in the Laboratory (Acell): A Model for Providing Professional and Personal Development and Facilitating Improved Student Laboratory Learning Outcomes. Chem Educ Res Pract 8:232–254.
52. Felder, RM (1996) Active-Inductive-Cooperative Learning: An Instructional Model for Chemistry? J Chem Educ 73:832–836.
53. Ramsay, A, Hanlon, D and Smith, D (2000) The Association between Cognitive Style and Accounting Students' Preference for Cooperative Learning: An Empirical Investigation. J Acc Ed 18:215–228.
54. Mohammed, S and Angell, LC (2003) Personality Heterogeneity in Teams: Which Differences Make a Difference of Team Performance? Small Group Research 34:651–677.
55. Huddle, PA (2000) A Poster Session in Organic Chemistry That Markedly Enhanced Student Learning. J Chem Educ 77:1154–1157.
56. Wimpfheimer, T (2004) Peer-Evaluated Poster Sessions: An Alternative Method to Grading General Chemistry Laboratory Work. J Chem Educ 81:1775–1776.
57. Mills, PA, Sweeney, WV, DeMeo, S et al. (2000) Using Poster Sessions as an Alternative to Written Examinations – the Poster Exam. J Chem Educ 77:1158–1161.
58. Barrie, SC (2006) Understanding What We Mean by the Generic Attributes of Graduates. High Educ 51:215–241.

Using Two-Dimensional Molecular Drawings to Evaluate Teachers' Conceptual Change in Chemistry

E. Steenberg and J. D. Bradley

Abstract Students are usually able to give adequate descriptions of macroscopic phenomena they observe and to develop suitable mental strategies to cope with the demands of symbolic chemistry language. Visualization skills or the skills necessary to describe microscopic phenomena need to be developed as an integral part of chemistry teaching and learning.

This paper discusses how use of the RADMASTE Molecular Stencil® could assist practising senior high school teachers to represent their microscopic views. Analysis of responses led to the identification of possible misconceptions or instances where the stencil was not used effectively. An exercise was designed to provide scaffolding in dealing with the chemistry concepts as well as skills in using the stencil. Subsequent responses could be analysed to ascertain to what extent use of the RADMASTE Molecular Stencil assisted the teachers in producing more complex microscopic representations and whether use of the stencil over a period of time could contribute to conceptual change.

1 Introduction

To understand and practice chemistry, students and teachers need to be able to link macroscopic properties of substances with microscopic phenomena and symbolic chemistry language [1,2,4,7–10]. These different kinds of descriptions often confuse beginners. This "chemists triangle" is often not explained, but it is assumed that students and teachers of chemistry have equally well-developed skills in all three aspects.

Macroscopic phenomena are observable, and students and teachers are usually able to give adequate descriptions of what they observe. Most students and teachers also develop suitable mental strategies to cope with the demands of symbolic chemistry

E. Steenberg (✉) and J.D. Bradley
RADMASTE Centre, University of the Witwatersrand, Private Bag 3, WITS, 2050
Johannesburg, South Africa
e-mail: erica.steenberg@wits.ac.za

M. Gupta-Bhowon et al. (eds.), *Chemistry Education in the ICT Age*,
DOI 10.1007/978-1-4020-9732-4_24, © Springer Science+Business Media, B.V. 2009

language [14], although its meaning may be imperfectly understood. Chemists need to develop representations and tools to mediate between observable physical substances and phenomena and aperceptual (unobservable) chemical entities and processes that account for physical characteristics of substances [8].

Visualization skills, or the skills necessary to describe or explain microscopic phenomena, require the student or the teacher to engage in abstract thinking [3,5,6,13]. It is thought that the demands of formal operational thinking are often the cause of poor performance in chemistry [3,6]. Poor visualization skills will also result if students or teachers are unable to operate on this cognitive level.

Providing personal experience in modeling of microscopic phenomena is an ideal way to develop visualization skills for students and teachers alike. Students and teachers should be given the opportunity to make their own drawings and build their own models of microscopic entities [8,12]. This needs drawing skills and modeling materials. Two low-cost tools have been developed at the RADMASTE Centre to meet these needs: the RADMASTE Molecular Stencil and the RADMASTE Molecular Modelling Kit.

In this paper it is described how the conceptions of practicing senior high school teachers could be evaluated from two-dimensional drawings of microscopic phenomena. Teachers were given the opportunity to use the RADMASTE Molecular Stencil in their drawings. Although still producing a two-dimensional model, it was envisaged that manipulation of the stencil would assist those teachers who experience difficulty with abstract thought [5], would encourage students to construct their own knowledge [12] and would equip teachers better to deal with the demands of the current curriculum (in South Africa).

2 Theoretical Background

Literature outlining the effect that use of molecular models could have on chemistry achievement emerged in the 1980s. A great variety of different molecular models have been developed and produced, including models that can be assembled into a three-dimensional structure. More recent developments include the use of Computer Aided Instruction (CAI) in virtual molecular visualization, to give but two of the numerous publications available in the field [8,13]. Modeling used in the present study was confined to the use of a molecular stencil (the RADMASTE Molecular Stencil), supported by the use of a three-dimensional molecular modeling kit (the RADMASTE Molecular Modelling Kit), and only a brief discussion of literature referring to the use of concrete modeling kits will be presented in this paper.

Gabel and Sherwood [5] published one of the first papers on using modeling kits to facilitate the transition from concrete-operational to formal-operational reasoning in chemistry. The authors established that American students who could manipulate space-filling molecular models were more successful in the study of chemistry than students who had only seen the models demonstrated. It was noted that students needed some time to become familiar with what the models represented.

It was also established that both concrete-operational and formal-operational students benefited from the use of models. One of the interesting observations made by these authors is that the use of models may be unrelated to Piagetian theory and that the use of models simply requires students to pay closer attention to the material being presented [5]. If models are only used in demonstrations and the students are not required to manipulate the models themselves, students may not be actively engaging with the material being taught [5,9].

Ben-Zvi and Hofstein [2] asked Grade 10 Israeli students to draw diagrams of molecules of elements and compounds. They found that the students could represent one molecule of an element with relative ease, but that students had difficulties in representing a molecule of a compound or representing an element in the gaseous or solid state. The difficulties were ascribed to the following four factors:

1. Inadequate understanding of the atomic model
2. Misleading use of models in textbooks, with only single units being shown
3. Misunderstanding of chemical equations, with students reading equations as representing single units and not moles of units
4. Students experiencing information overload

It should be borne in mind that multiple representations in chemistry place a considerable cognitive demand on students [7,10]. If the representations used in the teaching of chemistry lie within the students' Zone of Proximal Development (according to Vygotski's theory of learning), students will find representations meaningful, but if representations lie outside the Zone of Proximal Development, the representations will be largely meaningless [10]. Three levels of representational understanding could be distinguished by Madden and Jones [10].

1. Students who thoroughly understood representations found the models rich in meaning and could associate one type of representation with other types of representations.
2. Some students showed appreciation of surface features of representations only.
3. Some students experienced so much difficulty in understanding representations that they often ignored some features in the representations entirely.

The use of models could potentially reduce the demands on short term memory [9]. It has further been suggested that if science content is presented in visual form, learners can benefit from using the human visual sensory system to appreciate spatial representations and transformations [7]. Concept formation will be occurring simultaneously. Difficulties learners experience in understanding molecular visualizations could be classified into four areas [7]:

1. *Visual subtlety* – students should realize that models include visually subtle information.
2. *Complexity* – molecular models are complex with respect to the depth and amount of information they represent. This will be particularly problematic for novices, whereas experts can easily decode the information in a model and understand interrelations between different types of representations.

3. *Abstractness and conceptual depth* – students must understand the symbolism of the models they are using. Those new to the study of chemistry may not fully appreciate the links between theory and the models and the more uncertain the state of a student's conceptual knowledge, the less useful a model becomes. The introduction of scaffolding knowledge is therefore important.
4. *Differences in individual learning* – students who have appreciable visual skills to start with can progress to deeper understanding of information encoded in a model, whereas students with weaker visual skills might only be able to appreciate the three-dimensional nature of molecular structures for the first time.

In their study, Kozma et al. [8] examined representations produced spontaneously by practicing chemists, as well as representations generated by electronic modeling tools and how use of both these types of representations can be coordinated to promote understanding in chemistry. They found that teachers and students could use electronic or spontaneously created models to

1. Express their understanding of observed phenomena in terms of underlying aperceptual (unobservable) entities and processes
2. Test their understanding with experiments on physical materials and
3. Create a knowledge-building community around the common use of representations and tools to investigate and explain phenomena

In a more recent study by Merritt et al. [11], the authors are of the opinion that students in the USA find it difficult to understand the particle nature of matter because traditional curriculum materials present the concepts in a manner that does not help students to develop their own understanding. The particle nature of matter is often described in a short paragraph or chapter that outlines the history of the atom. Students don't develop appropriate ideas because they never apply and reapply these ideas to explain phenomena [11] and they do not internalize ideas related to the intrinsic motion of particles or the interaction between particles during a chemical change if they are not given additional opportunity to use modeling [4]. Merritt et al. [11] also address the very important issue of contextualizing learning in the particle nature of matter. They used a "driving question" to produce a context for students to learn about scientific phenomena. This "driving question" served to anchor students' learning and according to the authors "*shows students the legitimacy of their implicit knowledge and its availability as scaffolding in apparently unfamiliar tasks*". In another valuable contribution, the authors give examples of rubrics they have created to analyze students' molecular drawings [11].

In a recent study, students' manipulation of physical models was compared to their use of a technology-mediated modeling tool called "Chemation" [13]. The computer visualization tool described in the study helped students to model dynamic aspects of microscopic representations, since it allowed a build-up of "frame-by-frame" animations. Students had access to a "palette" of 21 atoms they could manipulate electronically. Sections of the curriculum that were studied included properties of substances, pure substances and mixtures, chemical reactions, and conservation of mass (conservation of atoms in a chemical reaction). Pre-tests and post-tests for computer-based as well as physical models showed significant gains in student knowledge [13]. The authors do not ascribe the gain in knowledge to the modeling tools alone, but rather to the overall learning environment that allows for improvement. In this particular study, most students in both groups showed a basic conceptual accuracy in their knowledge of two-dimensional representations, and the authors could not establish any differences between physical models and models created with the technology-mediated program [13].

The computer-based tool was superior in the case of dynamic representations [13]. It is proposed that use of a computer-based animation tool lowered the cognitive demands imposed on students, since during animation, students could use the animation to illustrate actual rearrangements of atoms, while they could focus on the chemistry content embedded in the representation. Students using physical models had to deal with both issues (rearrangement of atoms and chemistry content) and hence the use of static models might have led to cognitive overload [13]. This cognitive overload manifested in students providing inferior explanations for the phenomena under investigation [13].

It is interesting to note that students who used the computer-based tool did not do as well as students using physical models when mass conservation and the differences between pure substances and mixtures were considered [13]. In the case of mixtures, students using physical models had to build models of two pure substances and store these models in separate plastic bags. To model a mixture, molecules were taken from the two bags and placed together in a third bag. This manipulative exercise leads to meaningful learning, and is ascribed by the authors [13] to the tangibility of the physical models. In the case of students using the computer-based tool, the pure substances were represented separately in two frames, and the mixture was constructed as a third frame. Students presumably did not link the contents of the frames effectively. In the case where mass had to be conserved, students using physical models were very aware of the fact that they had to count and note the number of atoms in the beginning and had to keep the number of atoms the same throughout. Students using the computer-based tool often deleted single atoms formed during chemical reactions the models represented. The computer-based tool had the facility for students to add labels and chemical symbols to their representations, and this was noticed more extensively in computer-generated models.

In a second part of the study [13], the authors mention results they obtained when students had to design, interpret, and evaluate animations. Simply viewing animations did not lead to the development of stronger conceptual frameworks, but viewing had to be augmented with discussion and peer evaluation of each other's models [13].

A study in Taiwan [4] focused on students' conceptions in chemistry and although the study did not focus on modeling, some interesting aspects are relevant to the present study. The methodology of the study consisted of a two-tier test, in which the students had to state a fact in the first tier, but give their reasons for their answer in a second tier. Test items often consisted of two-dimensional molecular representations. The focus would therefore be on the interpretation rather than the creation of molecular representations. The results of the study fall outside the scope of this discussion, but the conclusion that models must be considered as essential and necessary in the design of a chemistry curriculum [4] supports the viewpoints of other authors cited, as well as findings in the present study.

3 Research Methodology

The study was conducted with a group of 23 practicing senior high school teachers. As full-time teachers, teaching Grade 10, 11, or 12 in a number of South African high schools, the teachers were enrolled in an Advanced Certificate of Education (ACE) course at the University of the Witwatersrand in Johannesburg, South Africa. The ACE certificate has been designed for in-service training of teachers and is delivered in mixed mode: teachers attend face-to-face contact sessions four times per year and work through assignments and portfolio activities on their own. The latter component has features of distance education - a comprehensive Course Workbook is used, but teachers have no direct access to lecturers when answering questions. At the start of the particular course, each teacher received written course materials, a RADMASTE Advanced Microchemistry Kit, a RADMASTE Molecular Modelling Kit, and a RADMASTE Molecular Stencil. The responses obtained with the latter item will be discussed. Students' drawings submitted as part of the course work over a period of 4 months were collected and analyzed.

The RADMASTE Molecular Stencil is a rigid, transparent stencil which allows two-dimensional drawing of a number of microscopic entities (Fig. 1), all to the same scale.

Teachers had to use the RADMASTE Molecular Stencil to answer the following questions:

Question 1 - Use a RADMASTE Molecular Stencil to represent the molecular changes corresponding to

1. *The boiling of liquid water and*
2. *The electrolysis of liquid water*

Show that the traditional classification of these changes as physical and chemical, respectively, is in agreement with the "micro" classification of these particular changes.

Question 2 – Use the RADMASTE Molecular Stencil to show the reaction of nitrogen and oxygen to form nitrogen monoxide in the air. Remember that there are approximately 4 molecules of nitrogen for every one molecule of oxygen. Draw a diagram to show the initial mixture (no reaction) and the final mixture (reaction complete).

Question 3 – Use the RADMASTE Molecular Stencil to represent the combustion of methane at different extents of reaction. Assume that you start with a reaction mixture comprising 3 molecules of methane and an excess of oxygen molecules. Represent the reaction mixture at extents of reaction (i) 0.00, (ii) 0.33, (iii) 0.67, and (iv)1.00.

(Note you will need to use separate pieces of paper for your drawings.)

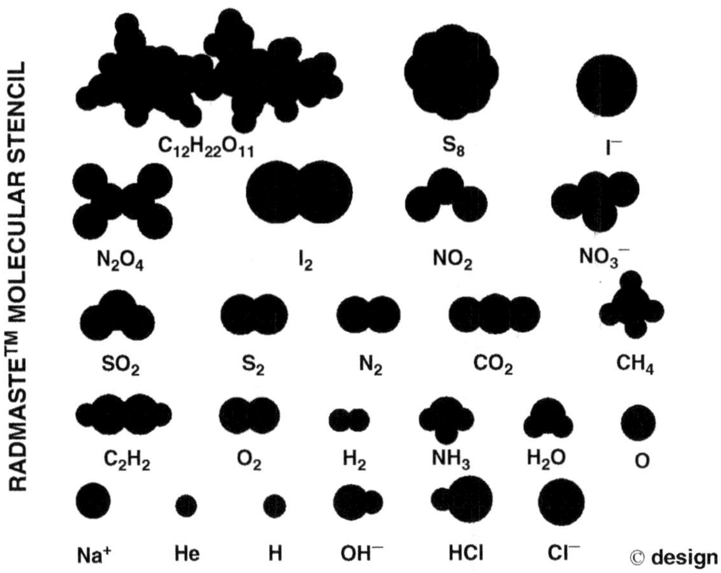

Fig. 1 The RADMASTE Molecular Stencil

4 Discussion of Results

4.1 Responses Obtained Prior to Working Through Scaffolding Material

Question 1.1 – Use a RADMASTE Molecular Stencil to represent molecular changes corresponding to the boiling of water.

This question formed part one of a question students had to complete individually as a portfolio assignment, working on their own in the absence of a tutor. If some students had spontaneously formed study groups and were working together, they would have had the opportunity to discuss their approaches and to compare draft drawings. Other students might have worked on their own and might not have interacted with fellow students at all before submitting their drawing.

From the group of 23 students, four students did not submit a drawing in response to the question, and the results will be expressed as a percentage of the remaining 19 students. In a situation where a portfolio assignment has to be submitted, the lack of submission should not be regarded as an indication that the student was unable to answer the question – other factors (e.g., missing deadlines, other commitments, lack of motivation, etc.) can also contribute to students in distance-education courses not submitting work.

There were instances of responses that could not be coded using pre-selected categories. The percentages of such responses have also been included in the table of results and the nature of aspects that made coding unfeasible will be discussed briefly.

The responses to this question were analyzed by considering:

1.1.1 *Whether students had used the RADMASTE Molecular Stencil*
1.1.2 *Whether there was any evidence of students scaffolding the modeling*
1.1.3 *The orientation of the molecules*
 (a) *Random or uniformly oriented*
 (b) *The spatial distance between the molecules in the liquid and the gas*
 (c) *Any indication of hydrogen bonding*
1.1.4 *Conservation of molecules in the liquid and the gas state*

Results for question 1.1 are summarized in Table 1a

1.1.1 *Did students use the RADMASTE Molecular Stencil?*

A total of 84.2%, or 16 of the 19 students, used the RADMASTE Molecular Stencil to create their drawings. In the case of the remaining 3 students, the stencil could possibly have been misplaced at the time, since the particular students completed subsequent questions with the stencil.

1.1.2 *Was there any evidence of scaffolding?*

For purposes of discussion, scaffolding was regarded as any additional writing, apart from labeling, e.g., "boiling water" or "liquid water." It could take the form of chemical symbols, a partial chemical equation or a balanced chemical equation.

Table 1a Results of molecular drawings representing the boiling of liquid water, using the RADMASTE Molecular Stencil

		Number of responses ($n = 19$)			Percentage of responses, %		
Aspect analyzed		Yes	No	Not coded	Yes	No	Not coded
1.1.1	Used the RADMASTE Molecular Stencil	16	3	0	84.2	15.8	0
1.1.2	Evidence of scaffolding	3	16	0	15.8	84.2	0
1.1.3 (a)	Are molecules drawn with random orientation?	6	11	2	31.6	57.9	10.5
1.1.3 (b)	Are the liquid water molecules closer together than the gaseous water molecules?	12	1	6	63.2	5.2	31.6
1.1.3 (c)	Are the molecules positioned in a manner that could suggest awareness of hydrogen bonding?	7	9	3	36.8	47.4	15.8
1.1.4	Number of liquid water molecules equal to number of gaseous water molecules	9	4	6	47.4	21.0	31.6

As much as 15.8% of students' responses showed some form of scaffolding. In all cases the scaffolding consisted of students writing the symbols for water with state descriptions to show the change when liquid water changes to gaseous water, namely $H_2O(l) \rightarrow H_2O(g)$.

1.1.3 *Orientation of the molecules in the drawings*

(a) *Random or uniformly oriented?*

Only six students (31.6%) showed the molecules in a random orientation. But 57.9% of students drew the molecules without rotating the stencil, very much like a letter-type stencil would be used, where all the stencil shapes face in a particular (vertical) direction (see Fig. 2). These results are perhaps not surprising, since many textbooks illustrate the particle model of solids, liquids and gases using circles for particles. These "structure-less" particles cannot represent randomness of orientations of molecules with other shapes.

One of the two uncoded responses was that of a drawing where the stencil was not used, and only circles were used to show the molecules. In the second uncoded response, the physical change was depicted by showing liquid water as hydrogen and oxygen atoms with the atoms represented as circles.

(b) *The spatial distance between molecules in the liquid and the gas state*

Ideally, the students should draw molecules of liquid water touching each other and those of the gas further apart. If they considered drawing a boundary shape, such as a rectangle to represent a container, the molecules of liquid water could also have been placed close to the bottom of such a "container" and the gas molecules could have been distributed evenly throughout the container, to the extent of "filling the container."

The fact that 63.2% of the responses showed the increased intermolecular distances for water in the gaseous state indicates that students have some grasp of the concept.

A typical response is shown in Fig. 2. It is noticed that the student increased the distance between the molecules, but not sufficiently. This could be due to physical constraints (i.e., the size of the space suggested for the drawing).

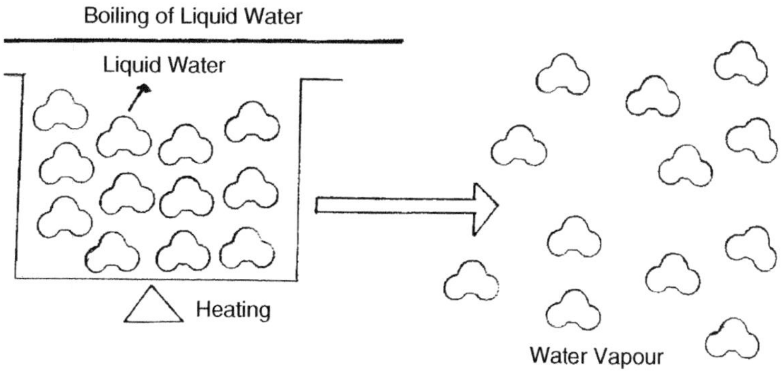

Fig. 2 Example 1 of student drawing to represent boiling of liquid water

Six responses could not be analyzed with respect to intermolecular distance, since four students only showed the product of the physical change (no comparison between liquid and gas possible) and two students showed a chemical change instead of a physical change.

(c) *Is there any indication of a special orientation between neighboring molecules that would suggest knowledge of the concept of hydrogen bonding?*

In 36.8% of the responses, the hydrogen atoms of a molecule were drawn in close proximity of the oxygen atoms of neighboring molecules. This could be fortuitous rather than deliberate, resulting from a slight lateral movement of the stencil during drawing of neighboring molecules, rather than a conscious effort to take hydrogen bonding into account. In cases where molecules were shown with a uniform orientation, the molecules were often drawn with the hydrogen atoms of neighboring water molecules closest together (see Fig. 2).

Uncoded responses were those drawings in which water molecules were shown as circles.

1.1.4 *Conservation of the number of molecules in the gas and the liquid state*

Perhaps intuitively, 47.4% of the students drew the same number of molecules in the liquid and the gas state. In four cases (21.0%) where the number of molecules in the two states were different, the number of molecules shown for gaseous water was always lower than the number of molecules shown for liquid water, irrespective of whether a container was shown or not. Indications are that misconceptions around mass changes were present.

However, 31.6% of responses could not be coded, again because students did not show both liquid water and gaseous water for comparison or because the change shown did not match the question.

Of the 19 drawings which could be used for analysis, only one drawing showed higher-order skills, with molecules being randomly orientated in a manner that could suggest hydrogen bonding in both the gas and the liquid; with the number of molecules being conserved during the physical change and the intermolecular distance increasing in the gas (see Fig. 3). The intermolecular distance in the gas did

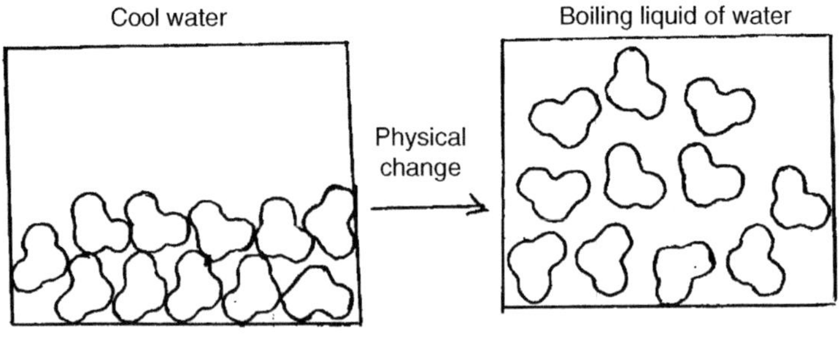

Fig. 3 Example 2 of student drawing to represent boiling of liquid water

not increase to the correct extent, but this could have been due to physical constraints of space the student used in the drawings.

Question 1.2 – Use *a RADMASTE Molecular Stencil to represent molecular changes corresponding to the electrolysis of water.*

This question formed part 2 of a question in a portfolio assignment exploring the differences between physical and chemical change.

Of the 23 students in the group, 7 students did not submit a drawing in response to the question, and the results will be discussed expressed as a percentage of the remaining 16 students. This would suggest that 3 students were prepared to attempt modeling of physical change, but not sufficiently confident to model chemical change.

Uncoded responses were those drawings that could not be analyzed by using the pre-selected categories. The percentages of such responses have also been included in the table of results and the nature of aspects that made coding unfeasible will be discussed briefly.

The responses to the question were analyzed by considering whether:

1.2.1 *Students used the RADMASTE Molecular Stencil*
1.2.2 *There was any evidence of scaffolding*
1.2.3 *The orientation of the molecules was*
  (a) *Random or uniform in the case of water molecules*
  (b) *Random or uniform in the case of hydrogen and oxygen molecules*
1.2.4 *Products of electrolysis were correctly shown as hydrogen and oxygen*
1.2.5 *Electrodes or an electrical circuit was shown*
1.2.6 *Atoms were conserved and electrolysis was represented by taking reaction stoichiometry into account*

Results for question 1.2 are shown in Table 1b.

1.2.1 *Did students use the RADMASTE Molecular Stencil?*

As many as 81.2% of students used the RADMASTE Molecular Stencil in question 1.2, compared to 84.2% in question 1.1. Three students did not use the stencil. Only one student did not use the stencil in both questions 1.1 and 1.2. Two students, who used the stencil in question 1.1, did not do so for question 1.2. The reason for the change in these students' approach to molecular drawings is not clear.

1.2.2 *Did students use scaffolding?*

A large number of students (68.8%) used scaffolding to help them draw a representation of the electrolysis of water. This percentage is more than four times the percentage of scaffolding recorded for question 1.1. Equations are rarely written for physical changes and students would possibly not regard a chemical equation as necessary scaffolding for a physical change. Writing the symbols for the products proved to be the most popular type of scaffolding (6 out of 11), followed by a partial or balanced chemical equation (3 out of 11); a descriptive paragraph (1 case) and a chemical equation for the reverse reaction (1 case). An example of scaffolding can be seen in Fig. 4.

Table 1b Results of molecular drawings representing the electrolysis of liquid water, using the RADMASTE Molecular Stencil

Aspect analyzed		Number of responses ($n = 16$)			Percentage of responses, %		
		Yes	No	Not coded	Yes	No	Not coded
1.2.1	Used the RADMASTE Molecular Stencil	13	3	0	81.2	18.8	0
1.2.2	Evidence of scaffolding	11	5	0	68.8	31.2	0
1.2.3 (a)	Water molecules oriented randomly	5	9	2	31.2	56.2	12.5
1.2.3 (b)	Hydrogen and oxygen molecules oriented randomly	2	7	7	12.5	43.75	43.75
1.2.4	Products correctly shown as hydrogen and oxygen	7	6	3	43.75	37.5	18.75
1.2.5	Electrodes shown	2	14	0	12.5	87.5	0
		Number of responses ($n = 16$)			Percentage of responses, %		
Aspect analyzed		Yes	Partly	No	Yes	Partly	No
1.2.6	Conservation of atoms and reaction stoichiometry	2	6	8	12.5	37.5	50.0

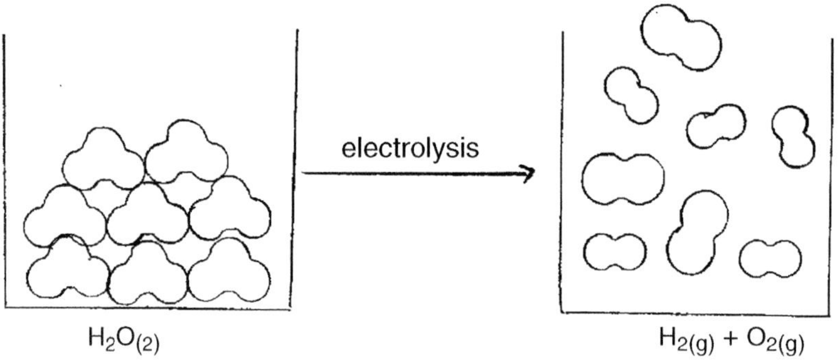

Fig. 4 Example of student drawing of electrolysis of water

1.2.3 *Orientation of molecules*

(a) *Orientation of the water molecules in the drawings*

Of the responses obtained, 56.2% showed the water molecules in a uniform orientation. This figure is very similar to the 57.9% reported for uniform orientation of molecules in question 1.1. Closer examination of students' drawings revealed that one student improved his drawing in question 1.2 in this respect, and another student only showed gas molecules with a random orientation, but liquid molecules with a uniform orientation (see Fig. 4).

(b) *Orientation of the product (hydrogen and oxygen) molecules in the drawings*

In 12.5% of the responses the orientation of the hydrogen and oxygen molecules appeared to be random. This aspect of the drawing was characterized by a large number of responses that could not be coded, since it appeared that students

(i) Did not know what the products of electrolysis of water were
(ii) Did not show any products
(iii) Showed product molecules as circles, presumably showing atoms of hydrogen and oxygen, in spite of using the stencil for reactants, or
(iv) Used formulae for the products instead of the stencil

1.2.4 *Products are correctly shown as hydrogen and oxygen*

Only 43.75% of students could identify the products of electrolysis of water correctly as hydrogen and oxygen. In three cases, students drew the reverse reaction (i.e., hydrogen + oxygen = water). Students had inadequate recollection of observations during the electrolysis of water and their lack of conceptual knowledge affected their ability to create suitable representations in this case.

Detailed scrutiny of students' drawings indicated that incorrect products were thought to be

(i) H^+ and OH^- (1 response)
(ii) Unreacted water molecules shown together with *atoms* of oxygen and hydrogen (1 response)
(iii) Water molecules described as 'jumbled' against each other (molecular shapes drawn to overlap (1 response)
(iv) Water molecules shown unchanged (1 response) or with deformed shapes (1 response)
(v) Atoms of oxygen and atoms of hydrogen (1 response)

1.2.5 *Showing the electrodes when representing electrolysis of water*

The majority of students (87.5%) did not show the electrodes in their representation of electrolysis of water. In the two cases where the electrodes were shown, this did not always lead to students being able to predict the products correctly. One student implied that a potential difference was applied across a liquid water sample during electrolysis, but this also did not help the student to identify the products as hydrogen and oxygen molecules.

1.2.6 *Conserving atoms and taking reaction stoichiometry into account*

Only two students (12.5% of the sample) represented both the nature and amount of products correctly. In cases where the students' responses could be classified as partly correct, their representations showed either the correct number of hydrogen molecules or oxygen molecules, but not both. In cases where students started with one molecule of water, they also showed two atoms of hydrogen and one atom of oxygen. Some students did not realize that to represent the stoichiometry of the electrolysis reaction, they would need to start with an even number of water molecules. Other drawings showed an equal number of hydrogen and oxygen molecules, indicating that students had not taken stoichiometry into account.

Responses in this question show that students did not know the concept of electrolysis well enough to draw a representation with any great deal of success. The fact that electrolysis was described in words in the course material just preceding the question did not seem to help the students to any great extent.

Question 2 – Use a RADMASTE Molecular Stencil to show the reaction of nitrogen and oxygen to form nitrogen monoxide in the air. Remember that there are approximately 4 molecules of nitrogen for every one molecule of oxygen. Draw a diagram to show the initial mixture (no reaction) and the final mixture (reaction complete).

Due to the nature of the course (mixed-mode delivery), students submitted the drawings for questions 1.1, 1.2 and 2 in a single portfolio assignment, so that there was no intervening face-to-face tutoring to discuss difficulties they had experienced during drawing of representations in questions 1.1 and 1.2 before they attempted question 2. The question posed an additional demand on students, namely that there was excess nitrogen.

The question differed from the previous two questions in that the chemical equation was given in words, and students had instructions to draw two "time frames," namely the initial reaction mixture and the mixture at the completion of the reaction. In asking students to draw the initial mixture and the final mixture, the cognitive load was somewhat decreased intentionally, allowing students to focus on the stoichiometry and nature of the product.

Responses to the question were analyzed by considering whether:

2.1 *Students used the RADMASTE Molecular Stencil*
2.2 *There was evidence of scaffolding*
2.3 *The correct ratio of molecules in the reactant mixture was shown*
2.4 *The correct number of product molecules was shown*
2.5 *The presence of excess nitrogen was shown at completion of the reaction*
2.6 *The shape of the product molecules corresponded to the formula NO*

A greater number of students attempted question 2, with 21 responses out of the sample of 23 students. In addition, all responses could be coded in pre-selected categories.

Results for question 2 are given in Table 2.

2.1 *Did students use the RADMASTE Molecular Stencil?*

All but one of the students (95.2%) used the stencil to generate their drawings. In the case of the one student who did not use the stencil, it would appear that the particular student had misplaced the stencil at the time of working on the portfolio assignment, since none of the drawings in questions 1.1, 1.2, or 2 were drawn with the stencil.

2.2 *Did students use a balanced chemical equation to scaffold their drawing?*

It is interesting to note that scaffolding for this question was only evident in 33.3% of the responses, approximately half of the percentage scaffolding recorded for the

Table 2 Using the RADMASTE Molecular Stencil to represent the formation of nitrogen monoxide in the presence of a nitrogen:oxygen molecular ratio of 4:1

		Number of responses ($n = 21$)			Percentage of responses, %		
Aspect analyzed		Yes	No	Not coded	Yes	No	Not coded
2.1	Used the RADMASTE Molecular Stencil	20	1	0	95.2	4.8	0
2.2	Evidence of scaffolding	7	14	0	33.3	66.7	0

		Number of responses ($n = 21$)			Percentage of responses, %		
Aspect analyzed		Yes	No	1 of each	Yes	No	1 of each
2.3	Number of reactant molecules correct	9	3	9	42.9	14.2	42.9

		Number of responses ($n = 21$)			Percentage of responses, %		
Aspect analyzed		Yes	No	No product shown	Yes	No	No product shown
2.4	Number of product molecules correct	14	6	1	66.7	28.6	4.7

		Number of responses ($n = 21$)		Percentage of responses, %	
Aspect analyzed		Yes	No	Yes	No
2.5	Excess nitrogen at completion shown	3	18	14.3	85.7

		Number of responses ($n = 21$)		Percentage of responses, %	
Aspect analyzed		Yes	No	Yes	No
2.6	Shape of product corresponding to nitrogen monoxide (NO)	8	13	38.1	61.9

question on electrolysis of water. The scaffolding was often in the form of a balanced chemical equation. It should be mentioned that the only artifacts available for analysis were the submitted portfolio assignments, and if students did rough drafts or scaffolding in rough on paper that was discarded, these would not have been recorded.

2.3 *Could students draw the reactant mixture using the given molecular ratios?*

Although a significant number of students (42.9%) represented the reactants by showing the ratio of nitrogen: oxygen molecules as 4:1, just as many students showed only one molecule of each reactant. This indicates that they could not internalize all the given information whilst attempting to draw a molecular representation. Students who showed other numbers of reactant molecules did not show any consistent pattern in their responses.

2.4 *Did students draw the correct number of product molecules?*

Two thirds of the students (66.7%) drew two product molecules when representing the product mixture, in spite of only one third of the group having used a balanced chemical equation in the proximity of their drawing. It must be borne in mind that a correct response of two product molecules would be possible whether students started with one molecule each of nitrogen and oxygen or whether they started with excess nitrogen.

2.5 *Did students show the excess (unreacted) nitrogen when the reaction was complete?*

Although nearly 43% students showed the reactants in the correct ratio, only 14.3% showed the unreacted nitrogen in the product mixture. This suggests that students had received limited instruction in the concept of excess reagent in their previous training and had most certainly not given any thought to representing such molecules in drawings. Balanced chemical equations also ignore this, e.g., reactions in air are frequently represented as O_2 in the equation.

2.6 *Did students represent the shape of the nitrogen monoxide molecule correctly?*

The RADMASTE Molecular Stencil does not have a template shape for the nitrogen monoxide molecule. This led to considerable cognitive disequilibrium and a large number of the students (61.9%) opted to use one of the existing shapes on the stencil. For these students, the shape on the stencil determined the structure of the product. The shape most often used was that of nitrogen dioxide, NO_2, with the shape of N_2O_4 being the second most popular. One student ingeniously solved the cognitive disequilibrium by using the stencil to represent the reactants and chemical symbols to represent the products.

At this stage of the course, it was realized that students needed to be guided more extensively before they could be expected to attempt more complex representations. The drawings students produced in response to questions 1.1, 1.2, and 2 were very useful to assess students' mental models, but students also had to develop some additional modeling and visualization skills in order for them to be able to progress to teaching aspects of modeling in their classrooms. During a face-to-face contact session, the first three questions were therefore revisited, but some scaffolding was done. Peer review of drawings and tutor intervention also occurred.

5 Introducing Scaffolding Material

Scaffolding of Question 1a

The question was divided into five subsections. In the first subsection, teachers had to write a chemical equation to show the boiling of water. This should emphasize that boiling of water is a physical and not a chemical change. In the second subsection, teachers had to reflect on the motion of water molecules in the liquid and the gas phase. This should encourage integration of aspects of the Kinetic Molecular Theory into the drawing. In the third subsection, teachers had to discuss the different effects of gravitational forces on water in the liquid phase and water in the gas phase. This led to discussions of the actual representations - e.g., if the drawing is shown in a box, does the bottom of the box necessarily represent the bottom of the container? In the fourth subsection, a hypothetical mass of water was given. To encourage consideration of the conservation of matter, prediction of the amount of water in the liquid phase and the gas phase was required (conditions which would ensure that conservation of mass will apply, were given). In the fifth subsection, teachers were given an example of a student's drawing, showing all the possible misconceptions, and had to identify misconceptions from the drawing.

Scaffolding of Question 1b

Scaffolding of this question was less extensive, with teachers being given a list of factors they had to take into account. The balanced chemical equation for the electrolysis of water and the stoichiometry of the reaction were mentioned specifically. Teachers had the opportunity to make their own drawing and the drawings were scrutinized and corrected if necessary during the face-to-face contact session. Peer evaluation of drawings was also permitted.

Scaffolding of Question 2

In scaffolding this question, the balanced chemical equation was inserted and the following paragraph was added to prompt teachers to consider the excess nitrogen: "This also means that there are nitrogen molecules that won't react – there is insufficient oxygen present. The excess nitrogen molecules will remain unchanged in the reaction mixture."

Some time spent in self-study elapsed before students were required to submit the last set of drawings.

6 Discussion of Results

6.1 *Responses Obtained After Introduction of Scaffolding Material*

Question 3. Show the combustion of 3 molecules of methane in excess oxygen at extent

(i) 0.00; (ii) 0,33; (iii) 0.67, and (iv) 1.00.

(Hint: You will have to make four separate drawings)

In giving students the hint to make four separate drawings an attempt was made to guide students to a "frame-by-frame" approach to represent the changes taking place during the course of a chemical reaction. In the analysis of responses, less attention was given to factors such as molecular orientation. The analysis focused on aspects of the representation that would illustrate students' understanding of the decrease in the number of reactant molecules and an increase in the number of product molecules during the course of a reaction. Each student could decide on the number of oxygen molecules to be represented at the start of the reaction (the excess reactant). This made each response unique, but the analysis tedious. The total number of responses out of the group of 23 decreased to 15. Of the 15 responses, only 13 could be considered for analysis in part 3.4 of the question. In the remaining two cases, there was no effort to change the number of reactant molecules and draw product molecules as the reaction proceeded and these responses were only coded in parts 3.1, 3.2, and 3.3.

Responses for question 3 are reported as follows:

3.1 *Whether students used the RADMASTE Molecular Stencil*
3.2 *Whether there was evidence of students using some scaffolding to help them cope with the demands of stoichiometry and varying extents of reaction*
3.3 *Whether the molecules in all drawings were oriented randomly*
3.4 *The number of reactant and product molecules at various extents of reaction (a) Extent 0.00, (b) Extent 0.33, (c) Extent 0.67, and (d) Extent 1.00.*

*Note on coding of responses**

There were three types of responses that could not be coded in all categories. Categorization was done where possible. In the first type of response, two students showed all reactant molecules at extent 0.00, but only product molecules at extent 0.33, 0.67 and 1.00. Coding could be done for 3.4 (a), but not for CH_4 and O_2 in 3.4 (b) to 3.4 (d). The products shown could be coded. In the second type of response, two of the representations showed the molecules reacting at the various extents (instead of molecules remaining), and the corresponding product molecules. Again coding could be done for 3.4 (a), but not for reactants in 3.4 (b) to 3.4 (d). The products shown could be coded. In the third type of response that could not be coded, there was one instance in which no products were shown at extent 0.33. Reactants could be coded. Because of these difficulties in coding, results for 3.4 (a) to 3.4 (d) are reported as percentages.

3.1 *Did students use the RADMASTE Molecular Stencil?*

In the last question, the stencil was used without exception. The marks allocated to molecular representations presumably served as sufficient incentive for all students to locate their stencils and to take note of guidelines given during the contact session.

3.2 *Was there evidence of scaffolding to support the drawing of the representation?*

In 60% of the responses to the question, there was evidence of students writing down chemical equations, calculating number of molecules of reactant remaining, etc.

An interesting case of scaffolding can be seen in Fig. 5. The student wrote down the balanced equation and drew the number of reactant molecules that would react at the various extents (only extent 0.33 shown). Thereafter the remaining reactant molecules and formed product molecules were calculated, but to cope with the cognitive load, the student also drew the reactants and products separately.

Another example of scaffolding is shown in Fig. 6. The student needed less scaffolding to reach the correct solution to the question.

3.3 *Orientation of molecules*

Two thirds of the students showed the molecules in random orientation. This represents a considerable increase when compared to earlier drawings. During the face-to- face session, the orientation of molecules was one of the points addressed.

(a) *Representing the number of molecules at extent 0.00*

As many as 69.2% of the students drew the correct number of methane molecules, namely three. In the remainder of the drawings, the number of methane molecules was most often only one or two. The error in the number of methane molecules could possibly be related to careless reading of the question.

Students experienced more difficulty in representing the excess oxygen molecules, with 53.8% of the recorded responses being correct. In cases where excess oxygen was not shown, the number of molecules varied from zero oxygen molecules in two cases, to six (the correct stoichiometric amount), with some values in between.

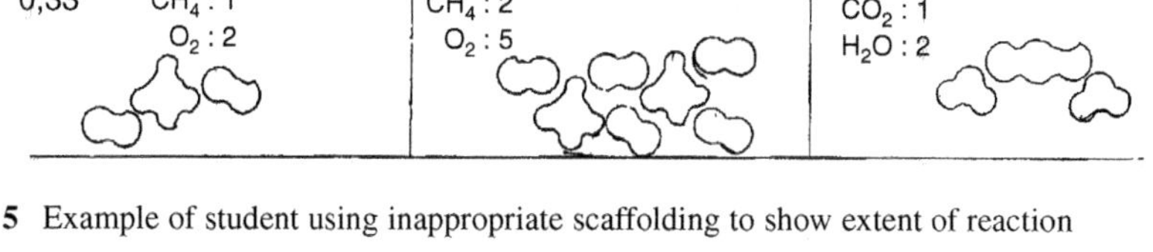

Fig. 5 Example of student using inappropriate scaffolding to show extent of reaction

Fig. 6 Example of student using appropriate scaffolding to show extent of reaction

Table 3 Using the RADMASTE Molecular Stencil to represent the reaction of 3 molecules of methane with excess oxygen at extent (i) 0.00, (ii) 0.33, (iii) 0.67, and (iv) 1.00

		Number of responses (n = 15)			Percentage of responses, %		
Aspect analyzed		Yes	No	Not coded	Yes	No	Not coded
3.1	Used the RADMASTE Molecular Stencil	15	0	0	100	0	0
3.2	Evidence of scaffolding	9	4	2	60.0	26.7	13.3
3.3	Molecules randomly oriented	10	3	2	66.7	20.0	13.3

	Percentage of correct responses, %			
3.4 Extent of reaction (n = 13)	CH_4	O_2	CO_2	H_2O
(a) Extent 0.00 Number of molecules	69.2	53.8	–	–
(b) Extent 0.33* Number of molecules	69.2	38.5	61.5	61.5
(c) Extent 0.67* Number of molecules	61.5	30.8	69.2	69.2
(d) Extent 1.00* Number of molecules	61.5	38.5	92.3	92.3

(b) *Representing the number of molecules at extent 0.33, 0.67, and 1.00*

The number of remaining methane molecules was represented with some success in all three "time-frames," with representation of the number of oxygen molecules remaining problematic throughout.

Students were more successful in representing the number of product molecules, and representation of the number of product molecules at extent 1.00 posed very few problems.

7 Conclusion

The aim of introducing the students to molecular visualization by using the RADMASTE Molecular Stencil was in the first instance not to evaluate their performance, but rather to collect data on the level of students' ability to answer questions on chemical concepts by using molecular visualization. The type of responses collected in the study could be coded to investigate aspects of two-dimensional molecular drawings, but since definitive rubrics, as suggested by various authors [11, 12] had not been developed at the time the study was initiated, it is difficult to compare students' modeling ability at the start and at the end of the course quantitatively. Perhaps, as suggested by Park et al. [12], one could at best expect a change in the distribution of responses. This is confirmed in the present study in aspects such as students' scaffolding of their drawings, students' representations of molecular orientation improving and coping with demands of stoichiometry at the same time.

The wide range of responses obtained in the present study is an indication of how students need to take different paths to develop concepts dealing with the particle nature of matter as suggested by Park et al. [12]. In that respect it was perhaps fortuitous that students in the present study were creating their first sets of drawings when working on their own and, apart from one instance of scaffolding, were given ample opportunity to develop chemical concepts as well as modeling skills independently and individually. Students were largely unfamiliar with the use of molecular modeling at the start of the course and this would have affected their ability to create molecular representations [11–13]. Representations created initially were very useful to show which aspects of physical and chemical change had to receive attention during the course and allowed the researchers in the present study to use the "*window into students' thinking*" [11]. Representations indicated that students experienced difficulty in conceptualizing molecular orientation, conservation of molecules in physical change, conservation of atoms in chemical change, reaction stoichiometry, excess and limiting reagents, and the changes in a reaction mixture during the course of a reaction.

When students make mistakes in a particular representational system, it could mean that they have not mastered the rules and semantic meanings associated with the representation or that they have not discovered the connection between a familiar and unfamiliar representational system [10]. This was possibly a contributing factor when teachers started to use the RADMASTE Molecular Stencil at the start of the course.

Providing a context (or driving question) is essential for meaningful learning to take place when molecular modeling is being used [11,13]. In that respect, the students in the present study were given an appropriate context throughout the course and the modeling exercises. The fact that drawings showed that certain concepts in chemistry were not well understood, contributed largely to inform tutors on aspects of chemistry that needed to be addressed.

Providing scaffolding for modeling exercises is another prerequisite if students are to benefit and construct their own knowledge as a result of modeling [8,13]. Intermediate representations used in the present study (such as the start and end of the reaction in question 2 and the various "frames" to show different extents of reaction in question 3) should have decreased the cognitive load students were experiencing [13].

In the present study, students were mainly required to create molecular drawings. During the scaffolding session, they had their first opportunity to comment critically on an existing drawing. This technique should be introduced earlier in the course, since other authors working in the same field recommend a combination of model building (or model drawing in the case of the RADMASTE Molecular Stencil), model interpretation, and peer model evaluation [11,13]. Findings by other authors [13] suggest that peer evaluation of models could lead to richer models and a stronger grasp of relevant chemistry concepts. Consensus model building [12] could also be introduced - this would entail some discussion, analysis of the chemical concept and negotiation of a most suitable model [12]. On a higher level, students should be encouraged to identify and analyze features of representations and use these features to explain, draw inferences, and make predictions about chemical phenomena or concepts [8]. It is suggested that as students' models develop closer towards a particle view of matter, the level of sophistication of their explanations will also increase [11].

There should be some advantages to giving instructions on how to use the RADMASTE Molecular Stencil at the start of a similar course in future. This will result in students being more confident to use the tool, but richness of data could be affected as a consequence. In the present study, students were perhaps cautious to submit drawings that were being assessed for credit, and this concern could be addressed by spending some time on creating, discussing, and explaining models as a non-credit assignment.

Results in the present study show changes in students' molecular drawings over a period of 4 months, but do not allow for definitive comment on conceptual changes that might have taken place. It is possible that a number of compartmentalized concepts, knowledge of which was formal (i.e., rote learned), were interlinked for the first time in students' minds and that this could be regarded as a type of conceptual change. The results are useful however, as a first step in analyzing and collecting data from students' drawings with the RADMASTE Molecular Stencil. In future research, it might be useful to combine molecular drawing tasks with other test items in order to evaluate conceptual change more clearly.

Questions in future assignments should also make provision for text augmentation, and students explaining their molecular drawings [13]. The latter could be extended to interviews with students or with follow-up questions on models students have created.

We are nevertheless optimistic that the experience students had in the present study to use the RADMASTE Molecular Stencil not only provided them with hands-on experience in creating molecular representations, but also developed skills in molecular visualization that they can introduce to their learners once they return to their classrooms.

References

1. Akoobhai B, Bradley JD, Steenberg E (2006) Macro, Micro, Symbolic: Tools for All to Construct the 'Chemists' Triangle'. Proceedings 19icce, Seoul, Korea
2. Ben-Zvi R, Hofstein A (1996) Strategies for remediating learning difficulties in chemistry. In: Treagust DF, Duit R, Fraser B (ed) Improving Teaching and Learning in Science and Mathematics, Teachers College Press, New York
3. Cantu LL, Herron JD (1978) Concrete and formal Piagetian stages and science concept attainment. J Res Sci Teaching 15(2):135–143
4. Chíu, M-H (2007) A National Survey of students' conceptions of chemistry in Taiwan. Int J Sci Ed 29(4):421–452
5. Gabel D, Sherwood R (1980) The effect of student manipulation of molecular models on chemistry achievement according to Piagetian level. J Res Sci Teaching 17(1):75–81
6. Herron JD (1976) Commentary on "Piagetian cognitive development and achievement in Science". J Res Sci Teaching 13(4):355–359

7. Jones L, Jordan K, Stillings N (2001) Molecular Visualization in Science Education, Report from the Molecular Visualization in Science Education Workshop, NCSA Access Centre, Arlington, VA, January 12–14, 2001 available at http://pro3.chem.pitt.edu/workshop/workshop_report_180701.pdf (accessed Nov 2005)
8. Kozma R, Chin E, Russell J, Marx N (2000) The roles of representations and tools in the chemistry laboratory and their implications for chemistry learning. J Learn Sci 9(2): 105–143
9. Kuo M-T, Jones LL, Pulos SM, Hyslop RM (2004) The relationship of molecular representations, complexity and orientation to difficulty of stereochemistry problems. Chem Educator 9:1–7
10. Madden SP, Jones LL (2002) Multiple representations and their role in solving ideal gas problems, presented at the Annual Meeting of the National Association for Research in Science Teaching. April New Orleans LA. Forwarded electronically by one of the authors (LL Jones) November 2005.
11. Merritt J, Shwartz Y, Krajcik J (2007) Middle School Students' Development of the particle Model of Matter.http://hi-ce.org/presentations/documents/MerritteetalNARST2007.pdf (accessed May 2008)
12. Park S, Park YS, Chae HK (2007) Comparison study of concept test with molecular modelling in the undergraduate inorganic chemistry of stereoisomerism. Proceedings of 2nd NICE Symposium,2007,Taipei.http://dochoonho.sunchon.ac.kr/NICE2/2ndNICE-Papers/2ndNICE-oral/Sohye Park et al. (pdf accessed May 2008)
13. Quintana C, Chang H-Y, Krajcik J (2007) Technology-mediated visualization and interpretation of chemical phenomena by middle school students http://www.hice.org/papers/documents/Chemation_ARA2007.pdf (accessed May 2008)
14. Steenberg E (2006) Investigation of the development of bridging students' cognitive skills relevant to the use and understanding of chemical formulae and equations. Unpublished MSc (Science Ed) thesis, University of South Africa.

Using the Personalization Hypothesis to Design E-Learning Environments

A. Yeung, S. Schmid, A. V. George, and M. M. King

Abstract This paper reports a study that set out to test the personalization hypothesis in the domain of chemistry. E-learning environments were used in this study that were compulsory pre-laboratory activities, used to prepare students to carry out experiments in chemistry. Retention and transfer tests were conducted immediately after students had completed the pre-laboratory work and were used to investigate students' academic performance. Approximately 600 students took part in the project in semester 2, 2005 and about 800 participated in semester 1, 2008, as well as a small bridging course group in 2007. Each study covered a range of levels of prior knowledge within the domain. Characteristics of language background and gender were also taken into account.

Our findings revealed that the personalized group does not in general perform significantly differently from the non-personalized group. However, it appears that the different performance of personalized or non-personalized groups is dependent on prior knowledge of participants in the domain under study. If prior knowledge is weak, significant improvements have been found for personalized over non-personalized instruction.

Abbreviations BC: Bridging Course; CHEM1001: Fundamentals of Chemistry 1A; CHEM1101: Chemistry 1A; CHEM1901: Chemistry 1A (Advanced) ; CHEM1002: Fundamentals of Chemistry 1B; CHEM1102: Chemistry 1B; CHEM1902: Chemistry 1B (Advanced); ESB: English-speaking background; HSC: Higher School Certificate; N: Non-personalized; NESB: Non-English-speaking background; P: Personalized; SID: Student Identification Number

A. Yeung, S. Schmid (✉), and A.V. George
School of Chemistry, The University of Sydney, NSW 2006
e-mail: s.schmid@chem.usyd.edu.au

M.M. King
Faculty of Education and Social Work, The University of Sydney, NSW 2006

M. Gupta-Bhowon et al. (eds.), *Chemistry Education in the ICT Age*,
DOI 10.1007/978-1-4020-9732-4_25,

1 Introduction

Student learning and understanding has the potential to improve with the use of computer-based multimedia environments. Such environments have been expected to offer a powerful effective means of delivering material and enhancing the learning situation [1]. Currently, there is much research activity investigating best practices for the design of multimedia instructional materials for creating effective e-learning environments. Several principles have already been established and tested such as the *multiple representation principle* that says that it is better to present explanations as words and pictures rather than solely words and the *coherence principle* stating that people learn better when extraneous pictures and sounds are excluded rather than included [1,2].

An area of particular interest concerns the personalization hypothesis. Moreno and Mayer [3] first asked the question "Does the use of personalized messages improve students' understanding of a multimedia science lesson?" (p 724). It was predicted that students who learn from a personalized computer program would remember more factual information and solve problems better than students who learn by means of neutral/non-personalized messages. Moreno and Mayer [3] conducted experiments using a short multimedia explanation of how lightning storms develop. Explanations included a 140-second-long animation accompanied by personalized speech and/or text (first person, conversational style) or non-personalized speech and/or text (third person, more formal style). The experiments found that students in the personalized (P) group generated more creative and correct solutions than students in the non-personalized (N) group in a follow-up transfer test (i.e., tests used to determine students' understanding) with an effect size of 1.00 in an experiment involving speech and 1.60 in an experiment using text.

More recently, Moreno and Mayer [4] studied the personalization hypothesis in the domain of biology for a plant design activity. This research involved a computer-based science simulation game, where students learn about the biological features of plants. Students interact with a virtual environment that has certain environmental conditions and are asked to design a plant that has root, stem, and leaf properties that would survive in given conditions. Guidance is provided by an on-screen agent, Herman the Bug, who not only guides students but also poses questions. Two different versions of the program were used, a personalized version and a non-personalized one. Examples are given below.

> **Introduction to the Program** ([5], p 173)
> P version: You are about to start a journey where you will be visiting different planets. For each planet, you will need to design a plant. Your mission is to learn what type of roots, stem, and leaves will allow your plant to survive in each environment. I will be guiding you through by giving out some hints.
> N version: This program is about what types of plants survive on different planets. For each planet, a plant will be designed. The goal is to learn what type of roots, stem, and leaves allow plants to survive in each environment. Some hints are provided throughout the program.

Findings revealed that students studying the P version obtained significantly better outcomes in learning as measured by retention and transfer tests and found the program friendlier. Students in the P group obtained 28% more marks on the transfer test than students in the N group resulting in a large effect size of 1.64 [5].

Given such a large effect we were interested to investigate whether the personalization effect could also be observed in the domain of Chemistry. The examination of differences in academic performance between students who completed personalized versions of online chemistry material and students who completed non-personalized versions will allow a better understanding of the way information should be phrased in online materials to best promote learning and improve academic performance of first-year chemistry students. That understanding is of significant interest since present chemistry teaching materials are mainly written in a non-personalized way. If personalized messages were found to have a positive effect on academic performance, the information is likely to have a significant impact on the design of such materials in the future.

Consequently this study set out to test the personalization hypothesis using first-year university students studying chemistry. The e-learning environments chosen involved compulsory pre-laboratory work activities used to prepare students for experimental laboratory work in chemistry. In addition to simply investigating whether personalization has an effect on academic performance in the domain of chemistry, this project also considers the importance of factors such as gender, and language background – English-speaking (ESB) or non-English-speaking (NESB).

2 Methodology

2.1 Participants

Two cohorts of three different groups of first-year students participated in this study at the University of Sydney. The first cohort consisted of students enrolled in semester 2 courses available to students undertaking mainstream science qualifications in 2005. These courses were Fundamentals of Chemistry 1B (CHEM1002), Chemistry 1B (CHEM1102), and Chemistry 1B – Advanced (CHEM1902). The second cohort consisted of students enrolled in semester 1 courses available to students undertaking mainstream science qualifications in semester 1, 2008. These courses were Fundamentals of Chemistry 1A (CHEM1001), Chemistry 1A (CHEM1101), and Chemistry 1A – Advanced (CHEM1901). All three units in each semester cover similar material, but differ in the level of assumed prior knowledge and the level at which material is presented. CHEM1001 students have either not completed chemistry for the Higher School Certificate (HSC), i.e., university entry level, or achieved comparatively poor results. CHEM1101 students have satisfactorily completed HSC chemistry, whilst CHEM1901 students have achieved a HSC chemistry mark above 80%. The units CHEM1002, CHEM1102, and CHEM1902 assume successful completion of a corresponding semester 1 unit, and include both general and organic chemistry content. In addition a third group of students participated in the study in

2007. These students were enrolled in the chemistry bridging course (BC) at the University of Sydney. It is an intensive course seven-day-long designed for students with a weak or no chemistry background to prepare them for studying first-year chemistry at university.

2.2 *Study Design – Large Group Versus Small Group*

Students were assigned to P and N groups and these again split into a small and a large group. Students in the large group completed the online modules in their own time wherever they wanted. Students in the small group completed the modules in a supervised session. These students also undertook retention and transfer tests. Academic performance of all students in the large group and the small group was examined in the same way by comparing the performance of students who completed the different versions of the modules. Methodologically, gathering data from a large group as well as focusing on a small group allows conclusions of acceptable generality to be drawn, whilst still ensuring that results authentically reflect the experiences of students. Figure 1 shows a schematic diagram of the methodology and the different aspects of the project each group completed.

2.2.1 Participant Information Survey

A survey was distributed to about 900 students during a laboratory session in semester 2, 2005, about 1,100 students in semester 1, 2008, and approximately 200 students enrolled in the 2007 bridging course. The purpose of this survey was to obtain information on particular student characteristics, including gender and language background (ESB or NESB).

2.2.2 Online Pre-laboratory Work Modules

Students in semester 2, 2005 were asked to complete two online pre-laboratory work modules. Students must complete such pre-laboratory work modules before each laboratory session as part of their assessment. The investigated modules were part of the regular pre-laboratory work program: E18 – *Percentage ionization of a weak acid* (predominantly involves calculations) and E21 – *Buffered and unbuffered solutions* (a more concept-based module). In semester 1, 2008 all pre-laboratory work modules used for assessment, were also used in the study. All students completed E1 (photochemistry), E2 (solubility), E6 (gas properties), E10 (thermodynamics), and E11 (molecular models). In addition students in CHEM1001 and CHEM1101 also completed W1 (stoichiometry) and students in CHEM1901 also completed E12 (inorganic solids). The BC students completed one module on stoichiometry only. Completing the modules involved students reading through content presented on the screen and answering some quiz questions. Students were allocated to either a P or N version of the modules depending on the last digit of their student

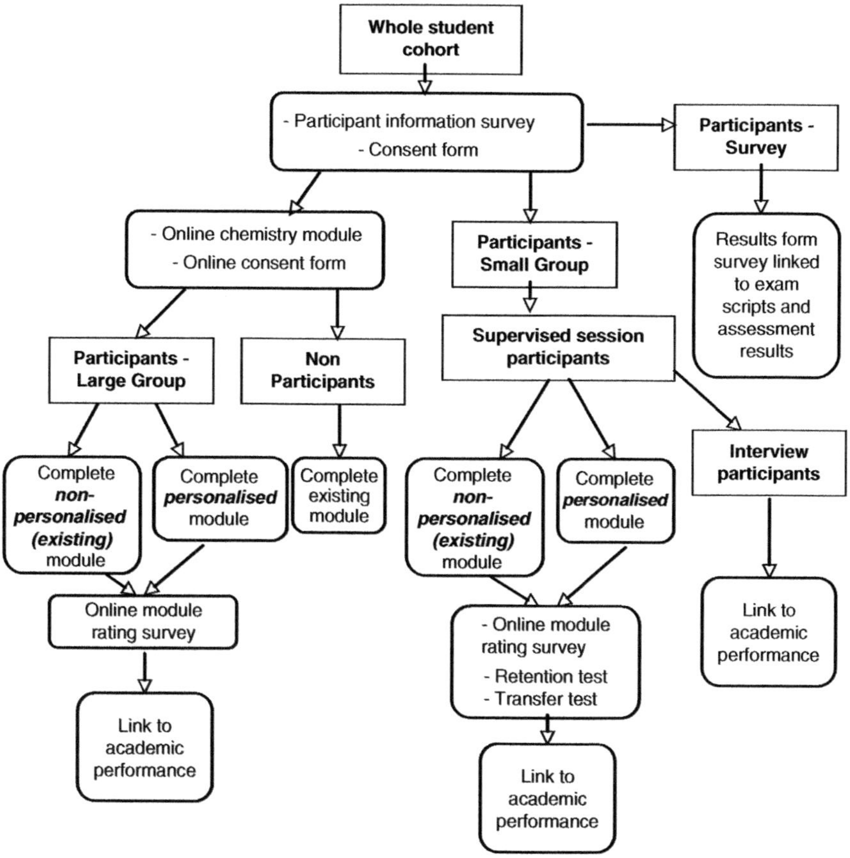

Fig. 1 Schematic representation of methodology

identification number (SID) in semester 2, 2005 or receipt number for the BC students. In semester 1, 2008 the student administration software allowed participants to be randomly assigned to either the P or N group.

2.2.3 Small Group – Additional Retention and Transfer Questions

In order to gain a better understanding of the effect of personalization on academic performance, students in the small group (34 students in semester 2, 2005 and 14 students in semester 1, 2008) also completed a retention test and a transfer test following the completion of the online pre-laboratory modules. The purpose of the retention test was to determine what students remember from the modules while the transfer test was used to determine students' understanding. Making students complete these further tests immediately after completing the modules ensures that no additional external factors can influence the learning outcomes of these students and thus give us a better indication of the personalization effect.

2.2.4 Module Survey

In semester 2, 2005 students were also asked to complete an online module rating survey after they finished both online modules in an attempt to determine whether personalization affects students' attitudes towards the modules. Other questions the survey sought to answer were questions related to student's motivational level, student's perception of the modules helpfulness and user friendliness, and how students use the modules to assist their learning.

2.2.5 In-Depth Interviews

Interviews were conducted with eight students of the small group in semester 2, 2005. Students were asked to provide more detailed responses for the module rating survey. During the interview students were also shown their retention/transfer test paper and were asked questions concerning different aspects of the paper. This allowed the interviewer to gain an insight into the methods students used to solve particular questions, especially the differences between students in the P and N groups. General questions about the modules were also asked with particular attention paid to the language used in the modules.

3 Results

Academic performance for all students was assessed by marks obtained in the online quizzes, which immediately followed the presentation of content in the online chemistry modules in semester 2, 2005. A total of 630 (67%) students gave consent to participate in the study and their results were used in the analysis of the large group. In semester 1, 2008 a total of 6 online chemistry modules was used for participants in all courses. Marks obtained in the quizzes associated with all modules were used as a measure of academic performance. A total of 824 (87%) students gave consent to participate in the study and their results were used in the analysis of the large group for semester 1, 2008. A total of 9 students (5%) from the 2007 bridging course gave consent to participate in the study.

3.1 Large Group

3.1.1 Semester 2, 2005

Independent samples t-tests were conducted to determine whether there was a difference in academic performance of students who completed the P or N versions of the online chemistry modules in semester 2, 2005. No significant difference in quiz performance was found for E18 ($t_{628} = -1.794$, $p = 0.073$) and E21 ($t_{219} = -0.256$, $p = 0.798$). The quiz marks of students who completed both online modules were

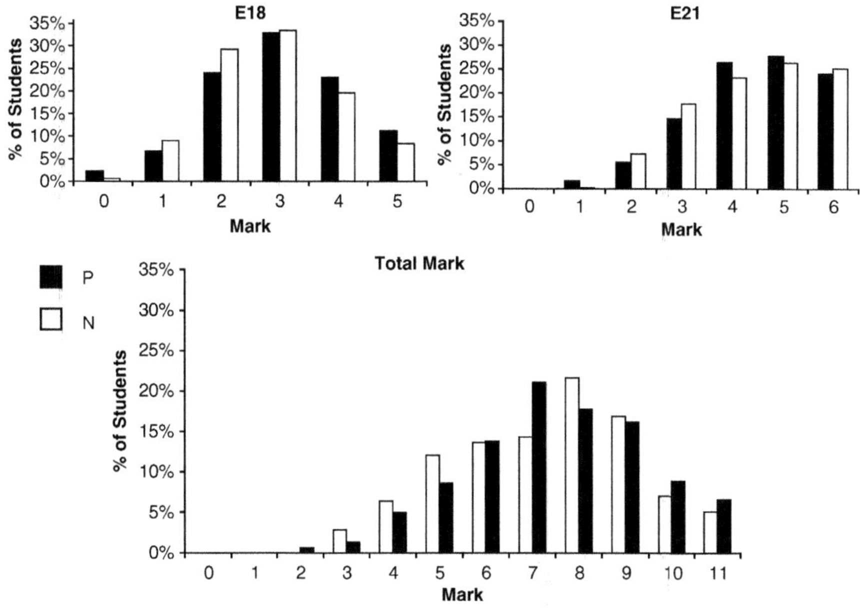

Fig. 2 Distribution of marks between the P and N group for E18 and E21

then summed to produce a *total mark*, and a further t-test showed no significant difference between the P and N group ($t_{615} = -1.299$, $p = 0.195$). These results were consistent for each of the courses. Since no differences were found between the overall marks of students in each group, x^2 analyses were conducted to determine whether there were differences in the distribution of marks for E18, E21 and total mark between the P and N groups. As shown in Fig. 2 it was found that there were no significant differences between the P and N groups for both E18 ($x^2 = 8.877$, df = 5, $p = 0.114$), E21 ($x^2 = 5.480$, df = 5, $p = 0.360$), and total mark ($x^2 = 12.113$, df = 9, $p = 0.207$).

Individual Questions from Pre-work

Individual quiz questions were looked at in detail in semester 2, 2005 to determine whether there were differences in performance between the P and N groups for particular questions. After conducting a series of independent samples t-tests, no significant differences were found between the P and N groups.

Gender

Independent samples t-tests were conducted to determine whether there was a difference in mean online quiz marks between students of each gender who completed the P or N versions of the modules in semester 2, 2005. It was found that there was no significant difference in quiz performance between the P or N group of males for E18 ($t_{276} = -1.370$, $p = 0.172$), or E21 ($t_{274} = -0.546$, $p = 0.585$), as well as females for E18 ($t_{347} = -1.179$, $p = 0.239$), or E21 ($t_{340} = 0.348$, $p = 0.728$).

Language Background

In semester 2, 2005, independent samples t-tests confirmed that NESB students in the P group performed significantly better than students in the N group for E18 ($t_{108} = -2.350$, $p = 0.021$, es = 0.45). This was the only significant difference found when investigating language background.

3.1.2 Semester 1, 2008

In semester 1, 2008, the quiz marks of all students in a course who completed either all P or all N online modules were summed to produce a *total mark* for the common modules that students in all courses completed. A further *total mark_1901,* included the marks for E12 that only the Chem1901 students completed, while *total mark_1001* and *total mark_1101* included the marks for W1 that only the CHEM1001 and Chem1101 students completed. Independent samples t-tests were conducted to determine whether there was a difference in academic performance between students who completed the P or N versions of the online chemistry modules. The results are summarized in Table 1.

No significant differences were found for *total mark* for students in CHEM1101 or students in CHEM1901. For Chem1001, however, a significant difference in performance was found between the P and N groups. When including the E12 exercise for the Chem1901 students (*total mark_1901*) there was also a significant difference for that group, which stems from the different performance for that particular module (*mark E12*). A separate independent samples t-test was conducted for E12, showing that students in the P group performed significantly better than

Table 1 (a) Summary of differences (t-tests results) between the P and N groups in semester 1, 2008 and (b) summary of differences (Mann Whitney results) between the P and N groups in semester 1, 2008

(a)		*T*	*df*	*p*
CHEM1001	*Total mark*	−2.400	229	0.0170
CHEM1101	*Total mark*	−1.341	332	0.1810
CHEM1901	*Total mark*	−1.685	86	0.0960
	Total mark_1901	4.631	83	0.0000
	Mark E12	−2.324	85	0.0230

(b)		*U*	*p*
CHEM1001	*Total mark_1001*	5187.5	0.043
CHEM1101	*Total mark_1101*	11420.5	0.302

those in the N group for E12 as shown in Table 1a. The data for *total mark_1001* or *total mark_1101* (i.e., including W1) did not show a normal distribution so a Mann Whitney test was conducted to determine if there were any differences in academic performance for students in CHEM1001 or CHEM1101. A summary of the results is shown in Table 1b. Students in the P group still performed significantly better than those in the N group for CHEM1001 while there was no significant difference between students in the P and N groups in CHEM1101.

3.1.3 Bridging Course

Independent samples t-tests were conducted to determine whether there was a difference in academic performance of BC students who completed the P or N versions of the stoichiometry online chemistry module. No significant difference in quiz performance was found ($t_7 = -1.40$, $p = 0.206$).

3.2 *Small Group*

As for the large group, independent samples t-tests were conducted to determine whether there were differences in quiz performance between students in the small group who completed the P or N version for the modules. No significant differences were found for E18 ($t_{32} = -1.393$, $p = 0.173$) or E21 ($t_{32} = 0.000$, $p = 1.000$). Furthermore, independent samples t-tests also showed that there were no differences in performance of students in the P and N groups for the additional retention ($t_{32} = 1.173$, $p = 0.249$) and transfer ($t_{32} = 0.793$, $p = 0.434$) tests.

3.2.1 Student Characteristics – Gender and Language Background

Further investigation of the personalization effect on academic performance with respect to student characteristics found a significant difference in that NESB students in the P group performed significantly better than the N group for E21 ($t_8 = -3.667$, $p = 0.001$, es = 1.95). However, this result was not seen in the large group of about 100 students. Generally, there were no significant differences between the P and N groups.

3.3 *Module Survey and Interviews*

Chi-square analyses were conducted to determine whether there were differences in the distribution of answers for the module survey between the P and N groups in semester 2, 2005. No significant differences were found, indicating that students in both groups had the same perceptions of the online modules despite the different versions they completed. These results were further verified in the interviews where students in both the P and N groups made the similar comments about different aspects of the modules, even though they completed two different versions.

However, although the number of students who completed the survey in the BC was small, there is a clear trend that emerges. Students who completed the P version found the module more helpful and more engaging than the N group, as shown in Fig. 3. The P group also felt that less effort was required to learn the material, as well as finding it easier to follow the explanations compared to the N group.

4 Discussion

The results of our study show that personalization of online material in chemistry does not seem to have an effect on academic performance of chemistry students in general. These results seem quite different and contrary to those obtained by previous research (e.g., [3]). There are, however, a number of distinct differences between these studies. The previous research involved psychology students who did not study in the domain under investigation [3, 5]. These students were given material in the domain of biology and physics, which they would not have been very familiar with. Our study involved chemistry students with different levels of prior knowledge who were presented with chemistry content. In addition, the present study was conducted on a much larger scale (~600 students in semester 2, 2005 and ~800 students in semester 1, 2008) than that of the previous research (~30 students). Furthermore the majority of students in this study completed the online modules in an authentic learning environment, i.e., at their own pace, time, and space. In contrast, previous research was conducted in a controlled setting, similar to that used for the small group in this study. In our initial study in 2005 virtually no evidence was found for a difference in performance for students that used either personalized or non-personalized material. In trying to understand that result we concluded that the explanation may be the expertise reversal effect.

It has been shown in empirical studies [6–8] that instructional design that is highly effective for inexperienced learners can lose its effectiveness or even have negative consequences for learners with more experience [9]. According to Sweller et al. [10] and Sweller [11] the level of learner expertise in a domain primarily influences the extent to which a learner can organize schemas and information in working memory. For inexperienced learners who receive no guidance on how to organize schemas, instructional material can act as a substitute for missing schema, and if it is effective, can act as a means of constructing schemas. Effective instruction provides the necessary instructional guidance while also minimizing cognitive load [10, 11]. However, instructional material that guides novices may be redundant to more experienced learners, and thus places excessive unnecessary load on limited working memory [9].

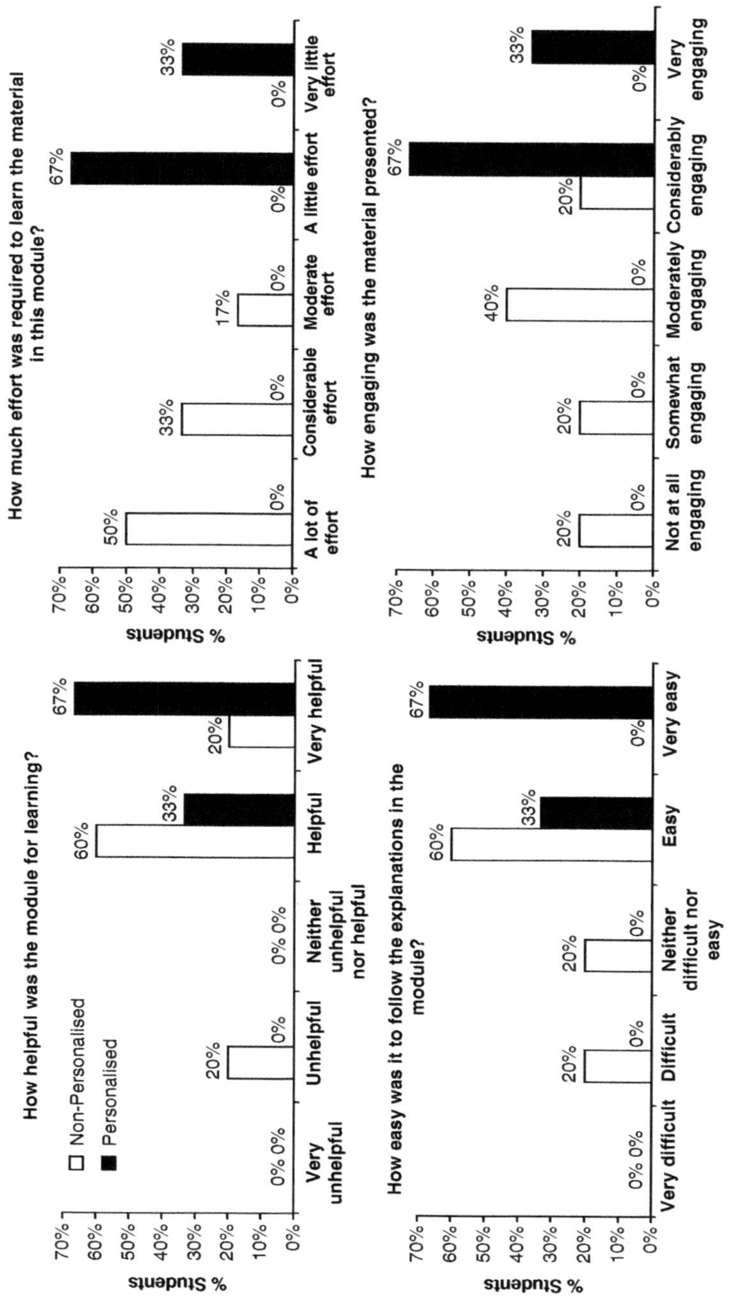

Fig. 3 Student responses from the BC for the module rating survey for each version of the module

So, personalization might only improve academic performance of students who are studying in an unfamiliar area. The participants in the 2005 study were already in semester 2 of first-year chemistry and therefore may have been too advanced in their chemistry knowledge to show an effect. Consequently, we extended the study to bridging course students in 2007 and semester 1 students in 2008.

While there was a difference in performance for the bridging course students (students with the P module had improved performance) the small number of participants did not allow any conclusions to be drawn from this study alone. It did, however, support the view that students with a weak background in a domain may benefit from personalized teaching material. Despite the small number of BC students who participated in the study, responses in the module rating survey show that students in the P group found the module more helpful and engaging.

In semester 1, 2008, it was found that for the students with the weakest background, i.e., CHEM1001, the P group performed better than the N group. For the students in Chem1101 and Chem1901 no difference was found for all common modules. Interestingly, the performance in the E12 module was better for the P group than the N group, which may seem contrary to our other findings, since Chem1901 students have the best chemistry background and are generally very good students. The E12 module, however, is an activity that includes new material on inorganic solids that has not been covered anywhere else. It therefore appears that even very good students benefit from personalized instruction material when studying in an unfamiliar area.

In semester 2, 2005, investigations of the large group found NESB students in the P group performed significantly better than NESB students in the N group for E18. Although this difference was not observed in the small group, investigations of the small group showed NESB students in the P group performed significantly better than NESB students in the N group for E21. Despite these conflicting results, which may be attributed to the small number of participants in the small group, it is likely that personalization has a positive influence on academic performance of NESB students. Such a difference is not seen for the ESB students. NESB students often have learning difficulties because of the language barrier that they face and personalization of text may assist them in their learning.

5 Conclusion

Our research has shown that, the positive effect of using personalized text in instruction material for students is not as general as first assumed. Contrary to previous findings in the domains of biology and physics, personalization of online chemistry material does not influence academic performance of all chemistry students. However, students with a weak background in chemistry or unfamiliarity in certain areas do improve their academic performance when exposed to personalized text rather than non-personalized text. Also non-English-speaking background students in the P group were found to perform significantly better than those in the

N group. These observed benefits of personalization suggest that consideration needs to be taken in the design of online materials in order to effectively promote student understanding. If certain phrasing of online material is able to promote learning and improve performance, then any new research or information in this area is likely to have a significant impact on the design of such materials in the future, since current materials are mainly written in a non-personalized way. Further investigation in various domains and with large sample sizes is required to further validate the personalization hypothesis. Moreover, investigations into the mechanism by which personalization can affect student learning, e.g., changes in interest or motivational levels, can provide a greater insight into the way students use online materials in their learning. Consideration of this information is worthwhile before making any major changes to the design and delivery of teaching materials in the future.

Acknowledgments The authors would like to thank all University of Sydney students who agreed to participate in this study. Data collections described in this work were authorized by the University of Sydney Human Research Ethics Committee, project reference numbers 08–2005/4/8461.

References

1. Mayer, RE and Moreno, R (2002) Aids to Computer-Based Multimedia Learning. Learn Instr 12:107–119.
2. Harp, SF and Mayer, RE (1998) How Seductive Details Do Their Damage: A Theory of Cognitive Interest in Science Learning. J Educ Psychol 90:414–434.
3. Moreno, R and Mayer, RE (2000) Engaging Students in Active Learning: The Case for Personalised Multimedia Messages. J Educ Psychol 92:724–733.
4. Moreno, R and Mayer, RE (2004) Personalised Messages That Promote Science Learning in Virtual Environments. J Educ Psychol 96:165–173.
5. Mayer, RE, Fennell, S, Farmer, L et al. (2004) A Personalisation Effect in Multimedia Learning: Students Learn Better When Words are in Conversational Styles rather than Formal Style. J Educ Psychol 92:389–395.
6. Kalyuga, S, Chandler, P and Sweller, J (1998) Levels of Expertise and Instructional Design. Hum Factors 40:1–17.
7. McNamara, D, Kintsch, E, Songer, NB et al. (1996) Are Good Texts Always Better? Interactions of Coherence, Background Knowledge, and Levels of Understanding in Learning from Text. Cognition Instruct 14:1–43.
8. Yeung, A, Jin, P and Sweller, J (1998) Cognitive Load and Learner Expertise: Split-Attention and Redundancy Effects in Reading with Explanatory Notes. Contemp Educ Psychol 23:1–21.
9. Kalyuga, S, Chandler, P and Sweller, J (2003) The Expertise Reversal Effect. Educ Psychol 38:23–31.
10. Sweller, J, van Merrienboer, JG and Paas, F (1998) Cognitive Architecture and Instructional Design. Educ Psychol 10:251–296.
11. Sweller, J (1999) Instructional Design in Technical Areas. ACER Press, Melbourne, Australia.

Calculation of Potential Energy in the Reaction of "F + HCl → HF + Cl" and its Visualization

A. Ikuo, H. Nagashima, Y. Yoshinaga, and H. Ogawa

Abstract Geometries and potential energies (PE) for intermediates of "F-H-Cl" around the transition state in the reaction of "F + HCl → HF + Cl" were calculated, and computer graphics (CGs) of PE in two-dimension (2-D) and PE surface in three-dimension (3-D), skeletal structures of intermediates on the way of the reaction, and the reaction profile were created. CG animation composed of four CGs was proposed in order to express images of the reaction.

1 Introduction

The reaction of simple molecule such as hydrogen halide and related compounds plays a fundamental role in the development of chemical kinetics and theoretical chemistry [1]. For example, changes of potential energy (PE) in the reaction of "F + HCl" have been reported experimentally [2] and theoretically [3]. However, the reaction has not been clarified enough in details such as PE surface in three-dimension (3-D) and so on.

Visualization of computer graphics (CG) is of great help to realize not only images of molecules but also images for dynamical reaction mechanism. The visualization have been tried and reported, for example, the interactive animation of hydrogen atomic orbital based on the atomic orbital calculation [4] and molecular structures of 215 organic compounds in the field of life science based on the molecular mechanics calculation [5]. It is our aim to produce programs, which provide images of chemical reactions. CG animation of chemical reaction of carbon dioxide and water based on molecular orbital calculations was reported [6]. The reaction was energetically favored than that with one molecule of water.

A. Ikuo (✉), H. Nagashima, Y. Yoshinaga, and H. Ogawa
Department of Chemistry, Faculty of Education, Tokyo Gakugei University
4-1-1 Nukuikita-machi, Koganei-shi, Tokyo, 184-8501, Japan
e-mail: ikuo@u-gakugei.ac.jp

M. Gupta-Bhowon et al. (eds.), *Chemistry Education in the ICT Age*,
DOI 10.1007/978-1-4020-9732-4_26,

The CG with two molecules of water has been demonstrated. Recently, the CG animation of esterification of acetic acid and ethyl alcohol has been reported [7].

Reaction profile is generally used to represent relationship between PE and reaction coordinate. This is often used in high school chemistry textbooks [8]. It is sometimes difficult for the learner to realize the meaning of reaction coordinate. Visualization of PE surface in 3-D could clearly provide images of reaction coordinate from the standpoint of subject on energy. A figure of PE surface in 2-D or contour diagram is often used in physical chemistry textbook of university [9]. Limited number of images on PE surface in 3-D is used in these textbooks. In this paper, we report here the calculation of the PE in the reaction of "F + HCl" and propose its CG animation combined with skeletal structures of intermediates and the reaction profile in order to express images of the reaction.

2 Procedure

2.1 Calculation Based on Quantum Chemistry

The semi-empirical molecular orbital calculation software *MOPAC* in the *CAChe Work System for Windows ver. 6.01* (Fujitsu, Inc.) was used in all of calculations for optimization of geometry by the *Eigenvector Following method*, for search of potential energies of various geometries of intermediates by use of the program with *Optimized map*, for search of the reaction path from the reactants to the products *via* the transition state by calculation of the intrinsic reaction coordinate (IRC) [10].

2.2 Production of Potential Surface

Data of potential energies and geometries of intermediates were extracted from the "energy.map" file from the calculation of *Optimized map,* and these were pasted on the worksheet of *Excel 2007* (Microsoft). Potential energy surface in 3-D was drawn by the *Excel.*

2.3 Production of CG Software

CG was drawn by the *CAChe* (Fujitsu, Inc.) and *Power point 2007* (Microsoft), and CG animation was produced by the *Flash 8* software (Macromedia, Inc.). Web-based CG software was produced by the *Homepage builder* software (IBM, Inc).

3 Results and Discussion

3.1 Optimization of States of Reactant and Product

The reaction of "F + HCl → HF + Cl" shown in Scheme 1 ***a*** is well known [9]. An attack of a fluorine (F) atom to hydrogen chloride (HCl) molecule leads to form hydrogen fluoride (HF).

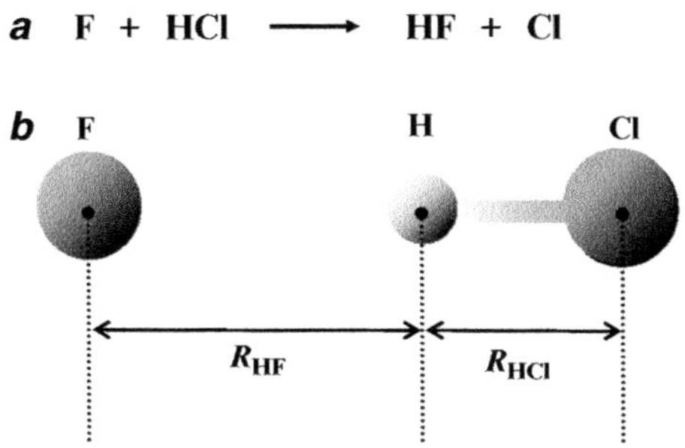

Scheme 1 Reaction of (***a***) "F + HCl" and (***b***) image of inter-atomic distance

Appropriate geometries of both HCl and HF molecules were fixed by calculation with the *Eigenvector Following method* in *MOPAC* with various Hamiltonians of AM1 [11], PM3 [12], and PM5. The optimization of the state of each molecule was started at the point of initially defaulted value of inter-atomic distance. The calculation was carried out until the cutoff value of less than 1.000 in gradient by root-mean-square (RMS) where the value less than 1.000 means to achieve the self-consistent field (SCF). Tentative heat of formation, ΔH_f, was obtained by MOPAC calculation. Results are listed in Table 1. In the case of HCl, the cutoff value by AM1 reached to the value of less than 1.000 in gradient only in 3 cycles of optimization, and the value of ΔH_f was −24.61233 kcal mol^{-1} with the value of 1.2842 Å of inter-atomic distance. Values of ΔH_f were obtained as −20.46808 and −30.41903 kcal mol^{-1} by PM3 and PM5, respectively. In the case of HF, the value by AM1 reached to −74.28070 kcal mol^{-1} in 6 cycles of optimization with the value of 0.8265 Å of inter-atomic distance. ΔH_f values were obtained as −62.75007 and −67.15007 kcal mol^{-1} by PM3 and PM5, respectively. Geometries of both HCl and HF by three Hamiltonians were determined by these optimizations.

From the standpoint of the value of subtraction by use of empirical values [13] in parentheses, the value by PM5 was larger than those by AM1 and PM3. Relatively large error of bond length concerning hydrogen molecule has been

Table 1 Calculations with AM1, PM3, and PM5 Hamiltonians

Sample	Hamil-tonian	ΔH_f[a]/kcal mol^{-1} Before optimization	After optimization	RMS gradient[b]	Number of cycles	Inter-atomic distance/Å Calcula-tion	(Experi-mental)[c]	Subtrac-tion[d]
HCl	AM1	–24.41511	–24.61233	0.487	3	1.2842	(1.2746)	0.0096
	PM3	–19.96390	–20.46808	0.079	4	1.2677		–0.0069
	PM5	–28.76089	–30.41903	0.036	5	1.2327		–0.0419
HF	AM1	–51.66330	–74.28070	0.454	6	0.8265	(0.9169)	–0.0904
	PM3	–51.16533	–62.75007	0.119	5	0.9377		0.0208
	PM5	–63.84668	–67.15007	0.132	5	0.9631		0.0462

[a]Tentative heat of formation ΔH_f was defined by MOPAC. Calculation was started from initially defaulted value of inter-atomic distance in each sample
[b]Value of gradient by root-mean-square method
[c]Microwave spectroscopy cited in [13]
[d]Subtraction of the value of experimental from that of calculation

reported by PM5 [14]. This is due to difference in approximation between by PM5 and by AM1 or PM3.

3.2 *Potential Energy (PE) Surface*

Image of an attack of F atom to HCl molecule is shown in Scheme 1b. Bond angle of F-H-Cl was adjusted to 180°. Intermediates of F-H-Cl on the way of the reaction were set for calculation of PEs as follows. Inter-atomic distance of F-H (R_{HF}) was changed from 0.80 to 4.50 Å at intervals of 0.05 Å, and that of H-Cl (R_{HCl}) was changed from 1.10 to 2.05 Å at intervals of 0.05 Å. PE of each intermediate on the way of the reaction was calculated by *MOPAC* with AM1 [11], PM3 [12], and PM5 Hamiltonians. Output file from the calculation by PM3 indicated eight fractures on the PE surface, and output on the way of the calculation by PM5 ceased due to an error. Only by AM1 was a smooth PE surface given. PE by AM1 in two-dimension (2-D) is shown in Fig. 1 along with figure legend of color boundaries. Inter-atomic distance of F-H (R_{HF}) is drawn as a horizontal axis and that of H-Cl (R_{HCl}) is as a vertical axis. The lowest value of PE is there around 0.8 Å of R_{HF} and 2.05 Å of R_{HCl}, and the secondarily lowest PE is around 2.2 Å of R_{HF} and 1.3 Å of R_{HCl}. The transition state is located near the point of 1.376 Å of R_{HF} and 1.354 Å of R_{HCl}. The intrinsic reaction coordinate (IRC) [10] method supported the transition state.

Figure 2 clearly shows these changes of PEs with display on PE surface in 3-D, which offers a bird-eye view of the reaction profile. Two Valleys of lower energies and hilltop on the transition state at the saddle point can be recognized boldly. Possible pathways of the reaction from the reactants of F and HCl to the products of HF and Cl *via* the transition state at saddle point can be readily traced.

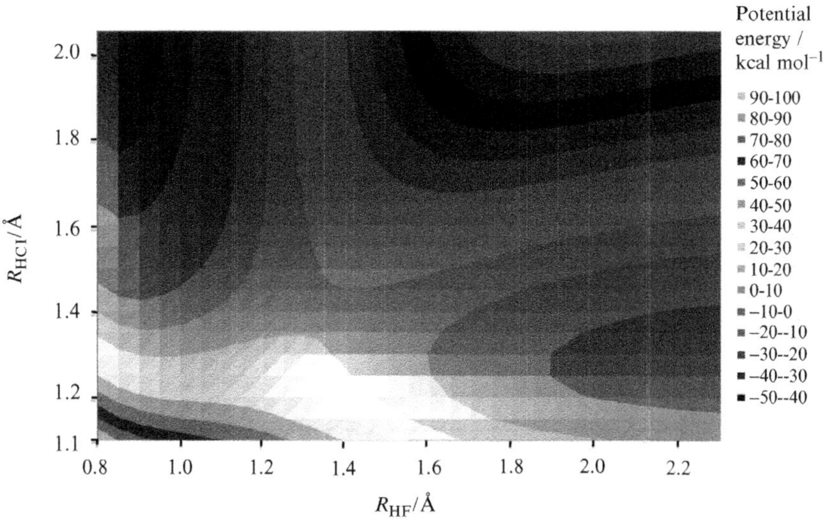

Fig. 1 PE in 2-D

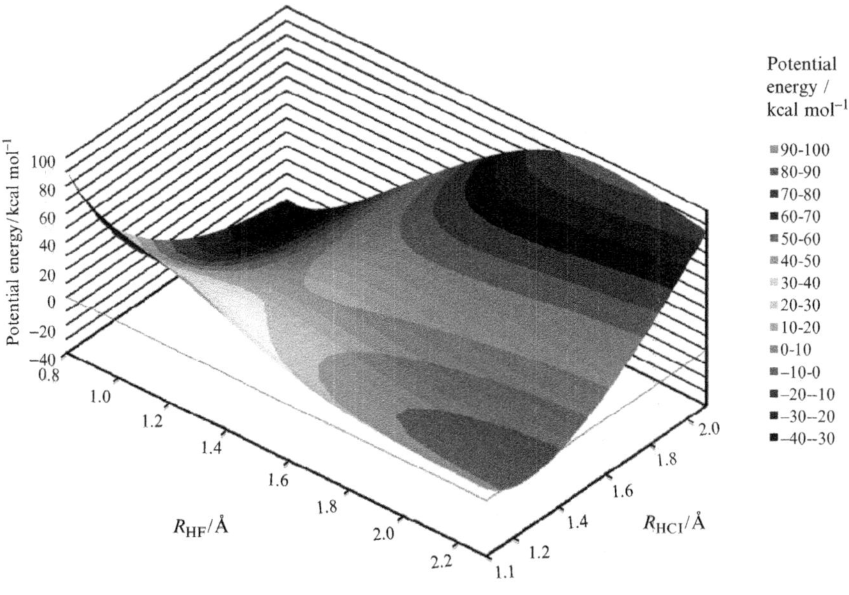

Fig. 2 PE surface in 3-D

The reaction profile is shown in Fig. 3. The profile was made by the calculation of IRC method with 35 steps. The reaction coordinate indicates the most probable pathway of the reaction according to the IRC theory [10]. Furthermore, the vibrational analysis of the intermediate was performed by use of the program *FORCE* in *MOPAC*.

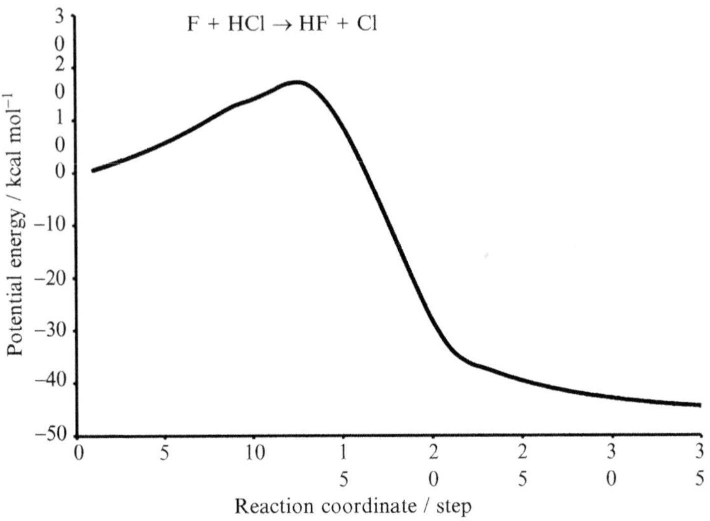

Fig. 3 Reaction profile

A single absorption peak in the negative region was found at −2,858 cm^{-1}. The result indicates vibrational mode due to the decrease of potential energy for the direction of only one path *via* a true transition state at the saddle point. Energy between the initial state of reactants and the final state of products was 38.96 kcal mol^{-1}. The value was in fairly good agreement with an experimental value of 33.06 kcal mol^{-1} [2].

3.3 Visualization with CG Animation

Four CGs of the PE in 2-D and PE surface in 3-D, structure of intermediate based on inter-atomic distance, and reaction profile were combined, and the combined CGs (abbreviated as combination CG) are shown in Fig. 4. The ball-and-stick model shows skeletal structure of the intermediate. The reaction profile demonstrates the degree of the reaction progress by the ball illustrated in the figure. The combination CG is able to provide information about PE and structure of intermediate of molecule in a certain state simultaneously.

CG animation was created by use of a number of the combination CGs. Figure 5 shows representative frames of the combination CGs on the way from the state of reactants (**a**) to that of products (**d**) *via* the transition state (**b**). The animation synchronizes information with the changes of PE, skeletal structure of the intermediate of the reaction, and the reaction profile in all stages. The ball on the reaction profile in the animation can move automatically along with the reaction coordinate. The animation demonstrates the degree of the reaction progress and structural change of the molecules of all stages with the movement of the ball, simultaneously.

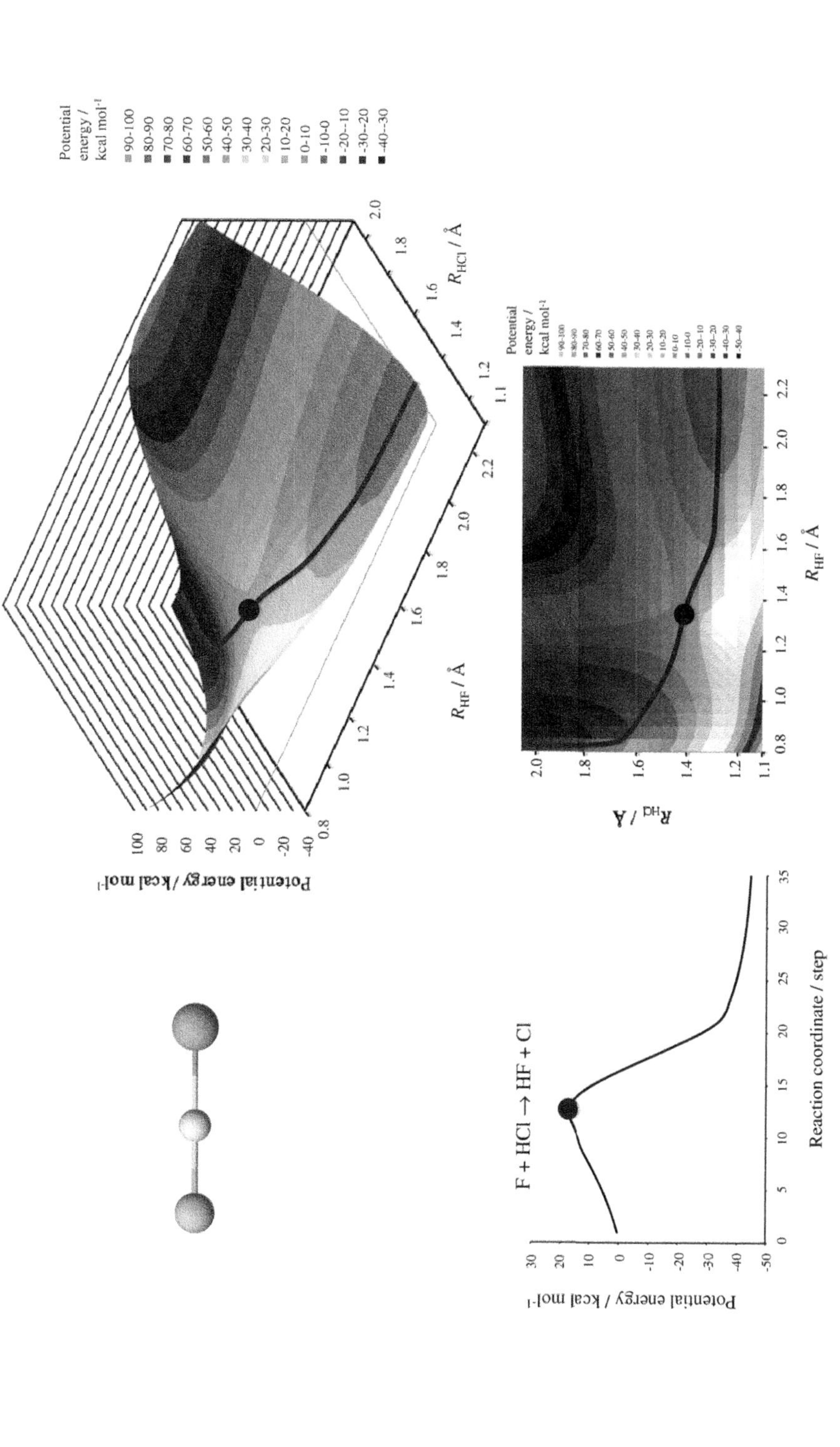

Fig. 4 Combination CG

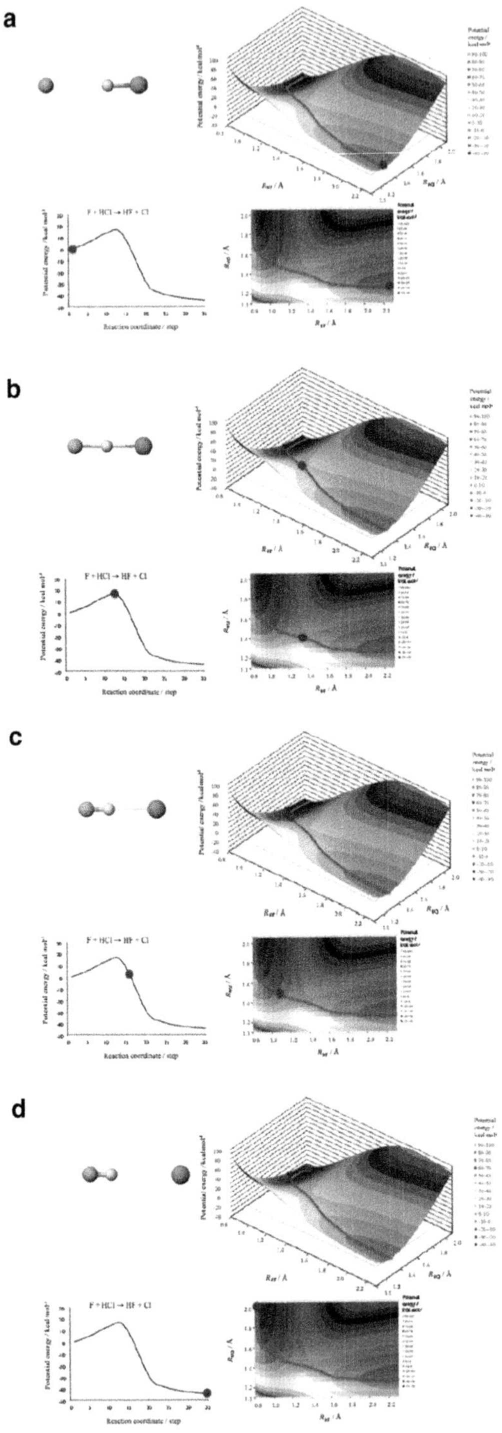

Fig. 5 CG animation

This leads up to pertinent acquisition of dynamical images of the chemical reaction. The animation could be applied in the high school, of course in the university, to teach the object of "structure and the chemical equilibrium of the material," which is a unit of senior high school chemistry [8].

4 Conclusion

Geometries and PE for intermediates of "F-H-Cl" around the transition state in the reaction of "F + HCl → HF + Cl" were calculated. The intermediate on the transition state was formed at the point of R_{HF} of 1.376 Å and R_{HCl} of 1.354 Å. This intermediate was supported by method of the IRC. CGs of PE in 2-D and PE surface in 3-D, skeletal structures of intermediates on the way of the reaction, and the reaction profile were created. CG animation composed of four CGs was made in order to express images of the reaction. The animation demonstrates the changes of PE and skeletal structure of the intermediate on the reaction profile in all stages, simultaneously.

References

1. Eyring, H. and Polanyi, M., Z. Phys. Chem., B12, 279 (1913); Sullivan, J. H., J. Chem. Phys., 46(1) 73–78 (1967)
2. Martin, J. D. D. and Hepburn, J. W., J. Chem. Phys., 109, 8139–8142 (1998)
3. Deskevich, M. P., Hayes, M. Y., Takahashi, K., Skodje, R. T. and Nesbitt, D. J., J. Chem. Phys., 124, 224303-1–224303-12 (2006)
4. Tokita, S., Sugiyama, T., Kondo, T. and Kikukawa, T., J. Chem. Software, 5, 129–136 (1999) (In Japanese); Tokita, S., Kido, F., Sugiyama, T., Watanabe, T., Tokita, N. and Azuma, C., J. Chem. Software, 8, 7–16 (2002) (In Japanese)
5. Shishibori, T. and Kurahashi, K., J. Chem. Software, 1, 183–186 (1993) (In Japanese); Shishibori, T., Naya, N., Ido, Y. and Kobayashi, R., J. Chem. Software, 2, 206 (1995) (In Japanese)
6. Ikuo, A., Ichikawa, T. and Teratani, S., J. Chem Software, 6, 45–54 (2000) (In Japanese)
7. Ikuo, A., Ikarashi, Y., Shishido, T. and Ogawa, H., J. Sci. Edu. Jpn., 30 (4), 210–215 (2006)
8. Textbooks of "Chemistry II" in Japanese high school: Daiichigakusyusya (2004); Jikkyosyuppan (2004); Keirinkan (2003); Sanseido (2004); Tokyosyoseki (2004) (In Japanese) Textbooks were listed in alphabetical order.
9. For example: Atkins, P. and Paula, Y.: Atkins Physical Chemistry 7th. Ed., pp. 966–969, Oxford University Press (2002); Moore, W. J.: Physical Chemistry, 4th. Ed., pp. 382–387, Tokyo Kagakudojin (1982) (In Japanese)
10. Fukui, K., J. Phys. Chem., 74, 4161–4163 (1970)
11. Dewar, M. J. S., Zoebisch, E. G., Healy, E. F. and Stewart, J. J. P., J Am. Chem. Soc., 107, 3902–3909 (1985)
12. Stewart, J. J. P., J Comp. Chem. 10, 209–220 (1989); Stewart, J. J. P., J Comp. Chem., 10, 221–264 (1989); Stewart, J. J. P., J Comp. Chem., 12, 320–341 (1991)
13. Maruzen (2004) Chem. Soc. Jpn. ed.: Kagakubinran Kisohen II (Ed), p. 798, (In Japanese)
14. Stewart, J. J. P., J Mol. Model., 13, 1173–1213 (2007)

Indian Chemistry Olympiad Programme: Outcomes of the Decade

S. Ladage

Abstract Indian Chemistry Olympiad programme is completing its first decade in year 2008. Since its inception, the programme has undergone academic and organizational changes. Today, the Indian Chemistry Olympiad programme is a significant educational activity within India and is contributing directly and indirectly towards chemistry education especially at tertiary level. The current paper reviews the growth of the programme and its tangible outcomes.

1 Indian Chemistry Olympiad Programme: Outline of the Programme

Homi Bhabha Centre for Science Education (HBCSE, TIFR) is the nodal centre responsible for implementation of Indian Chemistry Olympiad programme within the country. This programme is funded by Board of Research in Nuclear Sciences (BRNS, DAE), Department of Science and Technology (DST) and Ministry of Human Resource Development (MHRD).

Currently, the selection of the Indian team for forthcoming IChO goes through three stages (Table 1). On an average about 25,000 students appear for stage I examination that is conducted at about 900+ centres across India. The fee charged for this examination is notional and the examination is open to any student studying at higher secondary level (age 16–17 years). The prescribed syllabus for this examination is the National level chemistry syllabus of class XI and XII CBSE syllabus [1] (Central Board of Secondary Education). The thrust of the examination is more on topics from general chemistry. Stage II examination generally has seven to nine questions from areas of physical, analytical, inorganic, organic and introductory biochemistry and the syllabus for this examination is that prescribed for IIT-JEE

S. Ladage
Homi Bhabha Centre for Science Education (HBCSE), Tata Institute of Fundamental Research
V. N. Purav Marg, Mankhurd, Mumbai, 88, India
e-mail: savital@hbcse.tifr.res.in

M. Gupta-Bhowon et al. (eds.), *Chemistry Education in the ICT Age*,
DOI 10.1007/978-1-4020-9732-4_27,

Table 1 Indian Chemistry Olympiad Programme selection stages

Stage I: National Standard Examination in Chemistry (NSEC) (November last week) (duration 2 h)
First screening test consists of 100 multiple-choice questions with negative marking and is conducted at about 900+ centres across India. Any student from XI or XII is eligible to appear for this examination.
Stage II: Indian National Examination in Chemistry (INChO -Theory) (January last week) (duration 4 h).
Top 300 students selected from the NSEC qualify for this examination. It consists of seven to nine challenging problems in different areas of chemistry and is conducted at about 15 centres across the country.
Stage III: Orientation cum Selection Camp at HBCSE (after mid May)
Top 30–35 students selected from INChO (Theory) are invited to HBCSE for a week. Multiple theoretical and experimental examinations are administered to these students and at the end of the camp the Indian Team is selected to participate in the forthcoming IChO.
Stage IV: Pre-departure Training of the Team (after mid June)
Members of the Team are trained at HBCSE for 14 days prior to their departure for IChO.
For detailed information: http://www.hbcse.tifr.res.in/olympiads/

test (Indian Institute of Technology joint entrance test, the well established examination within the country with high academic standards). The CBSE and IIT-JEE syllabi [2] have considerable overlap with the theoretical syllabus of the final IChO examination [3]. Thus, most of the students who qualify for second stage of chemistry Olympiad are those who are studying for the IIT-JEE examination. The second stage is generally conducted at about 15 centres across India and this stage onwards no fee is charged from any student. The third selection round is conducted centrally at HBCSE and at this final selection round students face experimental examinations along with the theoretical examinations. In India, syllabi at secondary school and at higher secondary level are theoretical and thus, students hardly perform any experiments as part of their syllabi. Even the students who get selected for the final selection stage have little exposure to the experimental component of chemistry. The entrance examinations like IIT-JEE primarily test students regarding the theoretical components and thus overall, even though students are well prepared for the theoretical areas, they hardly have exposure to the experimental areas. Therefore, all students are familarized with glassware, instruments and general laboratory techniques prior to the experimental examinations at the final selection round. In fact, the experimental examinations are the characteristic feature of the Chemistry Olympiad programme and are often highly appreciated by participating students. The Chemistry Olympiad final selection tests are the only examinations within the country that assess the students on experimental domain in chemistry.

The selection examinations of Chemistry Olympiad programme are generating challenging theoretical and experimental questions in chemistry. The contextual theoretical questions set for Indian National Chemistry Olympiad (INChO) pertain to applications of concepts in chemistry to real situations. Exposure to such problems is one of the effective ways for developing students' interest in chemistry, especially when students at senior secondary level perceive chemistry as boring, abstract, sterile and non-relevant subject (as it is perceived not to have direct connection to engineering and medical streams). All these questions are available to any student or teacher as e-material and also in print format. As a part of Chemistry Olympiad activity, Chemistry Olympiad Awareness programmes are conducted in the regions that have low enrollments for Chemistry Olympiad stage I examination. In near future this material is expected to reach wider population of students and teachers across India through the educational portal that is an undertaking of the Government of India.

The experimental problems developed for the Chemistry Olympiad selection tests and training camps are a significant outcome of the Chemistry Olympiad programme. These experiments assess students for (i) experimental skills and (ii) comprehension of the experimental procedures. Attempts are made to keep the technical demands of the experiments designed for the selection tests as minimum as possible and also the experiments are standardized rigorously. This format of the experiments is adaptable to any regular undergraduate laboratory as it can be used for a large group and is useful to initiate discussion about the experiments. Thus, if used appropriately it will help in generating students' interest in the experimental component and also in better comprehension of the experiments. These experimental problems have considerably shaped the laboratory curricula of the new major initiatives set up within the country for attracting students towards careers in basic sciences. Attempts are also being made to implement these laboratory tasks (partly) in regular undergraduate chemistry laboratory so as to understand how students perceive these laboratory tasks as compared to their regular experiments.

2 Performance of Indian Teams at IChOs

Till date, all Indian students who have represented India at IChOs have come back with medals. Table 2 summarizes the performance of Indian Teams. Only few participating countries have such consistent performance at IChOs. The main reason why Indian students are not securing the top positions is the inadequate experience of handling the laboratory activities. Students selected in the Indian Team are exposed to rigorous experimental training only during the pre-departure camps. Since they are less used to working in the laboratory, overall Indian students are less confident regarding chemistry experiments. Thus, Indian students cannot take

Table 2 Indian Performance at IChOs

1999:	Two silver medals and two bronze medals (31st IChO, Bangkok, Thailand)
2000:	Two silver medals and two bronze medals (32nd IChO, Copenhagen, Denmark)
2001:	One gold medal and three silver medals (33rd IChO, Mumbai, India)
2002:	Two gold and two bronze medals (34th IChO, Groningen, The Netherlands) (One of the Indian Students stood first in experimental examination and secured overall 3rd rank)
2003:	Two gold and two silver medals (35th IChO, Athens, Greece) (One of the Indian Students secured overall 2nd rank)
2004:	One gold, one silver, two bronze medals (36th IChO, Kiel, Germany) (only year when female student was selected in the team and the student received gold medal)
2005:	Three silver, one bronze medals (37th IChO, Taipei Taiwan,)
2006:	One gold, two silver, one bronze medals (38th IChO, Gyoegnsan, Korea)
2007:	Two gold, one silver, one bronze (39th IChO, Moscow, Russian Federation)

quick decisions in chemistry laboratory especially under stressed situations. They often are less comfortable in handling the new type of glasswares or instruments (different than that handled by them) or when they have to perform a particular laboratory operation differently.

3 Resource Generation Camps (RGC) for Chemistry Teachers

Workshops for chemistry teachers are another important activity of the Chemistry Olympiad programme. This activity was initiated in year 2004. The camp introduces the participants to the theoretical and experimental Olympiad problems and presents an opportunity to teachers to generate problems through workshop mode.

The participants of the camp are from different states of the country and often most of them find this activity as an educative and challenging exercise. Tables 3 and 4 present the details of the RGCs conducted so far and number of questions generated during these RGCs respectively.

Generally the RGCs have duration of 4–5 days and the activity will be expanded further in near future. Some of the best questions set during RGCs are selected for the selection tests of Chemistry Olympiad and promising teachers accompany the Indian Teams for the final IChO event. Such teachers then become integral part of the Chemistry Olympiad programme. In future, attempts will be made to increase the number of RGCs at zonal level and to invite teachers from rural parts of the country. The teachers are also encouraged to organize such programmes at zonal level. These camps are helping in networking of chemistry teachers across India.

4 National Initiative on Undergraduate Sciences Programme (NIUS) in Chemistry

As an extension of Chemistry Olympiad activity, a programme called the National Initiative on Undergraduate Sciences (NIUS) is initiated by HBCSE. This programme is in its initial stages and still shaping up. NIUS is open to any promising undergraduate student including the students who qualify for the final selection round of Chemistry Olympiad. Currently the NIUS chemistry programme offers research projects in computational, interfacial and organic chemistry. From year 2009, the programme is expected to attain its full strength. The thrust of the programme is to encourage students to work as independent investigators and thus nurture their talents. Such opportunities are not an integral part of the existing undergraduate chemistry curricula within the country. These opportunities will expose students to some of the frontier areas and will be useful in sustaining their interest in basic sciences.

Table 3 Details of RGCs conducted

Year	Period	Type	Total participants
Camps Held At HBCSE			
2004/05	November 04	Theory	15
2004/05	February 05	Experimental	15
		Total	30
2005/06	September 05	Experimental	09
2005/06	November 05	Theory	15
		Total	24
2006/07	September 06	Theory	15
2006/07	March 07	Experimental	10
		Total	25
2007/08	October 07	Theory	6
Camps Held Outside HBCSE			
2005/06	January 06	Theory	25
2006/07	October 06	Theory	12

Table 4 Number of questions generated

Year	Period	Type	Total
2004/05	November 04	Theory	11
2004/05	February 05	Experimental	09
		Total	20
2005/06	September 05	Experimental	03
2005/06	November 05	Theory	14
		Total	17
2006/07	September 06	Theory	08
2006/07	March 07	Experimental	07
		Total	15
2007/08	October 07	Theory	08
		Total	08
2005/06	January 06	Theory	11
2006/07	October 06	Theory	04

Organizing seminars/workshops for chemistry teachers teaching at tertiary level is yet another activity that is planned under NIUS Chemistry Programme. These workshops will focus on content enrichment and pedagogical inputs and also expose teachers to the frontier areas of research. Generating quality materials for chemistry education at tertiary level is an important mandate of the NIUS programme. In other words, the NIUS programme is trying to create a platform where teachers, students and researchers will have fruitful interactions.

5 Indian Association of Chemistry Teachers

The most significant outcome of the Chemistry Olympiad Programme is the establishment of Indian Association of Chemistry Teachers (IACT) within the country. Currently IACT has more than 650 life members and holds annual conventions in different parts of the countries. It has six zones across the country and conducts and supports seminars related to research in chemistry or chemistry education organized at zonal levels. IACT is now responsible for setting paper for the first selection stage of the Chemistry Olympiad programme and also conducts the Chemistry Olympiad awareness programme both for teachers and students. It is keen on planning activities for chemistry education and is working towards this goal.

6 Summary

Over a decade, the Indian Chemistry Olympiad programme has become a significant educational activity within India. However, some of the goals are still not achieved by the programme. Currently even though the programme has spread across India, it is still confined to urban parts of the country. Also the number of female students appearing for selection tests is small as compared to male students. Till today, most of the students who have participated in IChOs have not taken careers in chemistry. In other words, the programme is contributing towards chemistry education especially at tertiary level, but is still not an integral part of the mainstream chemistry education at secondary school level and efforts in this direction are already on.

(The author of the paper is the coordinator of Indian Chemistry Olympiad programme)

References

1. Central Board of Secondary Education syllabus (CBSE). http://cbse.nic.in/senior curriculum 2010 vol-1.pdf. Accessed on May 2008
2. Indian Institute of Technology joint entrance test syllabus (IIT JEE). http://www.jee.iitb.ac.in/chemistry.html. Accessed on May 2008
3. International Chemistry Olympiad IChO theoretical syllabus. http://www.icho.sk. Accessed on May 2008

Indian National Chemistry Olympiad Examination: Implications for Teaching and Learning of Chemistry

S. Ladage and S. Narvekar

Abstract The selection process of Indian teams for the International Chemistry Olympiad (IChO) currently goes through three stages. Indian National Chemistry Olympiad (INChO) is the national level theoretical examination administered at the second stage of selection and around 300 students selected from initial screening examination qualify for INChO. The current paper reviews the INChO examinations for (i) the areas that are generally covered and the weightage allocated to them, (ii) its correlation with existing secondary school chemistry syllabi within India and with IChO theoretical syllabi, (iii) the prototypical performance of students on these examinations and (iv) its implications for teaching and learning of chemistry at senior secondary level.

1 Introduction

In year 1999, India started participating in the International Chemistry Olympiad (IChO). As compared to other Asian and European countries, India is a late entrant to IChO. From inception the Indian Chemistry Olympiad programme has undergone several changes, for both the academic and administrative aspects. Currently the selection process of Indian teams for IChO goes through three stages. Indian National Chemistry Olympiad (INChO) is the National level theoretical examination administered at the second stage of selection and around 300 students selected from initial screening examination qualify for INChO. This examination plays a crucial role in selection of the group (30–35 students) from which finally the Indian team is selected for forthcoming IChO.

In India, the Chemistry Olympiad examination can be taken by any student studying at higher secondary level (i.e. class XI and XII, age group 16–17 years). Most of the students that qualify for the INChO examination are from CBSE stream [1]

S. Ladage (✉)
Homi Bhabha Centre for Science Education Tata Institute of Fundamental Research
V. N. Purav Marg, Mankhurd, Mumbai 88, India
e-mail: savital@hbcse.tifr.res.in; saa@hbcse.tifr.res.in

M. Gupta-Bhowon et al. (eds.), *Chemistry Education in the ICT Age*,
DOI 10.1007/978-1-4020-9732-4_28, © Springer Science + Business Media B.V. 2009

(Central Board of Secondary Education syllabus). The CBSE is the National level syllabus and is more comprehensive as compared to that followed at the state levels. Most of these students are also preparing for the Indian Institute of Technology joint entrance test (IIT-JEE, a well established examination of high academic standards for engineering stream) or other equivalent tests. The syllabus for IIT-JEE [2] is more advanced as compared to CBSE. Further, there is good overlap between these syllabi and the theoretical syllabus for final IChO competition. Table 1 indicates only those theoretical topics that are part of the final IChO theoretical syllabus [3] but are partly covered or not covered by the CBSE and/or IIT-JEE syllabi.

Table 1 Comparison of theoretical syllabus of IChO with CBSE and IIT-JEE syllabi

Topics listed in theoretical syllabus of IChO*	CBSE	IIT
Rate law		
Differential rate law	No	No
Integrated rate law	No	No
Rate constant for second and third order reactions	No	No
Reaction mechanisms		
Opposing parallel and consecutive reactions	No	No
Spectroscopy		
UV/visible		
Identification of aromatic compound	No	No
Identification of chromophore	No	No
Dyes: colour vs structure	No	No
Beer's Law	No	No
Infrared		
Interpretation using a table of frequencies	No	No
Recognition of hydrogen bonds	No	No
X-ray		
Bragg's Law	No	No
NMR		
General concepts	No	No
Chemical shift	No	No
Spin–spin coupling and coupling constants	No	No
Integration	No	No
Interpretation of a simple ^1H spectrum (like ethanol)	No	No
Identification of *o*- and *p*-disubstituted benzene	No	No
Interpretation of simple spectra of ^{13}C (proton decoupled)	No	No
Other spin nuclei	No	No
Mass spectrometry		
Recognition of molecular ion	No	No
Recognition of fragments with the help of a table	No	No
Recognition of typical isotope distribution	No	No
Organic Chemistry		

(continued)

Table 1 (continued)

Topics listed in theoretical syllabus of IChO*	CBSE	IIT
Cis-Trans	No	Yes
E/Z	No	Yes
Enantiomers	No	Yes
R/S nomenclature	No	Yes
Cycloalkanes		
Mechanism involving carbo-cation intermediates	No	No
Relative stability of carbo-cations	No	Yes
1,4 addition to dienes	No	No
Halogen compounds		
Elimination reactions	No	Yes
Competition of elimination and substitution	No	Yes
Aldehydes and ketones		
Keto/enol tautomerism	No	Yes
Enolate anions (aldol condensation)	No	Yes
Alcohols to form acetals/ketals	No	Yes
Carboxylic acids and their derivatives		
Preparation of carboxylic acids by hydrolysis of Amide and nitriles	No	Yes
Reaction of acid chlorides to form amides	No	Yes
Mechanism of esterification	No	Yes
Multifunctional acids (hydroxyacids, ketoacids)	No	No
Polycarboxylic acids	No	No
Polymers		
Polyethene	No	No
Silicones	No	Yes
Biochemistry		
Carbohydrates		
Chain formulae	No	No
Fischer projections	No	No
Haworth formulae	No	No
Difference between α- and β-D-glucose	No	No
Fats		
Structure of fats in relationship to properties	No	No
Formula of glycerol	No	No
Amino acids		
Ionic structure	No	No
Isoelectric point	No	No
Sequence analysis	No	No
Enzymes		
Nomenclature, kinetics, coenzymes, function of ATP	No	No
Analytical chemistry		
Chromatographic methods of separation	No	No

*Currently the IChO theory syllabus is undergoing revision

Thus, the spectroscopy topics and biochemistry topics such as carbohydrates, fats and part of amino acids are not covered by both CBSE and IIT-JEE chemistry syllabi. For topics in organic chemistry, cycloalkanes and mechanisms involving carbo-cation intermediates are only two topics that are not taught both by CBSE and IIT-JEE. Under Analytical chemistry, the chromatographic techniques are not covered by CBSE and IIT-JEE chemistry syllabi.

2 Analysis of the INChO Papers

From its inception, the INChO has steadily evolved to its current pattern and today, on an average it covers seven to nine questions. These questions cover different branches of chemistry and sometimes a small question related to introductory biochemistry is also included in INChO paper. The total marks are 120 (sometimes the marks marginally deviate from 120). For the last 2–3 years, occasionally small subparts related to areas that are not taught at higher secondary level are also included in the questions, for example, questions related to spectroscopy topic in organic chemistry. Students appearing for this examination are aware about these subparts as past INChO papers are available to students online. Wherever applicable attempts are made to design context-based questions and often INChO questions cover interrelated concepts and their applications to unknown situations. Each question has subparts that vary in difficulty level and the expected answers are often devoid of prose. The duration of the INChO test is 4 h and the examination is conducted at about 15 centres across India. Generally this examination is scheduled in last week of January as by this time, students have practically completed their academic calendar. As a prototype, the theoretical areas of INChO 2006 are discussed in Table 2 and this list will exemplify the areas covered for INChO in general.

Generally, the INChO papers cover topics such as chemical equilibrium, chemical kinetics, chemical bonding and chemical thermodynamics as part of physical and analytical chemistry. The organic chemistry questions test students regarding functional group transformations, reactivity of compounds, stereochemistry (assigning stereo descriptors and drawing of different projection formulae), IUPAC naming and reaction mechanisms. As stated earlier, the questions regarding spectroscopy are sometimes included but often have less weightage and whenever asked all the required information is generally provided as part of the question. Inorganic chemistry either covers chemistry of main group elements and/or transition elements. Students are assessed for concepts such as chemical bonding, drawing of Lewis dot structures, co-ordination complexes, their geometries and isomerism, characteristic reactions of main group and transition elements and reactions involved in extractions of metals. Overall the questions at INChO comprehensively represent the core concepts of physical, analytical, organic and inorganic chemistry.

Figure 1 indicates the distribution of marks for INChO theoretical papers of academic year 2006, 2007 and 2008. Every year, the pattern of marks obtained is more or less similar and generally a Gaussian distribution of the scores is observed.

Table 2 Theoretical areas of INChO 2006

2006	Details of question/s	Concepts covered
Organic 2 questions 21 + 17 = 38 marks	Chemistry of Bombykol an insect pheromone Chemistry of carbon acids	Identification of intermediates and final products for ozonolysis and Wittig reactions, stereochemistry, IUPAC naming of organic compounds, application of NMR and IR spectra for structure elucidation, acid-base concepts, acidity of carbon acids and their reactions (condensation and cycloaddition), concept of resonance and Tautomerism
Physical 14 marks	Chemical kinetics of Nitrogen Oxides	Expression for rate laws and rate constant (using data), calculations of rate constant and use of steady state approximations to identify correct reaction mechanism/s
Physical 18 marks	Molecular Hydrogen	Chemical equilibrium constant calculations and its qualitative interpretation, equilibrium compositions and factors affecting these compositions, use of C_p, enthalpy and entropy data for equilibrium constant calculation at different temperatures
Analytical 13 marks	Solubility equilibrium	Equilibria of sparingly soluble salts, precipitation titrations and calculations of cation concentration during the course of titrations and indicators used in precipitation titrations
Analytical 10 marks	Electrochemistry	Emf calculations, Faraday Law, cell reactions involved in corrosion
Inorganic 10 marks	Chemistry of Boron	Trends in Acidic character of boron halides, Lewis structures, bonding in diboranes, borazine and other borane entities, reactions of diboranes
Inorganic 23 marks	Chemistry of Chromium and Nickel	Coordination complexes and their IUPAC naming, use of VBT for drawing orbital energy diagram, isomerism in co-ordination complexes, colours of the complexes and crystal field splitting energy, reactions involved in extraction of metals from chromite ore

On an average, the marks of the students who are selected for the final selection stage (top 30–35 students) are in the range 80–105 (66–87%).

To have better understanding of general errors committed by students, detailed analysis of each question for INChO 2006 is presented in Fig. 2. This figure indicates the distribution of marks obtained by students for each question. Overall, distribution of marks for question 1 (Bombykol, organic), question 4 (solubility equilibrium, analytical) and question 5 (molecular hydrogen, physical) are skewed

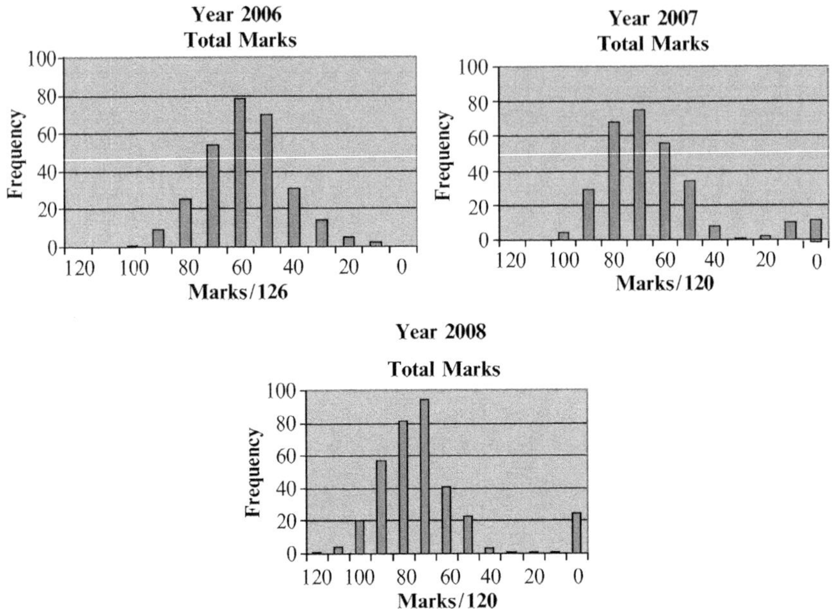

Fig. 1 Distribution of Marks for INCho (2006, 2007, 2008)

towards lower side indicating that these questions are more difficult than other questions. For questions 2 (carbon acid, organic), 3 (chemical kinetics) and 8 (co-ordination chemistry, inorganic), Gaussian distribution is obtained. Bimodal distribution is obtained for questions 6 (electrochemistry, analytical) and 7 (boron chemistry). The bimodal distribution indicates that either student secures full marks or 50% of the total indicating that the subparts of the question are not designed properly. For question related to Bombykol, particularly the subparts related to Wittig reaction could not be answered completely by students. This reaction is not part of higher secondary syllabus and thus it is understood that students could not solve all the subparts related to Wittig reactions. Same was observed for subparts related to use of NMR data. For question related to boron chemistry, students failed to answer the subparts related to reactions of diborane, structure of borazine and its reaction with HBr. For question 5, students could not use the Cp data for calculation of equilibrium constant at different temperature and since further subparts were dependent on this calculation and thus were left blank. Often care is taken to avoid questions with completely interdependent subparts but sometimes it is unavoidable. For question related to solubility equilibrium (question 4), students could not calculate the concentration of different entities present in reaction mixture during the course of titration.

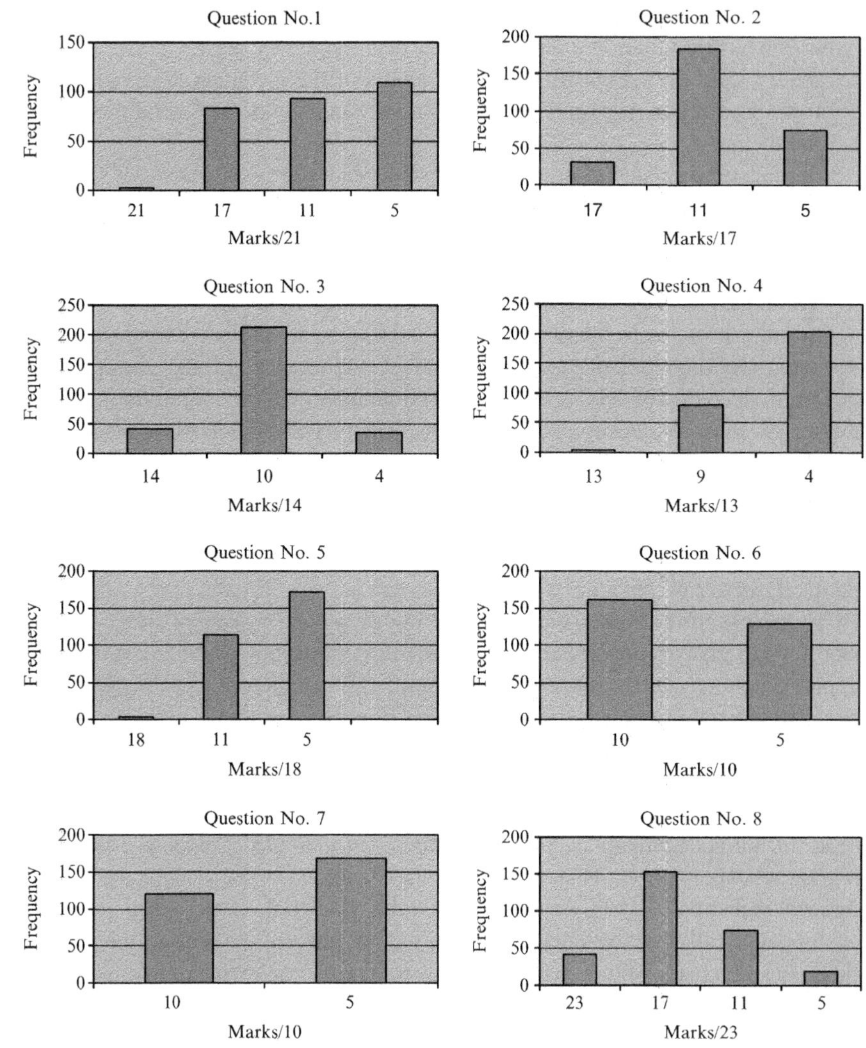

Fig. 2 Analysis of questions (INChO 2006)

3 Conclusions

In the final IChO, performance of Indian students in theoretical examination is good as compared to their performance in the experimental examination. On an average, students secure 45–55 marks out of 60.

Overall, Indian students find the area of multiple equilibria as one of the most difficult topic. In fact over the years it is observed that Indian students often secure less mark on questions related to multiple equilibria. This area demands good conceptual understanding and students fumble with the approximations to be made under the given situations. Generally the discussions in the textbooks at higher secondary level do not clarify the exact details of the given equilibria and further there are no discussions regarding the approximations to be made and the rationale behind them.

For questions related to organic chemistry, generally performance of students is good. However, Indian students perform relatively poor on the topics listed below: stereochemical representations (Wedge and Fischer projections), functional group transformation (mostly the carbonyl chemistry) and spectroscopy. The stereochemical representation demands orientation and rotations of molecules in space and their appropriate projections in two-dimensional frame. At the pre-departure training sessions, maximum practice is given to students in this area, still the concepts cannot be internalized completely over such short time interval. Also, sometimes the information given in the question is not analyzed properly and this leads to trivial errors. The biochemistry questions when asked (at IChO) often involve carbohydrate and protein chemistry. In this area, students make representational errors (e.g. Haworth projections, hydrogen bonding interactions in protein /nucleic acids). Despite the fact that students have least exposure to spectroscopy topics, Indian students fare relatively better in this area.

Thus, at least for Indian situations, the higher secondary school chemistry syllabus should give more practice to solve the equilibrium problems without any approximations and then the equilibria are altered in different ways so as to understand whether the problem can be solved with approximations. Multiple examples (involving acid-base and precipitation reactions) need to be included in discussions so as to develop comprehensive understanding of the topic. Organic chemistry requires more emphasis on areas of stereochemistry and reaction mechanisms. The topic of stereochemistry is normally taught without using models especially at higher secondary level. Hence, students face difficulties in visualizing the three dimensional structures of the molecules and thus the whole exercise becomes totally abstract. Overall, Indian students perform well in most of the areas of physical and inorganic chemistry. While studying chemistry, use of contextual problems is important as they explain the applications of chemical concepts to real situations. Thus, such problems should become an integral part of the chemistry syllabi. The chemistry education within the country also needs to emphasise the experimental areas and must engage students in experiments for meaningful learning of chemistry.

(Savita Ladage is the coordinator of Indian Chemistry Olympiad programme)

References

1. Central Board of Secondary Education syllabus (CBSE). http://cbse.nic.in/senior curriculum 2010 vol-1.pdf. Accessed on May 2008
2. Indian Institute of Technology joint entrance test syllabus (IIT JEE). http://www.jee.iitb.ac.in/chemistry.html. Accessed on May 2008
3. International Chemistry Olympiad IChO theoretical syllabus. http://www.icho.sk. Accessed on May 2008

Laboratory Courses in Organic Chemistry: A Case Study

L. Ravishankar and S. Ladage

Abstract The laboratory courses in chemistry at tertiary level are expected to (i) expose students to experimental techniques and skills, (ii) allow sufficient practice to master these techniques and (iii) also provide opportunities to students as investigators of the experimental work. The present paper discusses and analyses the laboratory courses in organic chemistry, prescribed by Mumbai University, one of the oldest Universities of India for (i) the content, (ii) skills/technical demands of the experiments, (iii) weightage with respect to time and marks, (iv) evaluation criteria used for assessment of students, and (v) the changes it has undergone over time.

1 Introduction

Today, with constructivist philosophy, it is well accepted that learners are active constructors of knowledge. The active construction of knowledge by the learners is through interactions with the environment. Thus, both learner and the environment are important for meaningful learning. The word *environment* refers to those external influences that interact with the learner during the learning process [1]. In the current study the environment will refer to laboratory courses in organic chemistry and the general ways in which they are implemented in the laboratory.

A good practical course in chemistry at tertiary level should help in development of the core skills such as problem solving, planning of experiment (including risk and safety of glassware, instruments, chemicals) and evaluation skills (observation, data collection, interpretation and reporting). Along with the core skills, a practical course should also familiarize the students with transferable skills such as time management, group work, literature search and use of chemical databases, use of computers and common software packages [2].

L. Ravishankar (✉)
Department of Chemistry, V.G.Vaze College of Arts, Science and Commerce, Mumbai–400081
e-mail: lakshravi@yahoo.com

S. Ladage
Homi Bhabha Centre for Science Education, Mumbai, 400088
e-mails: savital@hbcse.tifr.res.in; savitaladage@gmail.com

M. Gupta-Bhowon et al. (eds.), *Chemistry Education in the ICT Age*, 325
DOI 10.1007/978-1-4020-9732-4_29, © Springer Science+Business Media B.V. 2009

The adopted styles for laboratory instructions/implementation can be either expository, inquiry-based, discovery-based, or problem-based [3].

The expository style, also known as the 'cookbook recipe' style, is the most commonly used style of laboratory instruction [4]. It has been criticized heavily as it places very little emphasis on the students' thinking and creativity. Though this style helps students in exercising manipulative and data gathering skills, it fails to provide opportunities for independent design and planning of the experiment (Table 1). Also, such a style offers little motivation and stimulus to students. The other three styles present more opportunities to students as independent investigators. Nevertheless, the expository style is still widely implemented, as it can be used for large numbers, is economical with respect to infrastructure, resources, time, chemicals and easy to supervise and to grade [5]. However, attempts are being made to modify the expository style so as to incorporate the advantages of the other styles, at least partly [6–9].

In the current study, the organic chemistry laboratory courses of Mumbai University at the undergraduate and postgraduate level are analysed for various parameters. Mumbai University is one of the oldest Universities of India with 55 affiliated science colleges, having diverse conditions of infrastructure, students and teachers. At the postgraduate level, organic chemistry is the most preferred branch of chemistry and is offered in 38 affiliated colleges with more than 50% seats allocated to it. In others words, almost 50% of students at postgraduate level major in organic chemistry. The long-term objective that authors are interested in is to study feasible changes in the existing laboratory courses and the instructional practices which will lead to meaningful learning.

2 Analysis of Organic Chemistry Laboratory Courses

As a part of the undergraduate (UG) and postgraduate (PG) programme, students who major in organic chemistry take the laboratory courses for 5 continuous years. The University provides the syllabi and a common evaluation scheme for both UG and PG laboratory courses. However, the objectives of the syllabus are not mentioned by the university. The syllabi are regularly revised every 5 years. The current study analyses the organic chemistry laboratory courses for (i) the content, (ii) skills/technical demands of the experiments, (iii) weightage allocated to different type of experiments with respect to time and marks, (iv) evaluation criteria used for assessment of students, and (v) major changes the syllabi have undergone over time.

Table 1 Laboratory instruction styles

Style	Outcome	Approach	Procedure
Expository	Predetermined	Deductive	Given
Inquiry	Undetermined	Inductive	Student generated
Discovery	Predetermined	Inductive	Given
Problem-based	Predetermined	Deductive	Student generated

2.1 Content

The practical syllabus in organic chemistry as prescribed by Mumbai University introduces various experiments in a phased manner. Table 2 lists the type of experiments included in undergraduate (UG) and postgraduate (PG) syllabi.

At the first year of UG, only two types of experiments, namely (i) qualitative analysis and (ii) purification by recrystallisation followed by determination of melting point of the purified product are part of the syllabus. From the third year of UG onwards, recrystallisation is a part of synthesis or qualitative analysis experiments. Compounds given for qualitative analysis cover a wide spectrum of mono-functional groups and students normally perform about 12 organic spottings in the first year. Most of these reactions covered are familiar to the students. Thus, by the end of the first year the students are well versed with (a) the systematic qualitative analysis of organic compounds, (b) the techniques of recrystallisation, and (c) recording of physical constant.

During the second year of UG, simple one step organic synthesis (e.g. preparation of bromo or nitro derivative) and quantitative estimations based on specific reactions (e.g. estimation of phenol/aniline by bromination) are introduced along with qualitative analysis. The qualitative analysis is at micro-scale level and is more complex in UG second year, as compounds with multifunctional groups are given for analysis. In addition to characterisation, students have to identify the compound by performing the confirmatory tests.

Both synthesis and estimation emphasise the quantification aspect along with the study of organic reactions. Generally, students associate titration with inorganic chemistry. Estimation of organic compounds using titration helps in overcoming this notion.

In the third year of UG, no new type of experiment is introduced except distillation technique. The complexity of qualitative experiments is increased further by providing a binary mixture that demands separation by physical or chemical methods. This exercise introduces students to the techniques of chemical and physical separation of an organic mixture along with quantification, purification and identification. Further, the concept of recovery of the product is emphasised as a part of this experiment.

Table 2 Type of experiments in UG and PG courses

	UG	UG	UG	PG	PG
Type	1st year	2nd year	3rd year	1st year	2nd year
Qualitative analysis	+	+	+ (binary mixture)	+ (binary mixture)	+ (ternary mixture)
Synthesis	–	+	+	+	+
Quantitative analysis	–	+	+	–	–
Recrystallisation	+	–	–	–	–
Advanced separation techniques	–	–	–	–	+

+ = included; – = excluded

Separation of a binary mixture by micro-analytical techniques and one-step preparations continues in the first year PG practical organic syllabus. Students purify both the isolated compounds of the binary mixture and check their purity by Thin Layer Chromatography (TLC). One of the components is identified and its identity confirmed by derivatisation. The single-step organic preparations are more comprehensive and involve various classes of reactions, including synthesis of heterocyclic compounds. About 10–12 preparations are performed by the students. The PG second year involves the separation of a ternary mixture. Further, single-step preparations involving purification by column chromatography, vacuum sublimation, steam distillation or vacuum distillation are also included in second year of PG syllabus.

Spectral analysis is an integral part of any organic synthesis. The facilities for recording the spectra of the synthesised compounds do not exist in colleges; nevertheless students are familiarised with the interpretation techniques through interpretation of at least ten unknown spectra. The spectra include IR, UV, Mass, PMR and CMR data. This exercise provides a hands on experience in spectral identification of organic compounds.

A 30-marks project component was introduced at the PG second year level in the last revision (2002–2007) so that students get opportunities to do small research projects (duration 6–8 weeks) either at their institution or in an industry. Students are expected to write the report of the work done and also present the same. Most often, students are placed for industrial training in the Research and Development or Quality Control department of a company. The exposure received by students is very limited as often the companies use students as data collectors without revealing the purpose for which the data are collected. While the students are demonstrated the working of various analytical instruments like GC, HPLC etc., normally they are not allowed to handle them. Often due to lack of advanced infrastructure and funding, it is rather difficult for students to do a research project at their parent institution.

Thus, the syllabus covers different types of experiments along with introduction to spectroscopy and its applications in structure elucidation. Throughout 5 years, students get enough opportunities to master the skills involved in various types of experiments.

2.2 *Weightage (w.r.t. time and marks)*

The total duration of practical for the first year, second year and third year of UG is 6, 9 and 16 periods per week respectively. Table 3 presents allocation of time per week for different types of organic experiments.

Table 3 Time allocation for experiments

Type	UG	UG	UG	PG	PG
	1st year	2nd year	3rd year	1st year	2nd year
Qualitative analysis	3	4.5	4	4	4–6
Organic synthesis	–	2	4	4	4–6
Quantitative analysis	–	4.5	4	–	–

The figures in the table indicate the periods allocated to the experiments. At UG, one period = 48 min and at PG one period = 60 min.

The practical examinations are conducted annually. The pattern of examination is prescribed by the University, so as to maintain uniformity in all the affiliated colleges. While these laboratory examinations test the students for the various skills that are learnt as a part of the course, they do not, in the true sense present any unknown or novel situations where the learnt skills or techniques can be applied by the students.

University provides details regarding the type of experiments to be covered in the examination along with allocation of time and marks. Table 4 presents the distribution of marks in the final practical examination. The total marks are further subdivided for the different important steps of an experiment. Currently, the viva component (15 min) is an integral part of the laboratory examination to test students' understanding of the theoretical component of the experiments. Details of the type of experiments that are included for examination along with their respective time duration are presented in Table 5.

The weightage for organic experimental examination varies from year to year. For first year of UG, out of total 80 marks, 20 marks are allocated to experiments in organic chemistry. At the second year of UG, 20 marks out of 120 are allocated

Table 4 Marks allocation for experiments

Type	UG	UG	UG	PG	PG
	1st year	2nd year	3rd year	1st year	2nd year
Qualitative analysis	20	15	25	20	60
Recrystallisation	20	–	–	–	–
Synthesis	–	5	15	20	60
Quantitative analysis	–	12	20	–	–
Spectral problems	–	–	–	–	30
Project	–	–	–	–	30
Viva	10*	10*	6	5	10

*General viva covering physical, organic and inorganic chemistry

Table 5 Type of experiment/s with allocation of time

	Type of experiment/s	Duration
UG 1st year	Qualitative analysis or recrystallisation	3 h
UG 2nd year	Qualitative analysis and synthesis	3 h
	Quantitative analysis*	3 h
UG 3rd year	Separation of binary mixture & qualitative analysis	3 h
	Synthesis	3 h
	Quantitative analysis*	3 h
PG 1st year	Separation of binary mixture & qualitative analysis	3 h
	Synthesis	3 h
PG 2nd year	Separation of ternary mixture & qualitative analysis	6 h
	Synthesis	6 h
	Spectral problem	3 h
	Project presentation	30 min

*Optional instead of volumetric analysis

for organic chemistry. In case a student gets quantitative analysis involving organic estimation in place of volumetric analysis then the weightage for organic increases from 20 to 32. In the third year of UG, 50 marks out of 200 are allocated to organic chemistry. Similarly, if a student gets quantitative analysis involving organic estimation then the weightage increases from 50 to 70 marks. In the first year of PG, students study physical, analytical, inorganic and organic chemistry. Thus the total of 200 marks for laboratory examination is equally divided among all the four branches. In the final year of PG, as students specialise in organic chemistry, 200 marks (i.e. 40% of the total marks for theory and experiments) are for experiments in organic chemistry.

2.3 *Skills Related to Experiments*

The word skill refers either to the actual technique that needs to be learnt and/or understanding of the sub-concepts/principles associated with each type of experiment. Different types of experiments included in organic chemistry practical syllabus are analysed for the skills and the list of skills is presented below.

Qualitative Analysis

- Test-tube reactions
- Use of micro-scale apparatus
- Melting point determination
- Understanding of specific reactions for identification of functional groups and organic compounds
- Separation techniques for mixtures
- Purification techniques (distillation and/or recrystallisation)

Organic Synthesis

- Heating with burners/hot plates and heating of liquids
- Handling inflammable/moisture sensitive substances
- Synthesis in round bottom flask or using micro-scale glassware
- Techniques of distillation (normal/vacuum/steam distillation)
- Filtration and washing techniques (including handling of suction pumps)
- Purification techniques and appropriate drying techniques
- Handling of analytical balance
- Chromatography techniques (TLC and column)

Quantitative Analysis

- Dilution technique
- Titration technique (blank and back titration)
- Colour transitions of indicators
- Acid–alkali and iodometric/iodimetric titration
- Pipetting technique
- Quantitative volumetric determinations

The last type of experiment is spectral analysis that demands basic knowledge of spectroscopy, comparison with standard data and logical interpretation for structure determination of the unknown compound.

2.4 Changes in the Syllabi Over Time

The **syllabus** at UG level has not seen any major changes for the contents except in the last revision (2002–2007), when micro-scale techniques were introduced at second and third year of UG. The syllabus that has undergone major change in this revision is the second year PG syllabus. Greater emphasis has been given to preparatory chemistry and separation using chromatography and other advanced distillation procedures. The project component which was introduced as part of the revised syllabus emphasises the experimental project or placement in an industry (6–8 weeks).

3 Summary

The organic chemistry experimental syllabus trains the students in the various areas of qualitative, quantitative and spectral analysis, besides giving them enough practice in preparatory chemistry. However, most of the experiments are close-ended as the procedure and the expected outcome are stated in laboratory manuals. Even though students on an average gain mastery for the manual skills, their role in the laboratory remains as verifiers of the facts. As a result, very little scope exists for the student to plan and design any experiment. This is true especially for UG syllabus. It is difficult to implement the open-ended activities in a system that is catering to a large number of students. Further, the laboratory space in most colleges is limited and is shared by various classes on a shift basis. Thus, the laboratory is available to students and teachers for limited period of time.

However, in the prevailing system some changes could be introduced to enhance students' participation in the laboratory activities, thus making the laboratory courses more student-centric. One of the changes that can be implemented in the current course is the post-lab discussions. The post-lab discussions are important especially for undergraduate students as it helps in sensitising students to the (i) common avoidable errors and their rectifications and (ii) data evaluation. These sessions will help instructors in understanding students' difficulties and questions regarding the laboratory experiments. Overall, the post-laboratory sessions if handled appropriately will help in building confidence of students. Some of the small activities such as choice of appropriate solvent for recrystallisation can be done in mini-project mode. Similarly some of the regular synthesis can be done either by varying the molar ratios/conditions or by changing the substrates. If few experiments are done in this manner, the syllabus will become more meaningful for learning.

Also digital technology can be used effectively to augment the concepts that are introduced in the wet labs. Today, with the advances in technology, excellent online resources are available to train the students on various practical skills [10–12]. Alternatively, students could be divided into smaller groups and simple experimental problems could be given to a group. The group works on the problem with the help of the instructors, plans the experiment using previous experience, analyses the data obtained and draw their conclusions. This kind of a learning process has been incorporated in some Universities for teaching Organic Chemistry and found to be beneficial [13].

In terms of content, the syllabus is quite exhaustive and comparable to any other University. Where it probably falls short is in the attempts to make it more guided –inquiry-based rather than the expository-based.

(The current study is the first phase of the research project undertaken by the authors. Further work (starting in September 2008) involves survey to understand instructors' and students' feedback regarding the content and evaluation pattern about the existing organic laboratory courses. Subsequently some of the changes discussed above will be tested for their effectiveness.)

References

1. Domin DS (1999) A Review of Laboratory Instruction Styles. J Chem Educ 76: 543–547.
2. Mitchell T, Bridge M, Ellis F et al. (1999) Core Practical Skills. www.cpe.fr/ectnassoc/archives/lib/1999/1999_WGR_TeachLearnPractSkills.pdf. Accessed in May 2008
3. Johnstone AH, Al-Shuaili A (2001) Learning in the laboratory: Some Thoughts from the Literature. U Chem Ed 5: 42–51.
4. Abraham MR, Cracolice MS, Graves AP et al. (1997) The Nature and State of General Chemistry Laboratory Courses Offered by Colleges and Universities in the United States. J Chem Educ 74: 591–594.
5. Montes LD, Rockley MG (2002) Teacher Perceptions in the Selection of Experiments. J Chem Educ 79: 244–247.
6. Gaddis BA, Schoffstall AM (2007) Incorporating Guided-Inquiry Learning into the Organic Chemistry Laboratory. J Chem Educ 84: 848–851.
7. Hollenback JJ, Wixson EN, Geske GD et al. (2006) A New Model for Transitioning Students from the Undergraduate Teaching Laboratory to the Research Laboratory. J Chem Educ 83: 1835–1843.
8. Mohrig JR, Hammond CN, Colby DA (2007) On the Successful Use of Inquiry-Driven Experiments in the Organic Chemistry Laboratory. J Chem Educ 84: 992–998.
9. Vianna JF, Sleet RJ, Johnstone AH (1999) The Use of Mini-Projects in an Undergraduate Laboratory Course in Chemistry. Quimica Nova 22: 138–142.
10. Chittleborough GD, Mocerino M, Treagust DF (2007) Achieving Greater Feedback and Flexibility Using Online Pre-Laboratory Exercises with Non-Major Chemistry Students. J Chem Educ 84: 884–888.
11. Garratt J (1997) Virtual Investigations: Ways to Accelerate Experience. U Chem Ed 1: 19–27.
12. MIT Open Course Ware - http://ocw.mit.edu. Accessed in May 2008
13. Bradley AZ, Ulrich SM, Jones M Jr, Jones SM (2002) Teaching the Sophomore Organic Chemistry Course without a lecture. Are you Crazy? J Chem Educ 79: 514 –519.

Green Chemistry Initiatives Across the Undergraduate Curriculum

H. Traore

Abstract The responsibility of solving the world's future environmental problems will fall into the hands of the world's future scientists. While not all today's chemistry training will focus on careers related to such issues as renewable energy and pollution prevention, it is imperative that all young chemists be prepared to be green chemists. Just as we train undergraduate chemists to understand basic principles of chemistry, we must also teach them to "think green" and keep an eye on such concepts as the 12 principles of green chemistry [1]. Because green chemistry pervades every field in the discipline, it is possible to infuse the concepts of green chemistry into the entire undergraduate chemistry curriculum. At the University of Wisconsin-Whitewater, we have taken some initiatives to create a Green Chemistry Initiative Center (GCIC). The GCIC will have a multidisciplinary approach and three objectives: teaching, research training, and community outreach. In the area of teaching, the GCIC will facilitate a change in curriculum through the incorporation of lab experiments based on the green chemistry innovations recognized by the Presidential Green Chemistry Challenge Awards between 2000 and 2007, into existing classes [2]. In the area of research training, collaborations with local industries and farms through student internships will enhance learning. Community outreach through arts and other activities with surrounding high schools will be used as an awareness tool and to promote good citizenship.

1 Introduction

Green chemistry in higher education implies an institutional mission that integrates this new chemistry concept in all aspects of campus life, from curriculum, research, and community outreach to governance, operations, and professional development. Green chemistry can be defined in many contexts such as sustainability: producing

H. Traore (✉)
Department of Chemistry, University of Wisconsin-Whitewater, WI, 53190
e-mail: traoreh@uww.edu

M. Gupta-Bhowon et al. (eds.), *Chemistry Education in the ICT Age*,
DOI 10.1007/978-1-4020-9732-4_30,

modern materials for society and meeting the needs of future generations through environmental stewardship and the production of better and safer products affordable to all people. This paper will focus on a proposal to integrate green and sustainable chemistry in University of Wisconsin-Whitewater through Green Chemistry Initiatives Center. This program will require a multidisciplinary approach to teaching, research training, and community outreach involving collaborations between the sciences, technology, engineering, and math (STEM) disciplines and economics, political science, sociology, and psychology. The areas of concentration will include "greening" our laboratory experiments, and among others, renewable energy, the use of chemical feedstock from renewable resources, and development of benign, nonpolluting processes. The outcome will be to produce innovative STEM teachers, scientists, and engineers who will produce and practice green chemistry and sustainability solutions.

2 Background

In an article titled "more campuses are practicing than teaching green" published online in the environmental news magazine *Plenty*, Ted O' Callahan [3] reports that 63% of students want to know about a school's commitment to the environment as part of their college decision. He also acknowledges that many college administrators believe environmental issues will be the key challenge facing this generation of students and that schools must prepare them to find solutions. Furthermore, according to O'Callahan, a recent study by the National Wildlife Federation [4] has indicated that overall attention to teaching environmental courses has lagged behind the greening of buildings and grounds. The study shows that only 53% of colleges offer environmental studies majors and minors. O'Callahan justifies this situation by insisting that it is much simpler to take on the tangible task of improving lighting or water efficiency than to wrangle faculty, a notoriously independent group, for a systematic greening of the curriculum. To illustrate his argument, O'Callahan takes a chemistry department as an example. He cites John Warner [5], one of the pioneers of green chemistry. After 10 years of effort to build up a program at the University of Massachusetts-Lowell, John says that academic inertia is probably the largest barrier. There is not so much opposition to green chemistry, per se, as there is resistance to any change at all. When faculty are faced with introducing a subject they themselves never had in their education, some immediately question the need. He added: "Younger untenured faculty typically see the opportunity, but they risk not getting tenure if they criticize the existing system." The number of students interested in the green chemistry program dwarfs the number interested in the rest of the chemistry department, according to Warner. O'Callahan recognizes that many schools are finding ways to introduce green facts to existing endeavors. He cites Julian Dautremont-Smith, Associate Director of the Association for the Advancement of Sustainability in Higher Education (AASHE), confirming some new programs offering grounding in a traditional field and an overlay of sustainability: environmental architecture, sustainable agriculture, or ecological economics. The B-school gut course operations management at Stanford University prepares MBA students to manage sustainability issues by teaching the way Starbucks encourages affair trade and improved farming techniques. Alaska Pacific University or any of the other schools in the Eco League, a consortium of colleges, offers experimental programs with environmental focuses. Goucher College is one of the 4% of schools that have made environmental literacy a mandatory part of the curriculum. While some colleges make it a required part of the first year, Goucher students must simply work one of many environmental courses into their studies before graduating. The University of Wisconsin-Whitewater is offering a forum in which green chemistry will be at the center of initiatives to reach out to the community, including industries and farmers, to address this problem through the Green Chemistry Initiatives Center.

3 Methodology

The implementation of this initiative will focus in three areas.

3.1 Community Outreach

This paper will examine the "three-legged stool" model to explore the potential for increased public awareness and outreach: Green Chemistry (ICE) – Information, Communication, and Education.

Community outreach-special projects:

- Providing an interactive Web-based resource of local case studies on sustainable business practices linked to directories of service providers
- Reaching out to particular sectors of small- and medium-sized businesses on energy efficiency and/or environmental management system projects
- Providing a forum for companies interested in waste minimization and recycling to share ideas and outlets
- Bringing together environmental representatives of institutional organizations such as hospitals, universities, schools, hotels, etc., to share problems and solutions to environmental compliance and issues management
- Providing regular meetings for those interested in green chemistry design

3.2 Teaching

We believe that it is very important that university students be exposed to real-world, state-of-the-art examples of green chemistry in the mainstream courses that they encounter in a typical college chemistry curriculum. Many industries are now practicing green chemical principles. Those students who are versed in green chemistry will be most attractive to these industries, and will be able to foster the practice of green chemistry in these industries, as well as initiate the practice and discussion of green chemistry throughout industry and academia. Our goal is to infuse green chemistry into chemistry courses. Each of the instruction units will have three parts:

- "The module" – this is where the green chemistry topic is discussed, and the instructor will have her/his students go to the Web page to read and study the material.
- "Notes to Instructors" – this suggests how and where the particular module could be used in a particular course, and other courses in which the module might also be used.
- "PowerPoint Presentation" – this can be downloaded by the instructor and the students. This can then be used by the instructor to present the material, and by the students to take notes.

We will develop and place on the Web an introduction to green chemistry and some green chemistry teaching modules. Over the academic years to come we will insert these modules into the courses that we teach in order to green the chemistry curriculum. The following courses will be affected:

- General chemistry
- Biochemistry
- Inorganic chemistry
- Organic chemistry
- Analytical chemistry
- Physical chemistry

Also, student training will include the economical and ethical development of safe chemical processes to produce commodity materials for all segments of society.

3.3 Research

We will work with industry and foster interdisciplinary research through collaboration with chemical engineering and other disciplines. We will promote the general concepts of clean technology towards an improved public understanding of science. Also, we will work at the frontiers of modern chemical research in the areas of clean synthesis, catalysis, novel materials and application of renewable resources and seek to do the following:

- Develop new cleaner chemical processes to replace environmentally unacceptable methods.
- Apply innovative catalyst technology to established industrial processes.
- Reduce waste through increased reagent and solvent efficiency.
- Develop environmentally acceptable routes to important organic products.
- Design new environmentally friendly materials including those based on renewable resources.

4 Results and Discussion

4.1 *Results*

The first year of implementation was highlighted with a series of seminars and lectures. Professor John Warner gave two lectures in the campus as a groundbreaking for green chemistry. The two-semester senior chemistry students' seminar themes focused on green chemistry. Following this event, students made a number of requests for research in green chemistry. In our research group, we had five students (including three interns from Marseilles (IUT)). Faculty participation at two international green chemistry conferences in South Africa and Ethiopia helped to establish collaborative work with different institutions.

In our non-chemistry major course (Consumer Chemistry), we have adopted a new book (*11th Ed. Chemistry for Changing Times*) by John W. Hill and Doris K. Kolb. This book has a topic in green chemistry at the end of each chapter that helps faculty to infuse and illustrate each chemistry topic with green chemistry example. In addition, a series of departmental discussions has led to the adoption of a new curriculum in regard to the new Auburn City Schools (ACS) curriculum guidelines. To fund this project, we have submitted a grant proposal to National Science Foundation-Course, Curriculum, and Laboratory Improvement (NSF-CCLI) titled "Establishing an Environmental/Green Chemistry degree track: A cross-disciplinary approach." If funded, we will start a new degree track in environmental/green chemistry in collaboration with the biological sciences and geography departments.

A new collaboration between Michael Fields Agriculture Institute and the College of Letters and Sciences has been established with a new committee to plan a lecture series and student internships. Michael Fields Agricultural Institute [6] is renowned for its green agriculture emphasis. A new collaboration with the school district is underway, starting with a two-day green chemistry show at a local high school. A grant was submitted to the Wisconsin Environmental Education Board (WEEB) for summer research experiments involving local high students.

4.2 *Discussion*

Promoting green chemistry within the University and to the public is critical for the growth of the field. However, incorporating green chemistry into the undergraduate curriculum does not imply that students will be automatically drawn to the idea. Green educators need to balance the excitement of the new curricula with the issue of what students want. The educators should overcome the perception that green chemistry is a niche topic rather than state-of-the-art science and therefore can attract and retain the best and brightest students.

5 Conclusion

Professor Michael Cann from University of Scranton, Pennsylvania in the United States, said: "Although a small and growing number of research students are focusing on green chemistry, in order to broaden this base we must bring green chemistry into the classroom. We must teach our students of today, our chemists of tomorrow, to view all chemistry work with pollution prevention in mind. It is time to turn a significant amount of our energies toward the greening of the chemistry curriculum" [7]. In response to the green requirement at Goucher college one senior said: "I think it is an intelligent move. I wouldn't mind being a little forced to learn more about the physical world we live in and what is happening to it". Dautremont-Smith concluded: "The green sector has taken off in the last few years, so there are more jobs than ever before. Employers want people who understand sustainability issues and students want to contribute to solving environmental problems." At the University of Wisconsin-Whitewater the creation of the Green Chemistry Initiatives Center will be used as a "catalyst" to implement and promote green chemistry in academia, industry and the community.

Acknowledgements The author would like to acknowledge Dr. Steven Sahyun, University of Wisconsin – Whitewater, for his assistance in preparing this manuscript.

References

1. Paul T. Anastas and John C. Warner, *Green Chemistry: Theory and Practice* (Oxford University Press, Oxford, 1998)
2. Presidential Green Chemistry Challenge Awards, *http://www.epa.gov/opptintr/greenchemistry/presgcc.htm*
3. Plenty Magazine-Environmental News and Commentary, *http://www.plentymag.com/features/2008/09/greening_higher_education*
4. National Wildlife Federation, *http://www.nwf.org*
5. John Warner, *http://www.warnerbabcock.com/about.html*
6. Michael Fields Agricultural Institute, *http://www.michaelfieldsaginst.org/*
7. Professor Michael Cann, University of Scranton, Pennsylvania, *http://academic.scranton.edu/faculty/CANNM1*

Education About the Use of Quantities, Units and Symbols in Chemistry: The Earlier the Better

J. Stohner and M. Quack

Abstract We focus on a collection of useful rules and recommendations also relevant for chemical education and exemplified in detail in the International Union of Pure and Applied Chemistry (IUPAC) publication *Quantities, Units and Symbols in Physical Chemistry*, well known as the IUPAC 'Green Book'. The 2007 new edition has jointly been published by IUPAC and The Royal Society of Chemistry (RSC).

1 Introduction

The IUPAC 'Green Book' has a long history going back to 1969 when the *Manual of Symbols and Terminology for Physicochemical Quantities and Units* was first published by M. L. McGlashan, the then Chairman of the IUPAC Physical Chemistry Division. The first edition of the Green Book as we know it now was published in 1988 and the third, revised and enlarged edition was published recently by IUPAC with RSC [1].

The intention of the Green Book ever since its appearance was not to present a list of recommendations as commandments, but rather its aim was and still is to help the user in what may be called a 'good practice of scientific language'. Many well-established conventions are used in science and technology, but mixing conventions can lead to misunderstandings or even cause severe errors. One of those errors, caused by confusion of metric and imperial units, led to the loss of the NASA Mars Climate Orbiter (MCO) in 1999, worth about 200 Million USD of equipment and a non-quantified loss of scientific data and work. The reason for the loss of MCO was that Lockheed Martin Astronautics (LMA) software used imperial units (lbf s) and Jet Propulsion Laboratory expected output from the LMA software

J. Stohner
ZHAW Zürich University for Applied Sciences, ICBC Institute for Chemistry and Biological Chemistry, Reidbach T, Einsiedlerstrasse 31, CH-8820, Wädenswil, Switzerland
e-mail: sthj@zhaw.ch

M. Quack
ETH Zürich, Physical Chemistry, Wolfgang-Pauli-Strasse 10, CH-8093, Zürich, Switzerland
e-mail: martin@quack.ch

M. Gupta-Bhowon et al. (eds.), *Chemistry Education in the ICT Age*,
DOI 10.1007/978-1-4020-9732-4_31,

Table 1 Various reference levels to measure heights with respect to sea level

Country	Description	Reference level (town)
Austria	Meter über Adria (m ü. Adria)	Trieste
France	mètres d'altitude (m)	Marseille
Germany	Meter über Normalnull (m ü. NN)	Amsterdam
Great Britain	metres above sea level (m ASL)	Newlyn
Italy	metri sul livello del mare (m s.l.m.)	Genua
Switzerland	Meter über Meer (m ü. M.)	Marseille

in metric units (N s), sending the satellite off course [2]. Another incident concerned the construction of a bridge in Laufenburg (Hochrheinbrücke, 2003/4) between Switzerland and the southern part of Germany. While Germany measures its altitude relative to North Sea level, Switzerland uses the Mediterranean Sea (see Table 1); both reference levels differ by 27 cm.

This was known to the engineers; unfortunately the correction was performed in the wrong way introducing a difference in height by 54 cm. It was stated that the correction could be made without financial adjustments (because the insurance of the engineering company covered the additional costs). One can estimate enormous daily economic losses due to the use of inadequate units and conventions worldwide under a variety of circumstances.

These examples show that it is worth caring about units as well as their proper usage.

2 Symbols and Units for Quantities and Operators

Symbols for physical quantities should be single-lettered using the Latin or Greek alphabet. The letters may be capital or lower case but should be printed in italic (slanted) type. Subscripts and superscripts may be added for clarity. All subscripts and superscripts are printed in Roman type (upright) except when these are symbols for physical quantities and therefore printed in italic type. Symbols for units should always be printed in Roman type. Similarly, symbols for chemical elements, elementary particles and mathematical operators (e.g. sin, exp, ln, d/d*x*, etc.) are also printed in Roman type (see sections 1.3 and 1.6 in [1]).

3 Base Quantities and the SI as a Coherent System of Units

The International System (SI) consists of seven base quantities and their corresponding units (see sections 1.2, 3.3 and 3.8 in [1]):

> metre (symbol: m), kilogram (symbol: kg), second (symbol: s), ampere (symbol: A), kelvin (symbol: K), mole (symbol: mol), and candela (symbol: cd).

A *coherent system* of units is such that equations between numerical values of physical quantities can be written exactly as the corresponding equations between the physical quantities themselves. This coherent system therefore avoids numerical factors between units and is especially useful for so-called dimensional checking. The SI is such a coherent system of units. The coherence is lost, however, when prefixes are used.

4 Physical Quantities and Quantity Calculus

The value of a *physical quantity* Q can be written as a product of a *numerical value* $\{Q\}$ and a unit $[Q]$

$$Q=\{Q\}[Q]$$

Equations between quantities do not depend on the choice of units; however, equations between numerical values do depend on the choice of units. Physical quantities, numerical values and units may be manipulated by algebraic rules ('quantity calculus'). The wavelength λ of one of the yellow sodium lines, for example, can be written in various equivalent ways

$$\lambda = 5.896 \times 10^{-7}\ \text{m} = 589.6\ \text{nm}$$
$$\lambda\ /\ \text{m} = 5.896 \times 10^{-7}$$
$$\lambda\ /\ \text{nm} = 589.6$$

Conversion between different units proportional to energy E can be achieved by using quantity calculus:

1 cm^{-1} is converted to eV (electronvolt)

$$(1\,\text{cm}^{-1})hc\left(\frac{e}{e}\right) \triangleq \frac{(1.986\,445\,501\times10^{-25}\,\text{J})\times10^{2}e}{1.602\,176\,487\times10^{-19}}$$
$$\triangleq 1.239\,842\times10^{-4}\,\text{eV}$$

and 1 kcal mol^{-1} is converted to cm^{-1}

$$\frac{(1\,\text{kcal}\,\text{mol}^{-1})}{hcN_{\text{A}}} \triangleq \frac{4.184(1\text{kJ}\,\text{mol}^{-1})}{hcN_{\text{A}}}$$
$$\triangleq \frac{4.184\times(10^{3}\,\text{J}\,\text{mol}^{-1})}{(1.986\,445\,501\times10^{-25}\,\text{J})\times10^{2}\,\text{cm}\times(6.022\,141\,79\times10^{23}\,\text{mol}^{-1})}$$
$$\triangleq 349.7551\,\text{cm}^{-1}$$

In a table or graphical representation, it is desirable to display numerical entries by dividing the quantities by their corresponding units (see also Fig. 1 and Table 2):

$$\ln\,(p/\text{MPa}) = a+b/\text{T} = a+b'(10^{3}\ \text{K}/T)$$

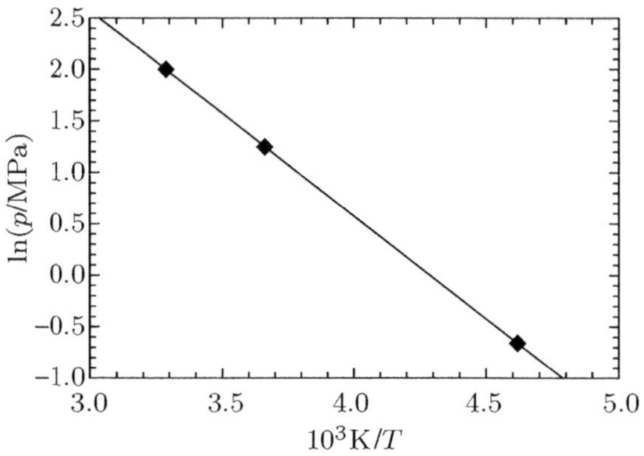

Fig. 1 Graphical representation of ln(p/MPa) against the reciprocal absolute temperature

Table 2 Data for the graphical representation of ln(p/MPa) against the reciprocal absolute temperature

T/K	10^3 K/T	p/MPa	ln(p/MPa)
216.55	4.6179	0.5180	−0.6578
273.15	3.6610	3.4853	1.2486
304.19	3.2874	7.3815	1.9990

Algebraically equivalent forms may be used in place of 10^3 K/T, such as kK/T or 10^3 (T/K)$^{-1}$. From the following equation one can see by writing the first few terms of a series expansion

$$\ln x \approx 2\left(\frac{x-1}{x+1}+\frac{(x-1)^3}{3(x+1)^3}+\cdots\right)$$

that omitting the unit of the logarithm's argument would imply that one subtracts a number from a pressure. It is therefore mandatory to write ln (p/MPa) instead of ln (p).

5 Units outside the SI

There are non-SI units which are accepted for use with the SI: min (minute), h (hour), d (day), eV (electronvolt), Da (Dalton; this is equivalent to u, 1 Da = 1 u), u (unified atomic mass unit) and some more (see section 3.7 in [1]). The minute (60 s), the hour (3600 s) and the year (86400 s) are defined here as exact multiples of the second. The year (symbol a) is not among those units, because there are different kinds of 'year', the Julian year being defined in terms of days as 365.25 d, the Gregorian year as 365.2425 d, and the Mayan year as 365.2420 d (see section 7.2 in [1]). The old definition of the second was based on an appropriate function of the tropical year 1900.

One should avoid using non-SI units from the following unit systems: the esu (electrostatic unit system), the emu (electromagnetic unit system) and the Gaussian unit system. However, equations relating these still widely used unit systems to the SI are listed in Chapter 7 of [1] which also makes extensive use of quantity calculus to help converting between those systems of units.

Sometimes, quantities of dimension one (frequently called 'dimensionless' quantities) are written in mathematical equivalent forms denoted by special symbols (e.g. % for percent, 10^{-2}) or abbreviations (ppm, parts per million, 10^{-6}). These are often ambiguous, for example, ppt could mean parts per thousand (10^{-3}) or parts per trillion. The latter is also ambiguous since a trillion can either be 10^{12} (American system of names) or 10^{18}. Similarly, the frequently used ppb (parts per billion) is ambiguous since a billion can either be 10^{9} (American system) or 10^{12} (European system). Since those quantities of dimension one can always be replaced by a proper fractional expression and ambiguities must be avoided, their use is deprecated (see section 3.10 in [1]).

Calories should not be used, because there are different calories: cal_{th} (thermochemical calorie, 4.184 J), cal_{IT} (international calorie, 4.1868 J), and the cal_{15} (15 °C calorie, approximately 4.1855 J). It is often not clear which conversion was used in a given context. Many more examples can be found in section 7.2 of [1].

6 Summary and Outlook

We have given here a very short overview of various topics covered in the 2007 edition of the Green Book which are relevant for everyday life of scientists because they facilitate cross-border communication among various disciplines in chemistry and physics. A 'Green Book Light' version is in preparation for use at high-school and undergraduate university level.

The present third edition has been substantially revised and extended with new sections (e.g. on uncertainty) compared to the second edition. The most accurate recent fundamental physical constants and atomic masses are tabulated. The symbol as well as the subject index has been extended considerably to facilitate the usage of the Green Book. A table of numerical energy conversion factors is given and the most recent IUPAC periodic table of the elements is given on the inside back cover.

The definitions of the seven SI base units are subject to changes as experimental methods lead to an increase in precision and accuracy. This led in the past to a redefinition of the second by counting the periods of the radiation corresponding to the transition between the two hyperfine levels of the ground state of the caesium 133 atom. In 1967/68 it replaced the previously adopted definition of the second as the fraction 1/31 556 925.9747 of the tropical year for 1900 January 0 at 12 h ephemeris time. Today, for example, one wishes to redefine the kilogram (along with the ampere, the kelvin and the mole) in terms of fundamental physical constants [3, 4]. Presently, the definition of the kilogram still relies on a definition, that goes back to the French Revolution using a prototype (see [5, 6] and references therein). We should refer here also to a related brief summary which we presented at the SASP 2008 conference [7].

Comments and suggestions on the IUPAC Green Book are welcome. Please send them to the email addresses as given above.

Acknowledgments We thank our dear colleagues and co-authors of the Green Book E.R. Cohen, T. Cvitas, J.G. Frey, B. Holmström, K. Kuchitsu, R. Marquardt, I. Mills, F. Pavese, H.L. Strauss, M. Takami, and A. J. Thor. We also thank numerous commentators and reviewers who have contributed to the monograph in one way or another. Discussions with Peter Atkins and Frih Schaefer are gratefully acknowledged.

References

1. Cohen, E.R., Cvitas, T., Frey, J.G., Holmström, B., Kuchitsu, K., Marquardt, R., Mills, I., Pavese, F., Quack, M., Stohner, J., Strauss, H.L., Takami, M., Thor, A. J.: Quantities, Units and Symbols in Physical Chemistry. 3rd Edition, IUPAC & The Royal Society of Chemistry, Cambridge (2007); ISBN 978-85404-433-7. http://www.rsc.org/Publishing/Books/0854044337.asp
2. The report can be found at ftp://ftp.hq.nasa.gov/pub/pao/reports/1999/MCO_report.pdf.
3. Mills, I.M.: On defining base units in terms of fundamental constants. Mol. Phys., 103, 2989–2999 (2005)
4. Mohr, P.J.: Defining units in the quantum based SI. Metrologia, 45, 129–133 (2008)
5. Feder, T.: To redefine kilogram, experiments must weigh in. Physics Today, 59, 32 (2006)
6. Mills, I., Mohr, P.J., Quinn, T.J., Taylor, B.N., Williams, E.R.: Redefinition of the kilogram: a decision whose time has come. Metrologia, 42, 71–80 (2005)
7. Stohner, J., Quack, M.: Quantities, Units and Symbols in Physical Chemistry. In: Beck, R.D., Drabbels, M., Rizzo, T.R. (eds.) Contributions 16th Symposium on Atomic and Surface Physics and Related Topics (SASP 2008), p. 219–222. Innsbruck Univ. Press, Innsbruck (2008)

ICT and Reporting Skills in Chemical Engineering Education

J. C. Reijenga and M. M. Roeling

Abstract Electronic submissions of reports from practical courses or group work and PowerPoint presentations have been a standard practice for many years now. A more recent development is reporting in the form of a web site. This web site is compulsory for all multidisciplinary projects in our Master Program and sometimes comes in place of a regular report. The web sites are a separate point in the grading system, and in addition to the contents, special attention should be given to the layout, ease of navigation, readability, text/figure ratio and other requirements of good web-practice. Subjects presented reflect current and ever-changing research topics and interests of society. Recent examples are illustrated.

As a result of our efforts, our students have been enabled to broaden the spectrum of communication tools. The resulting web sites have the additional advantage that they can serve the student's professional portfolio. Furthermore they may serve to illustrate new formats of cooperation between academia and industry.

1 Introduction

In the context of recent history of many ICT tools in relation to education (and of technical innovations in general), one can distinguish the following three phases:

1. Introduction of a new tool and use by early adapters
2. General acceptance regarding usefulness: educational institutions start teaching how to use the new tool
3. Ability to use the ICT tool is regarded as general knowledge in freshmen

Additional complications arise from the fact that in many cases, it is some of the students, rather than the teaching staff, that form the early adapters, and that the ICT tools show a very rapid development during the three phases indicated.

J.C. Reijenga (✉) and M.M. Roeling
Department of Chemical Engineering and Chemistry, Eindhoven University of Technology, 513
P.O. Box 5600, Eindhoven, MB, The Netherlands
e-mail: j.c.reijenga@tue.nl

M. Gupta-Bhowon et al. (eds.), *Chemistry Education in the ICT Age*,
DOI 10.1007/978-1-4020-9732-4_32,

Since the 1970s we have witnessed elementary use of computational hardware move from phase 1 to phase 3 (ranging from slide rule to pocket calculators, mainframe terminals and minicomputers to desktop and laptop computers). Hardware was followed by software. Use of text editors underwent similar development after the typewriter-era: from Wang and Wordstar in the 1970s, Wordperfect 4.1 and ChiWriter (first DOS-based WYSIWYG editor) in the 1980s to Word in the 1990s and LaTex. Our university went from phase 2 to phase 3 in the late 1990s with respect to text editors. The use of presentation software followed in the late 1990s: freshmen at present already know how to set up a simple PowerPoint presentation. Attention point remaining is of course the quality of the content of the report and of the presentation.

The need to be able to "program" declined however, after the use of Fortran (1950s), Cobol, Algol and Basic (1960s), Pascal (1970s), and C^{++} (1980s), because more sophisticated software for dedicated tasks became available. Best early examples are mathematical tools such as VisiCalc (1979), Lotus123 (1983), SlideWrite Plus (late 1980s), and Excel (early 1990s). Phase 2 at present includes programs such as Mathlab, Maple, and Aspen: introductions are incorporated within the framework of relevant courses.

During the same decades, the following general trends have been observed in university education: the increased need for developing competences (in addition to the specific knowledge of facts), and the requirement for academics to be able to communicate science and technology in a wider context: in multidisciplinary cooperation, and communicating with a broader, more general audience. Societal relevance and sustainability aspects of science and technology are not new phenomena, but the engineer of the future should also realize they are not one-way streets: communicating science for a broader audience has become increasingly important.

2 Methodology

At present the situation regarding ICT skills in our university is not unfavorable, as indicated above: our graduated master students are able, through repeated practice, to conduct a literature search, perform advanced experiments in computer-controlled equipment, write scientific reports, give oral and poster presentations, and perform modeling simulations using the appropriate ICT tools.

As already indicated, the need to communicate science and technology to a wider, more general audience increases. Important improvements in the corresponding skills are required. When trying to do so, one should realize that there is a trade-off between three priorities in the triangle of level of detail, interactivity, and general audience, as depicted in Fig. 1.

Communicating with direct colleagues and experts can be very well covered by the bottom-part of the triangle. Doing so with a more general audience, in an interactive way is quite a different matter. In general, posters are more effective if used interactively, but experience learns that authors are not always present during poster

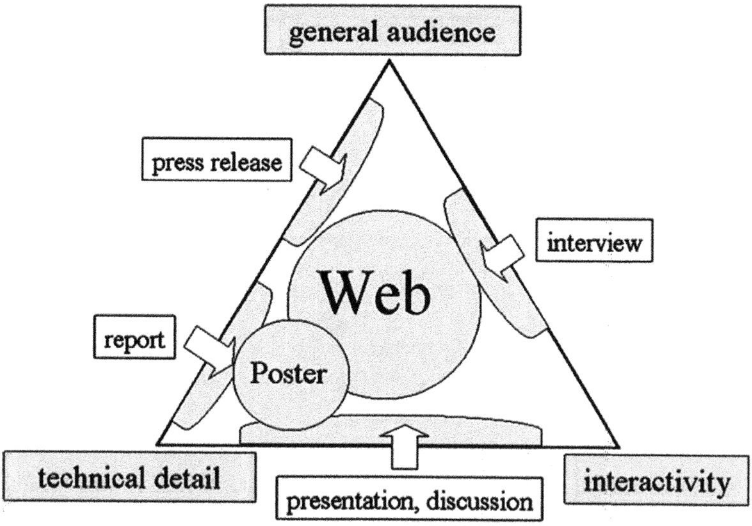

Fig. 1 Communication tools to be chosen, taking into account target audience, required degree of technical detail and interactivity

presentations. In addition, it only works if the text and the discussion are adjusted to the target audience. This is of course a universal requirement for communication. We are convinced that poster presentation and especially web-presentation should deserve more attention in science education. The latter is the subject of the present contribution.

In 1997 the so-called Notebook project was started on our campus: each student was supplied with a standard notebook computer, which was partly paid by the University and to be used in all educational activities. Electronic submission of results of practical courses was common since then, as all students are given a university e-mail address. Since 1998 we have been experimenting with web sites, replacing reports. At that time several practical courses in the second year each had different ways of finalizing: oral presentation, report, poster, or web site. The idea behind this was that we found it necessary that students are familiar with several communication and reporting tools, and to interact in different ways with different target audiences. For the latter (a practical course in analytical chemistry) students worked in pairs and made a web site on a special server. This web site was then graded as a report. This practice became a tradition and since 2002 all master students involved in a multidisciplinary project, report their findings by means of a web site. The multidisciplinary project[1] comprises a major subject in the master program in which 4–6 students from different departments work together on a 3-month project proposed by an external client, in many cases an executive from industry. Deliverables include a project plan, interim report, presentations, collected literature, sometimes a working prototype or a business plan, and a web site. The size of the course amounts to 8 ECTS units.[2] The multidisciplinary project has

Fig. 2 Screen-shot of Microsoft FrontPage with familiar folder structure and format/edit buttons

previously served as a platform for experimenting with the use of other ICT communication tools in the context of international cooperation.[3]

Initially, an afternoon course in web-design using Microsoft FrontPage was given. FrontPage is included in some editions of Microsoft Office, used on our campus. As seen from the screen-shot in Fig. 2 it is very user-friendly, especially for those familiar with other MS Office applications such as Word. At present, a 1-h introduction is given only on request for those students not familiar with making a web site.

We fully realize we are not the only university encouraging web site building by students. The Chemical Society in Britain for a couple of years organized Exemplarchem,[4] an Internet-based exhibition of exemplary students' work in Chemistry. It was a voluntary event, and students from any university could submit their entry through a local coordinator. Unfortunately the competition was discontinued after 2004. Many other universities worldwide enable students (and staff) to set up an individual web site for portfolio purposes, but contents vary widely, and in most cases these webs only serve as on-line CV. Our student project web sites on the other hand, are embedded in the educational goals of a compulsory course in the curriculum and highlight the results of a particular activity. The student web pages server of our Department at http://students.chem.tue.nl now contains 89 web sites, and new sub-webs are added each semester.

3 Results and Discussion

The web site has a substantial hit-count, and provides a valuable reference platform for both general audience, and secondary school students, as seen from e-mail response to the general contact address. Some of the popular sites are mentioned below.

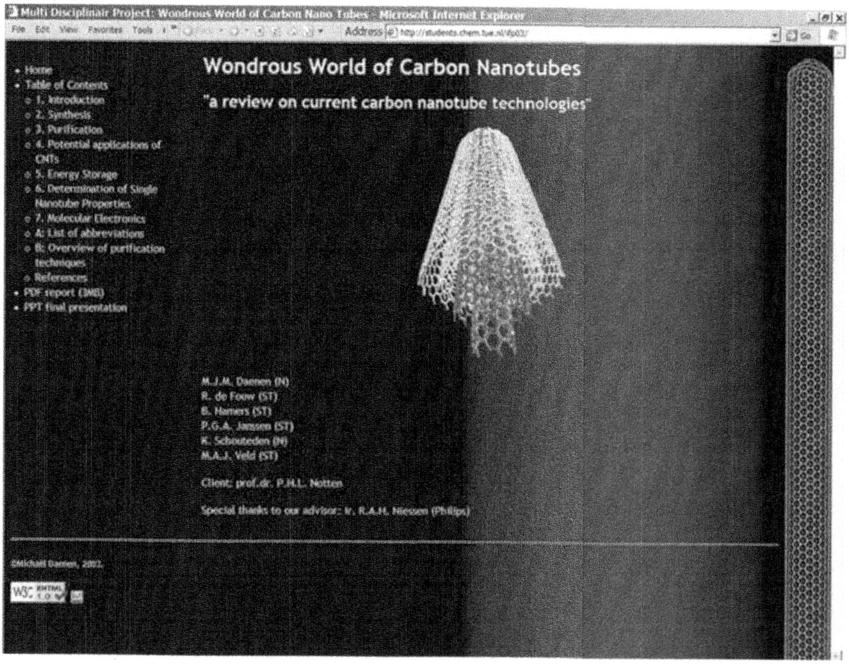

Fig. 3 Screen-shot of the most cited student web site http://students.chem.tue.nl/ifp03/

The web site of Wondrous World of Carbon Nanotubes[5] (Fig. 3), a project initiated by Professor Peter Notten of Philips Research Laboratories is cited most.

In this case, the full report, downloadable from the web site, was even translated into Farsi, after having obtained prior permission to do so. The reasons for popularity of any web site in general are threefold:

1. The site has an effective and attractive layout and navigation.
2. The content of the site is interesting from a scientific and societal point of view.
3. The site is easily found by search engines, by way of proper meta tag keywords in the html code.

Popularity of a web site, as determined by hit count is just one aspect. Naturally, students are not graded proportional to the hit count of their web site. There are many more-or-less obvious criteria, some of them subjective though, for a good web site. Guidelines, tips, and tricks are offered in excessive abundance, on paper and especially on the Internet. Some sources focus on a specific web-editing tool, others deal with a particular category of web sites, certain target audiences, or give general guidelines for good practice.[6–11] A general starting point is recently given in the ten Principles of Effective Web Design by Smashing Magazine.[12]

1. Don't make users think.
2. Don't squander user's patience.

3. Manage to focus user's attention.
4. Strive for feature exposure.
5. Make use of effective writing.
6. Strive for simplicity.
7. Don't be afraid of the white space.
8. Communicate effectively with a visible language.
9. Conventions are our friends.
10. Test early, test often.

Several other examples of student's web site are the following:

- *Chemagine 2030 – Back from the Future*[13] is an award winning site that presents a scenario study of the chemical industry in the industrial region of the Netherlands and Belgium in the year 2030.
- *All that glitters is not green – biofuel in the Netherlands*[14] addresses the controversy of increasing food prices due to the increased use of biofuels. The press release formulated a strong recommendation.
- *Clinical lab-on-a-chip*[15] gives an overview of miniaturization in the field of automated analysis of body fluids.
- *Quantitative determination of benzoic acid and sorbic acid in soft drinks*[16] describes a very popular experiment that is regularly carried out in our students' lab with classes of potential freshmen.

Criteria we particularly pay attention to are obvious and fast navigation, good readability, uniform layout and font, and a high figure-to-text ratio. We fully realize that not all sites meet all of these requirements in equal extent. Some of the earlier sites for example, are in Dutch. Others address subject of minor or only local importance. The students are also stimulated to find positive inspiration in some sites, and to recognize bad practice in others, and as a result be critical about the quality of their own communication towards others.

4 Conclusions and Recommendations

The system has been in operation successfully for several years now, and we have learned to particularly pay attention to the following points:

- Web sites should be in English, targeted towards an audience interested in societal implications of recent technological innovations.
- Some groups assign the task of making the web site to the group member most capable of such a task, with the results that the others did not learn from the experience. We try to solve this by having the students include it explicitly in the individual learning objectives at the start of the project.
- The web site should be graded separately from the report, the presentation, and the overall group performance. If not the students tend to make a sloppy job of the whole thing, just copy/pasting the report contents. A (non-zero) weighing

factor of the web site in the final grade is determined in collaboration with the students.

- Even more so than in the case of written reports, students nowadays seem to have a tendency to "forget to mention explicitly the sources of their information," especially if the information has been obtained from the Internet. For this reason, we warn the students that we use anti-plagiarism software[17] in order to check the material they submit. Naturally, we do this only after having instructed them how to deal with literature sources, citations, and references in a professional manner.[18]
- The importance of finding the right balance between quality of the form and of the content cannot be stressed enough, in addition to the requirement that the contents should be carefully tuned to the intended target audience. We all realize however, that this equally applies to reports, posters, and oral presentation.

By including web design as an inherent educational goal of a compulsory course in the Master curriculum, our students have been enabled to further broaden the spectrum of communication tools. The resulting web sites form a data base of past projects and activities and have the additional advantage that they comprise part of the student's professional portfolio. Furthermore they may serve to illustrate new formats of cooperation between academia and industry, and to promote science and engineering in general.

References

1. http://edu.chem.tue.nl/6z020/
2. http://en.wikipedia.org/wiki/European_Credit_Transfer_and_Accumulation_System
3. Reijenga, JC, Siepe, H, Yu, LE, Wang, CH (2003) Collaborative Learning and Cyber-Cooperation in Multidisciplinary Projects. Chem. Eng. Educ. 37:114–119
4. http://www.chemsoc.org/exemplarchem/
5. http://students.chem.tue.nl/ifp03/
6. Anderson, A (2004) Compact Guide to Web Page Creation and Design. Prentice Hall, Upper Saddle River, NJ
7. Donnelly, D (2000) WWW Design: Web Pages from around the World. Rockport Publishers, Beverley
8. Austin, B (2000) Web Page Design in Easy Steps. Computer Step, Southam
9. Mackinnon, M (2002) An Introduction to Web Site Design. Usborne Publishing Ldt, London
10. Hawkes, L (2001) A Writer's Guide to Document and Web Design. Prentice Hall, Upper Saddle River, NJ
11. Veen, J (2000) The Art and Science of Web Design. New Riders Press, Indianapolis
12. http://www.smashingmagazine.com/2008/01/31/10-principles-of-effective-web-design
13. http://students.chem.tue.nl/ifp16/
14. http://students.chem.tue.nl/ifp22/
15. http://students.chem.tue.nl/mdp12/
16. http://students.chem.tue.nl/6n220 - 23/
17. http://ephorus.com/
18. http://students.chem.tue.nl/guidelinescitation.pdf

Linear Free Energy Relationships (LFER) as a One-Hour Classroom Lecture for Postgraduate Students: Correlation of the Nature of the Transition States

V. Jagannadham

A linear correlation exists between the logarithm of a rate constant or equilibrium constant for one series of reactions and the logarithm of the rate constant or equilibrium constant for a related series of reactions respectively. Typical examples of such relations (also known as linear Gibbs energy relations) are the *Brønsted relation* and the *Hammett equation*. The name arises because the logarithm of an equilibrium constant (at constant temperature and pressure) is proportional to a standard free energy (Gibbs energy) change, and the logarithm of a rate constant is a linear function of the free energy (Gibbs energy) of activation. It has been suggested that this name should be replaced by linear Gibbs energy relation, but at present there is little sign of acceptance of this change. The area of Physical Organic Chemistry which deals with such relations is commonly referred to as 'Linear Free-Energy Relationships (LFER)'.

Gibbs energy of activation (standard free energy of activation) is $\Delta G_o^{\neq}$ the standard *Gibbs energy* difference between the *transition state* of a reaction (either an *elementary reaction* or a *stepwise reaction*) and the ground state of the reactants. It is calculated from the experimental rate constant k via the conventional form of the absolute rate equation: $\Delta G^{\neq} = RT\,[\ln (k_B/h) - \ln (k/T)]$ where k_B is the Boltzmann constant and h the Planck constant ($k_B/h = 2.08358 \times 10^{10}\ K^{-1}\ s^{-1}$). The values of the rate constants, and hence Gibbs energies of activation, depend upon the choice of concentration units (or of the thermodynamic standard state).

Although spontaneous transformations all have negative ΔG° s, not all exergonic processes are spontaneous, due to activation energy barriers to reaction. Treatment of activation energy was framed in terms of enthalpy or potential energy. It should now be clear that, if entropy factors are to be incorporated in the activation barrier, we should be thinking about *Free Energy of Activation*, $\Delta G^{\neq}$. The defining equation then becomes:

V. Jagannadham (✉)
Department of Chemistry, Osmania University, Hyderabad, 500 007, India
e-mail: vjchem@osmania.ac.in or jagannadham1950@yahoo.com

M. Gupta-Bhowon et al. (eds.), *Chemistry Education in the ICT Age*,
DOI 10.1007/978-1-4020-9732-4_33, © Springer Science+Business Media, B.V. 2009

$$\Delta G^{\neq} = \Delta H^{\neq} - T\Delta S^{\neq}$$

Transition state theory proposes an equilibrium between reactants and the transition state, so each of the functions in this equation may be viewed as a [$F_{\text{Transition State}}$ – $F_{\text{Reactants}}$] difference, where F represents H, S or G. The equations demonstrate the similar exponential relationship of ΔG° to K_{eq} and $\Delta G^{\neq}$ to k.

$$K = e^{-\Delta G^o / RT} and\, k = \frac{k_B T}{h} e^{-\Delta G^{\neq} / RT}$$

Since the rate constant equation incorporates the temperature variable twice and $\Delta G^{\neq}$ also changes with temperature, observed reaction rates are clearly temperature dependent. Physical Organic Chemists make general use of this relationship in two ways. First, it is a rule-of-thumb that a 10°C increase in reaction temperature will roughly double the rate of that reaction. Second, since this rule applies as well to accompanying reactions, the rates of such side reactions also increase with temperature, sometimes more than the desired reaction. Consequently, the practical yield of the desired product may actually decrease at higher temperatures. Thus, a cleaner (less contaminated) product is often obtained by running a reaction at the lowest effective temperature that gives the desired product. Because $\Delta G^{\neq}$ incorporates a temperature-dependent entropy factor and is related exponentially to the rate constant, k, reaction rate studies at different temperatures may be interpreted to provide the activation parameters.

Both the *Arrhenius* and the *Eyring equation* describe the temperature dependence of reaction rate. Strictly speaking, the *Arrhenius equation* can be applied only to gas reactions. The *Eyring equation* is used in the study of gas, condensed and mixed phase reactions - all places where the simple *collision model* is not very helpful. The Arrhenius equation is founded on the empirical observation that conducting a reaction at a higher temperature increases the reaction rate. The *Eyring equation* is a theoretical construct, based on *transition state* model.

Now let us derive a LFER on the basis of Eyring and Hammett equations: Taking Eyring equation

$$k = \frac{k_B T}{h} e^{-\Delta G^{\neq} / RT} \quad (1)$$

Taking logarithm

$$\log k = \log \frac{k_B T}{h} - \frac{\Delta G^{\neq}}{2.303RT} \quad (2)$$

and from Hammett equation

$$\log k = \log k_{\text{o}} + \rho\sigma \quad (3)$$

from Eqs. 2 and 3

$$\log\frac{k_B T}{k} - \frac{\Delta G^{\neq}}{2.303RT} = \log\frac{k_B T}{k} - \frac{\Delta G_o^{\neq}}{2.303RT} + \rho\sigma \tag{4}$$

Simplifying and rearranging Eq. 4 one will get

$$\Delta G^{\neq} = \Delta G_o^{\neq} - 2.303RT\rho\sigma \tag{5}$$

This equation with a particular value of ρ, applies to any reaction involving a reactant having a series of substituents. For a second series of homologous reactions, with reaction constant ρ′, one can show that

$$\Delta G^{\neq\prime} = \Delta G_o^{\neq\prime} - 2.303\mathrm{RT}\rho'\sigma \tag{6}$$

Dividing Eqs. 5 and 6 by ρ and ρ′ respectively and subtracting one from the other and on simplification, we get

$$\Delta G^{\neq} = (\rho / \rho')\ \Delta G^{\neq\prime} + \text{constant} \tag{7}$$

Thus there is a linear relationship between $\Delta G^{\neq}$ of one series and to $\Delta G^{\neq\prime}$ of another series. Now let us see the application of this relationship to the following putative examples.

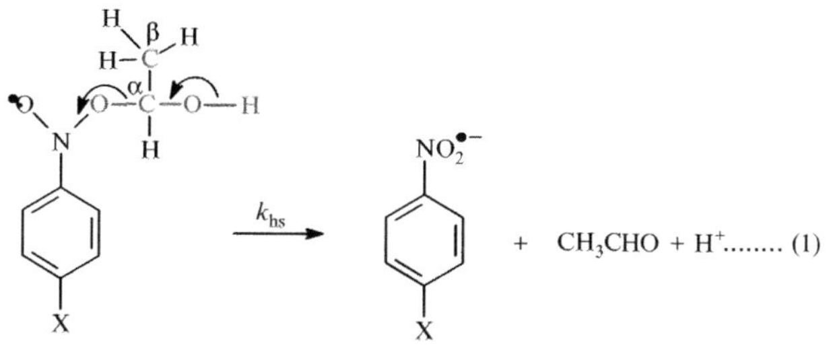

Nitroxyl adduct I

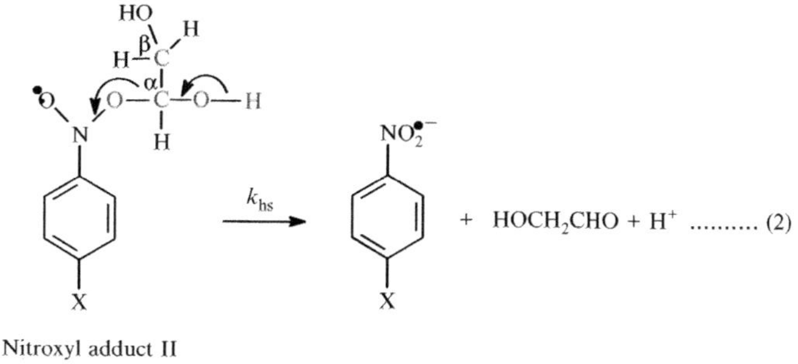

Nitroxyl adduct II

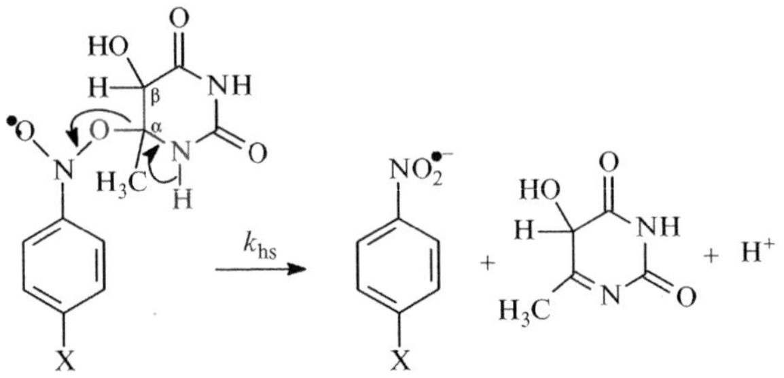

.... (3)

Nitroxyl adduct III

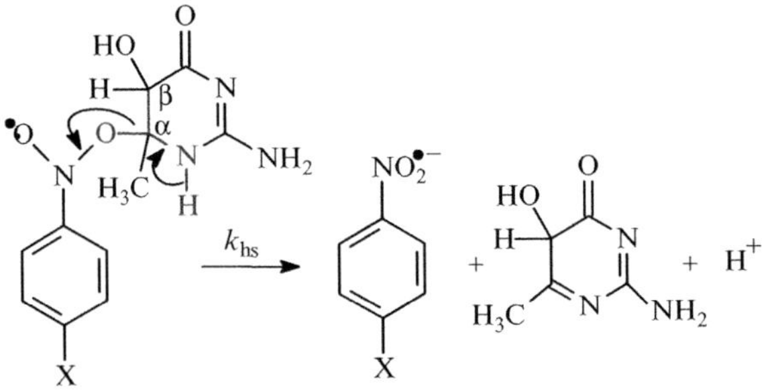

.... (4)

Nitroxyl adduct IV

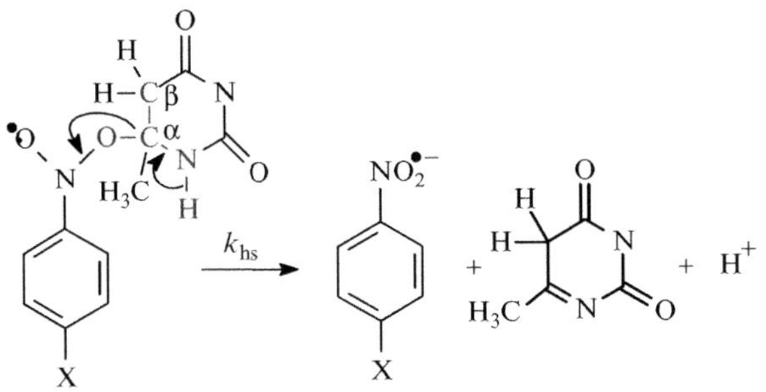

.... (5)

Nitroxyl adduct V

All the nitroxyl adducts I–V were generated [1–4] pulseradiolytically in aqueous solution from nitrobenzenes and the corresponding α-hydroxyethyl, α,β-dihydroxyethyl, 6-methyluracil-6-yl, 6-methyl-*i*-cytosine-6-yl and 6-methyldihydrouracil-6-yl radicals. Production of α-hydroxyethyl, α,β-dihydroxyethyl, 6-methyluracil-6-yl, 6-methyl-*i*-cytosine-6-yl and 6-methyldihydrouracil-6-yl radicals were described elsewhere [1–4]. The heterolysis rate constants, k_{hs}, and all activation free energy changes were taken from those references. The LFER plot for the heterolysis reaction of the nitroxyl adducts I and V is linear (Fig. 1) ($r = 0.99$). The slope of this plot, which should be equal to the ratio of the Hammett ρ values for the heterolysis of the two, adducts I and V, results as 1.0, in agreement with the individually determined ρ values (both 1.5). Also the solvent isotope effect (SIE) on k_{hs} of adduct V, that is

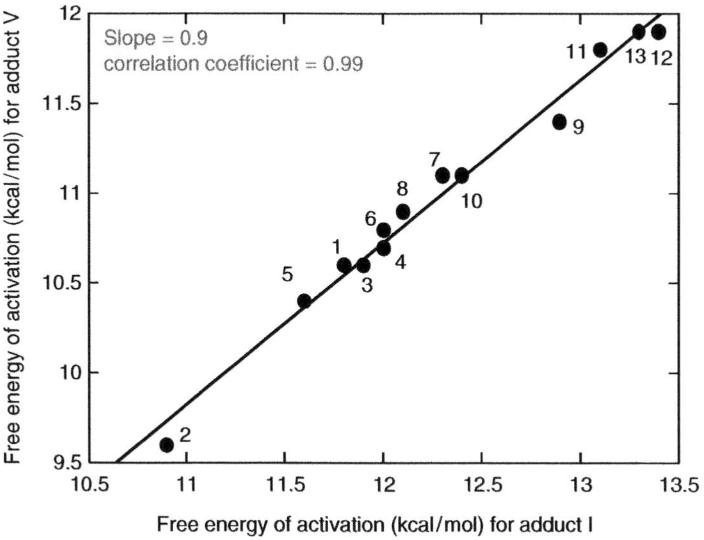

Fig. 1 Correlation of free energies of activation of Nitroxyl adducts I and V

$k_{hs}(H_2O)/k_{hs}(D_2O)$ found to be 1.7 – 2.3 is similar to that of 2.2 of the adduct I. All these observations led to conclude that the two transition states of the heterolysis of adduct I and adduct V were essentially similar in terms of synchronous O–H and C–O bond-breaking assisted by solvent water as shown below. This leads to immobilization of solvent water which was reflected in high negative $\Delta S^{\neq}$ [1, 4].

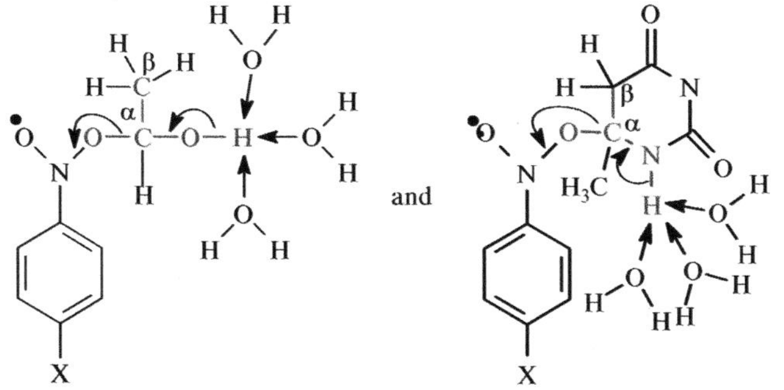

Hence these are all plenty of evidences for the close resemblance of the α-hydroxyethyl ($CH_3C^{\bullet}HOH$) and 6-methyldihydrouracil-6-yl radicals and which lead to the similar transition states of the adducts on their way to products.

In contrast, the LFER plot for the heterolysis reaction of the nitroxyl adducts III and IV were not perfectly linear when plotted against the nitroxyl adduct I and adduct V (Figs. 2, 3, 5, 6) (r = only 0.93, 0.92, 0.95 and 0.91 respectively), which indicates that the α-hydroxyethyl ($CH_3C^{\bullet}HOH$) is not a good model for the

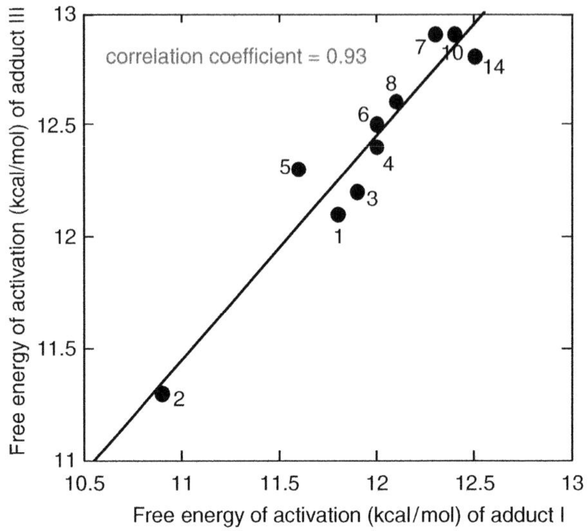

Fig. 2 Correlation of free energies of activation of Nitroxyl adducts I and III

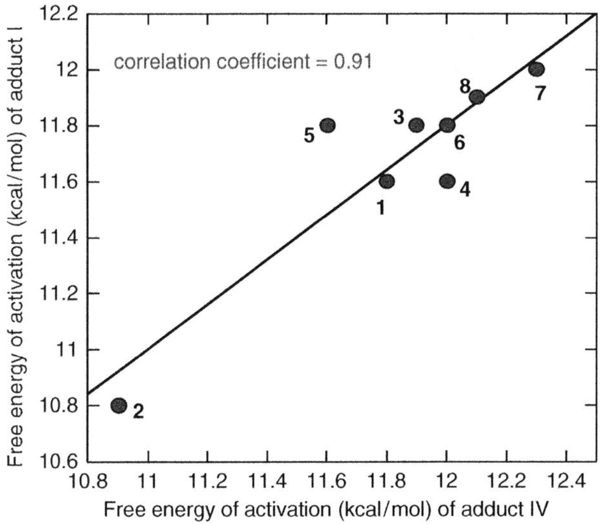

Fig. 3 Correlation of free energies of activation of Nitroxyl adducts I and IV

6-methyluracil-6-yl and 6-methyl-*i*-cytosine-6-yl radicals. That these belong to a class of their own is shown not only by similar solvent KIE (1.2 – 1.4) which is similar to that of (1.5) observed for the heterolysis of the adduct II but also by the good correlation (Fig. 4) (r = 0.98) between their own (for the heterolysis of nitroxyl adducts III and IV) corresponding $\Delta G^{\neq}$ values. To make an attempt to correlate the

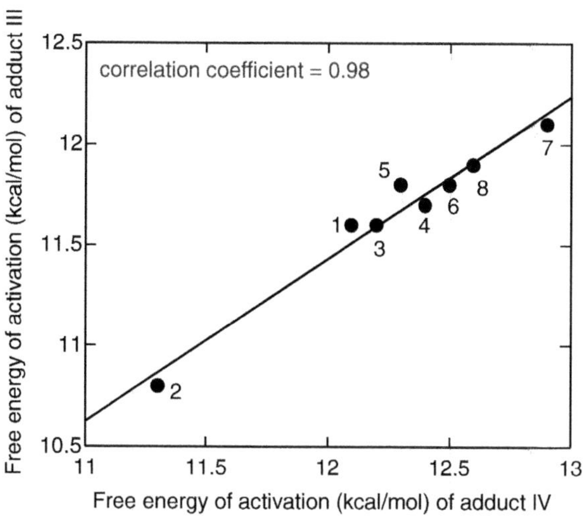

Fig. 4 Correlation of free energies of activation of Nitroxyl adducts III and IV

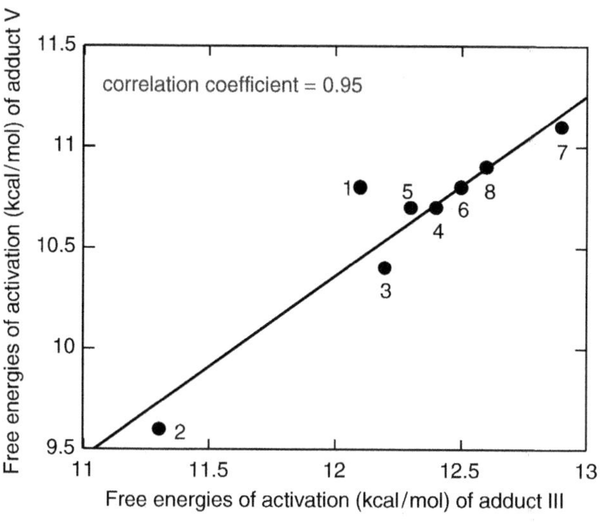

Fig. 5 Correlation of free energies of activation of Nitroxyl adducts III and V

$\Delta G^{\neq}$ values of adduct III and IV with that of adduct II, sufficient data for adduct II is not available. Yet the similar solvent KIE is a strong evidence to say that the α,β-dihydroxyethyl radical ($HO^{\bullet}CHCH_2OH$) is a good model for the 6-methyluracil-6-yl and 6-methyl-*i*-cytosine-6-yl radicals when their adducts III and IV lead to similar transition states on their way to the products. And adducts II, III and IV differ with adducts I and V where former set have an OH group in β position as shown in their structures above. The strong difference in KIE between the two sets of adducts I and

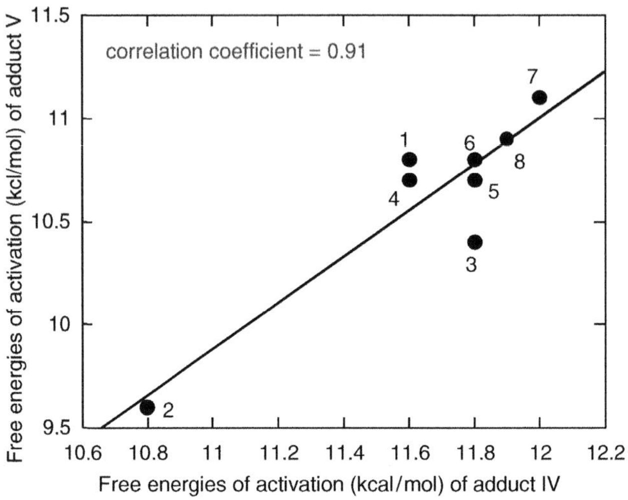

Fig. 6 Correlation of free energies of activation of Nitroxyl adducts IV and V

Serial number	**Substituent in the benzene ring**
1	SO_3CH_3
2	NO_2
3	CN
4	SO_2NH_2
5	$COCH_3$
6	CO_2CH_3
7	SO_3^-
8	$CONH_2$
9	CI
10	CH=NOH
11	H
12	CH_3
13	OCH_3
14	CO_2^-

The serial numbers from this table reflect on the figures for identification of substituents in benzene ring

V (1.7–2.3) and II, III, IV (1.2–1.5) could be understood in terms of the differences in stabilities between the transition states of the two different sets of adducts. The two sets of transition states differ by an OH group in β position to the deprotonatable OH/NH group. Due to the – I effect of OH group in the β position, the adducts II,

III, and IV are expected to be less stable than the adducts I and V. Besides in the reduction of k_{hs} (of C – O bond heterolysis, which is measured in terms of the build up kinetics of the nitrobenzene radical anion), this leads to an increased tendency to deprotonate at an early stage of the transition state, i.e. deprotonation tends to precede the heterolysis of C – O bond. This is symbolized as shown below:

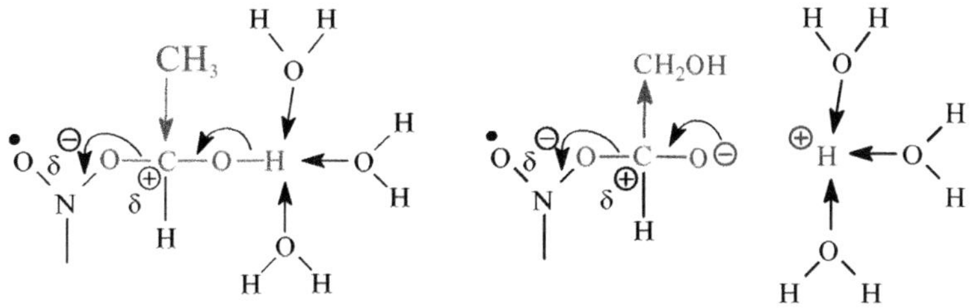

synchronous O-H and C-O bond breaking

O-H depotonation (partly) preceding C-O bond breaking

So in conclusion the two sets of adducts (I and V), and (II and III) behave differently and one can conclude that the α-hydroxyethyl radical is a good model for 6-methyldihydrouracil-6-yl radical and α,β-dihydroxyethyl radical is a good model for 6-methyluracil-6-yl and 6-methyl-*i*-cytosine-6-yl radicals.

Dedication: This article is dedicated to my mentor, Professor Dr. Steen Steenken (Retd), Max-Planck-Institute for Radiation Chemistry, Muelheim a.d. Ruhr, Germany.

References

1. Jagannadham, V. and Steenken, S. (1984), *J. Am. Chem. Soc.* 106: 6542–6551.
2. Steenken, S. and Jagannadham, V. (1985), *J. Am. Chem. Soc.* 107: 6818–6826.
3. Jagannadham, V. and Steenken, S. (1988), *J. Phys. Chem.* 92: 111–118.
4. Jagannadham, V. and Steenken, S. (1988), *J. Am. Chem. Soc.* 110: 2188–2192.

What Makes a Good Laboratory Learning Exercise? Student Feedback from the ACELL Project

A. V. George, J. R. Read, S. C. Barrie, R. B. Bucat, M. A. Buntine, G. T. Crisp, I. M. Jamie, and S. H. Kable

Abstract Over the last 7 years, a group of Australian universities have been collaboratively running a chemistry education project, now called ACELL (Advancing Chemistry by Enhanced Learning in the Laboratory). One of the key aims of ACELL is to facilitate the development and evaluation of educationally sound chemistry laboratory exercises with the goal of improving the quality of students' learning in the laboratory in Australia, New Zealand, and throughout the world. As part of this project, ACELL has developed an instrument for investigating students' perceptions of their laboratory learning experiences. To date, ACELL had collected data on 19 experiments from 972 students across 7 universities in Australia and New Zealand using this instrument, and this data collection is ongoing. As a consequence, ACELL is in an unusually good position to identify and discuss both procedural and cognitive factors that influence students' evaluation of their laboratory learning experiences, such as assessment, the quality of notes, interest, and the inclusion of opportunities for independent learning. Our results are both surprising and encouraging, and indicate that students can be highly cognitively engaged, even with traditionally "boring" content, provided a suitable learning environment is established. This paper will describe the research approach undertaken, discuss the range of factors which appear to significantly influence students' learning experiences, and consider the implications for the design of educationally sound chemistry laboratory exercises.

A.V. George (✉), S.C. Barrie, and S.H. Kable
The University of Sydney, Sydney, NSW, 2006, Australia
e-mail: george@chem.usyd.edu.au

J.R. Read, M.A. Buntine, and G.T. Crisp
The University of Adelaide, Adelaide, SA, 5005, Australia

R.B. Bucat
The University of Western Australia, Crawley, WA, 6009, Australia

I.M. Jamie
Macquarie University, Sydney, NSW, 2109, Australia

M. Gupta-Bhowon et al. (eds.), *Chemistry Education in the ICT Age*,
DOI 10.1007/978-1-4020-9732-4_34, © Springer Science + Business Media, B.V. 2009

1 Introduction

Thirty-five Australian Universities teach chemistry and over 20,000 students per year pass through these courses [1]. Students study chemistry as a discipline in its own right and as a central component of other degree programs. Chemistry is highly conceptual, and students can find it difficult to relate the molecular level of explanation to macroscopic properties of everyday substances. Understanding the language and symbolism of chemistry is critical for students to be able to engage with the concepts of the discipline [2, 3].

Laboratory activities are an integral part of chemistry education and according to the Royal Australian Chemical Institute report *The Future of Chemistry Study: Supply and Demand of Chemists* [4], almost 50% of student time is spent in laboratory work, and so it is imperative that the opportunities afforded by this substantial learning environment are realized. Notwithstanding an extensive literature describing the benefits of laboratory learning, the value of laboratory activities beyond developing technical skills (such as handling glassware) has been questioned, most recently by Hawkes [5]. Hawkes argues that laboratory activities are expensive and time consuming, and that the costs involved are not justified (particularly for non-science majors) by the technical skills developed. This position has been criticized [6–9], yet it reinforces the challenge to chemistry educators to provide compelling evidence that laboratory classes achieve more than Hawkes implies.

Laboratory work is integral to bridging the gap between the molecular and macroscopic levels of chemistry. Good laboratory programs provide a learning environment where students can forge links between theoretical concepts and experimental observations [10]. Moreover, learning goals that can be achieved through laboratory experiences include subject-matter mastery, improved scientific reasoning, an appreciation that experimental work is complex and can be ambiguous, and an enhanced understanding of how science works [11]. Skills that can be developed in high-quality laboratory exercises include: manipulation of equipment; experiment design; observation and interpretation; problem solving and critical thinking; communication and presentation; data collection, processing, and analysis; laboratory "know-how," including developing safe working practice and risk assessment skills; time management; ethical and professional behavior; application of new technologies; and team work [12, 13].

An extensive literature describes up-to-date chemistry laboratory exercises for students that extend beyond the traditional "follow the recipe" format [14]. Bennett and O'Neale [12, p. 59] have commented that students following a recipe "*are not 'doing an experiment', but 'carrying out an exercise'*." They argue that "recipe experiments" make limited intellectual demands on students, who "*often seem to go through the motions ... with their minds in neutral.*" By contrast, in a well-designed laboratory exercise students can experiment and engage, both individually and collaboratively [15], in open-ended labs [16] and inquiry-based learning activities [17] that apply theoretical concepts to relevant, real life problems. Equally, pure discovery approaches can be ineffective [18, 19], in part because they can lack sufficient structure necessary to support student autonomy [20], and in part because they can foster behavioral rather than cognitive engagement [21].

Students often indicate a preference for experimental learning methods [22, 23] and, within the chemistry curriculum, this is predominately found in the laboratory [12, 24]. In a well-designed laboratory students interact closely with teachers and peers, so learning can be enhanced, monitored, and assessed effectively [10,13,16, 24, 25]. It has been recognized that students find a well-designed laboratory program stimulating and motivating [26, 27]; moreover, they allow students to "scaffold" each other's learning [28]. Well-designed laboratories can be a popular component of science courses [10, 29] and can promote quality learning [30]. However, this literature also notes that the potential for "deep" learning is often not realized for reasons that include inappropriate experiments [12], poor educational design [10,13], and/or inadequate resources [31]. Moreover, an undergraduate laboratory setting is one that can induce anxiety in students [32] drawing undue attention to relatively simple activities, and reducing the available working memory needed for meaningful learning [33–37] by introducing extraneous cognitive load [38–41]. To some extent, such problems can be reduced by appropriate sequencing of activities [42].

Concerns such as these are certainly not new - in fact, according to Lock [43] and Hodson [44], discussions of the value of laboratory work have been occurring since the late nineteenth century - and others who have recently raised concerns about the value of some laboratory work include Marthie et al. [45] and Bennett [46]. Some research (such as [47–49]) has been undertaken in an attempt to address ways in which laboratory work can be made more effective for promoting student learning. Nevertheless, as Hofstein and Lunetta [50] note, "*researchers have not comprehensively examined the effects of laboratory instruction on student learning and growth in contrast to other modes of instruction, and there is insufficient data to confirm or reject convincingly many of the statements that have been made about the importance and the effect of laboratory teaching*" and that there "*is a real need to pursue vigorously research on learning through laboratory activities to capitalize on the uniqueness of this mode of instruction*" (p. 213). Despite the progress that has been made in developing our knowledge of learning and instruction in the 26 years since these statements were made, these comments remain true today [51–54].

1.1 ACELL: Promoting Effective Learning in the Laboratory

The challenge remains to provide students with laboratory programs that are relevant, engaging, and offer effective learning outcomes. The Advancing Chemistry by Enhancing Learning in the Laboratory (ACELL) project and its predecessors, the Australasian Chemistry Enhanced Laboratory Learning (ACELL) project [55–59] and the Australian Physical Chemistry Enhanced Laboratory Learning (APCELL) project [1, 60, 61] are examples of contemporary efforts designed to tackle this challenge.

ACELL covers all areas and levels of undergraduate chemistry. The ACELL project has three principal aims:

- To make available, via a database [62], materials relating to undergraduate chemistry experiments which are educationally sound and have been evaluated by both students and academic staff. These materials consist of everything needed to introduce the experiment into another institution, as well as evaluation data relating to both the chemical and the educational aspects of the experiment.
- To provide for the professional development of chemistry academic staff by expanding the understanding of issues surrounding student learning in the laboratory.
- To facilitate the development of a community of practice in chemistry education within the broader academic community of the Australasian region.

ACELL provides a single place, the web site [62], from which a complete, ready to implement experiment may be acquired. It provides a package containing student notes, technical, and demonstrator guides. In addition ACELL aims to add value to existing or new experiments by supplying the pedagogical analysis through an Educational Template, which provides the adopter with effective guidance on integrating the experiment into the educational objectives of a unit or course. The Template not only provides learning outcomes, but also evaluation data relating to the educational value of the experiment. Moreover, it is the principal tool used in achieving the second aim of ACELL. Most chemists engaged in teaching may not be particularly well read in education literature, and are not conversant with the language and methodologies of that discipline. The Template engages teaching chemists into thinking about their laboratory-based teaching and learning and at the same time provides an accessible entry point into educational concepts. The Template is divided into four sections, which present (i) a general summary, (ii) an analysis of the educational objectives, (iii) empirical data relating to student experiences, and (iv) relevant documentation. Beyond the Template itself, access to educational theory literature has been offered by the provision of further explanatory material on the web site.

These outcomes of the project have contributed to academic staff development by, for example, providing educators with a framework to identify and integrate intended student learning outcomes from the outset of designing and/or reviewing a laboratory exercise. Further staff development opportunities were initiated through the active involvement of staff delegates as "students" during the project workshops. All experiments accepted into the ACELL database are evaluated in a workshop. In these workshops an approximately equal mix of educators and undergraduate students participate as learners and both contribute "learner" evaluation feedback. During the workshop academic staff delegates were deliberately assigned to test experiments in areas both inside and outside their fields of sub-discipline expertise, forcing them to move beyond their comfort zone. In this way, the evaluation of each experiment drew on the specialist expertise of staff, whilst still allowing them plenty of opportunity to experience other experiments from the perspective of a student. Likewise, student delegates were mixed so that they were able to undertake experiments across a broad range of chemistry sub-discipline areas and undergraduate year levels. Many staff found that their conceptions of student approaches to experiments were challenged by the experience of working with, and as, students. This resulted in a new-found insight on the part of some staff delegates into the student perspective of learning.

2 Methodology – The Student Learning Experience

All experiments submitted to ACELL for evaluation are subject to a standard testing procedure described in the ACELL Guidelines and Procedures document [62]; the purpose of this testing is to demonstrate the transferability of the experiment to new institutions and to evaluate it from both chemical and educational perspectives. Laboratory testing of each experiment is carried out first at an ACELL workshop and then in an authentic undergraduate teaching environment. Following the testing, student feedback regarding their perceptions of the quality of the learning experience is sought.

Data are collected using the ACELL Student Learning Experience (ASLE) survey, which is distributed to all students who have undertaken each given experiment. Collection of data relating to the ACELL processes is authorized by the Human Research Ethics Committee at the University of Sydney, project number 12-2005/2/8807; the processes described in this ethics application are followed and thus completion of the survey is voluntary, and all responses are anonymous. Although the anonymity of the survey prevents any formal statistical testing to examine whether the respondents are a representative sample of the entire cohort, we ensure that responses are received from sufficient numbers of students to be able to draw meaningful conclusions; the response rates are also reported, allowing others to make informed judgments of the likelihood that conclusions can be applied to the entire cohort with confidence.

To date, ACELL has collected data on 19 experiments from 972 students across universities in Australia and New Zealand using this instrument, and this data collection is ongoing. As a consequence, ACELL is in an unusually good position to identify and discuss both procedural and cognitive factors that influence students' evaluation of their learning experiences, such as assessment, the quality of notes, interest, and the inclusion of opportunities for independent learning. Our results are both surprising and encouraging, and indicate that students can be highly cognitively engaged, even with traditionally "boring" content, provided a suitable learning environment is established.

The ASLE instrument includes 14 Likert scale items; a summary of the statements is presented in Table 1. Twelve of the statements probe students' perceptions of aspects of the experiment (such as interest, skill development, guidance from notes and demonstrators, and improved understanding of chemistry); the remaining two items concern the time available for the experiment, and ask for an overall rating of the experiment as a learning experience. In addition, the instrument includes five open-response items, which are:

- Did you enjoy doing the experiment? Why or why not?
- What did you think was the main lesson to be learnt from the experiment?
- What aspects of the experiment did you find the most enjoyable and interesting?
- What aspects of the experiment need improvement and what changes would you suggest?
- Please provide any additional comments on this experiment here.

Table 1 Summary of ASLE Likert scale items*

Number	Item
1	This experiment has helped me to develop my data interpretation skills
2	This experiment has helped me to develop my laboratory skills
3	I found this to be an interesting experiment
4	It was clear to me how this laboratory exercise would be assessed
5	It was clear to me what I was expected to learn from completing this experiment
6	Completing this experiment has increased my understanding of chemistry
7	Sufficient background information, of an appropriate standard, is provided in the introduction
8	The demonstrators offered effective support and guidance
9	The experimental procedure was clearly explained in the lab manual or notes
10	I can see the relevance of this experiment to my chemistry studies
11	Working in a team to complete this experiment was beneficial
12	The experiment provided me with the opportunity to take responsibility for my own learning
13	I found that the time available to complete this experiment was
14	Overall, as a learning experience, I would rate this experiment as

* For items 1–12, a + 2 (strongly agree) to −2 (strongly disagree) scale is used, with a 0 (neutral) midpoint - for these items, the ideal response is + 2. For item 13, a + 2 (way too much time) to −2 (nowhere near enough time) scale is used, with a 0 (about right) midpoint - for this item, the ideal response is 0. For item 14, a + 4 (outstanding) to 0 (worthless) scale is used, with a 2 (worthwhile) midpoint - for this item, the ideal response is + 4.

Data from the Likert items are examined looking at the histograms (for distribution) and numerically by calculating the mean and standard deviation of the responses, as well as the percentage of respondents in broad agreement (agree or strongly agree), in line with standard ACELL analysis practice [62]. Data from the open-response items are separated into thematically distinct comments, and then coded into categories as part of a content analysis, following the procedure outlined by Buntine and Read [63], which is broadly based on the approach of Miles and Huberman [64]. Thematic separation of comments was done with the aim of minimizing the number of comments which need to be coded as relating to more than one category.

3 Results and Discussion

3.1 Individual Experiments

Results from the evaluation of individual experiments provide useful insight into both the strengths and weaknesses of an individual experiment. As an example, consider the responses to items 3, 10, and 14 for one particular experiment, shown in Fig. 1. It is clear that the student response to item 10 (relating to relevance) is much stronger than is the response to item 3 (relating to interest). This provides a starting point for the implementation of incremental change for the experiment – try to

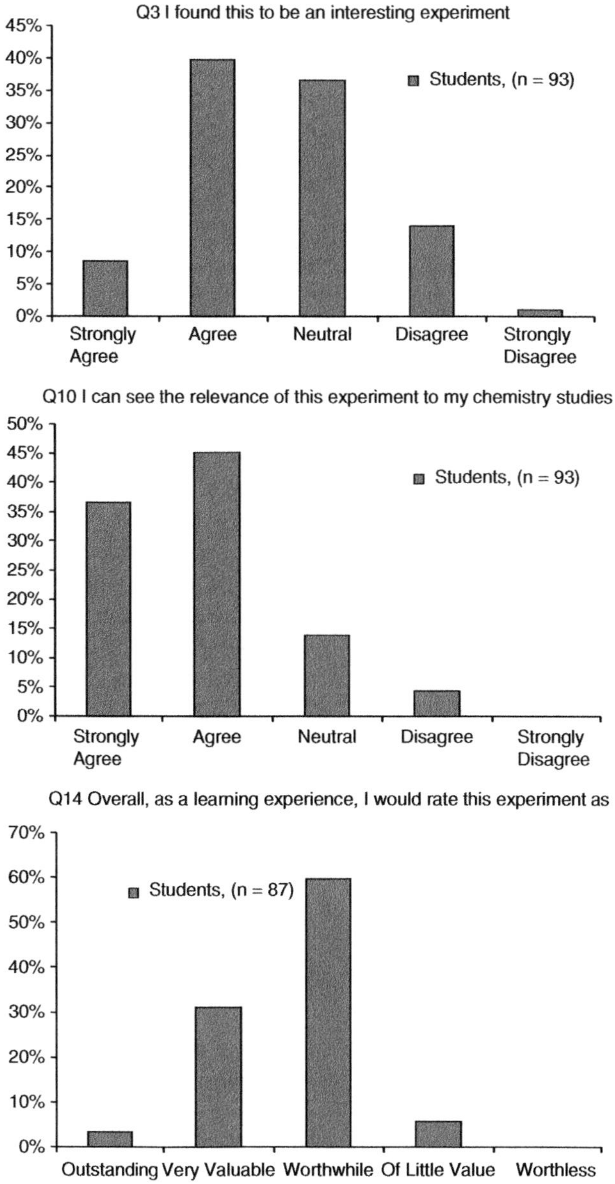

Fig. 1 Response patterns to ASLE survey items 3, 10, and 14 for one experiment

make the experiment more interesting. (Of course, that is more easily said than done!) In this particular case, an observation from the ACELL workshop process provides additional information. At the February 2006 workshop, the data collection relating to each experiment included an item where delegates were asked about their expectations for interest amongst students – responses to this item (and the aggregate data from all experiments) is shown in Fig. 2. This clearly shows that delegates' expectations were significantly higher than were students' actual interest levels in the case of this experiment.

The literature of educational psychology provides a possible interpretation for these observations. Distinctions have been drawn between individual interest, which reflects a fairly stable and enduring characteristic of an individual, and situational interest, which arises spontaneously due to characteristics of individual learning activities [65–68]. Situational interest can be sub-divided into triggered and maintained situational interest, with this sub-division effectively reflecting the difference between "caught" and "held" attention. Tasks that are involving and meaningful (and preferably related to students' goals) have been shown to maintain a situational interest once triggered [69], with maintained situational interest having been shown to be associated with a higher level of cognitive engagement than triggered situational interest. In practical terms, this means that it is desirable for an experiment (or sequence of experiments) to include a mix of activities to both trigger situational interest and to maintain it once triggered.

In this particular experiment, the differences in expected and realized interest levels can be rationalized by noting that the experiment includes activities to maintain situational interest and to promote individual interest, but little in the way of situational interest triggers. In the case of the workshop delegates, no such triggers were needed as they had chosen to participate in a chemistry experiment workshop, and thus can be expected to mostly express individual interest in the subject. Furthermore, one of the aims of the experiment (to de-mystify the "black box" which many modern instruments can appear to be by building a "home-made" version) was intriguing for those with a high level of chemical knowledge, but may have been beyond those with less knowledge. As a consequence, it might be hypothesized that interest levels could be improved by the inclusion of situational

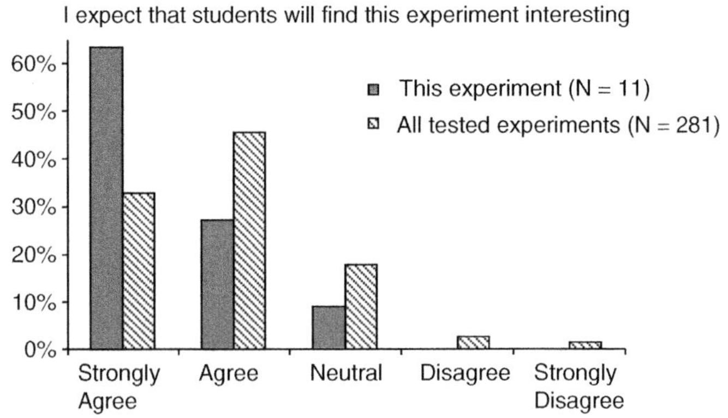

Fig. 2 Workshop delegates expectations for student interest for all experiments and the one shown in Fig. 1

interest triggers, such as by providing additional contextualization or by presenting the activity as some sort of game or competition [70].

Content analysis of the data from the open-ended items on the ASLE instrument provides evidence to support this interpretation of the data. Those who found the experiment interesting generally described reasons which pointed to engagement beyond the levels expected for merely triggered situational interest - this suggests that the experiment was effective at promoting engagement amongst those whose interest was triggered (or possibly, those who have more developed individual interests in chemistry). However, the low levels of interest in some parts of the cohort confirm that there were students whose situational interest was not triggered, and their engagement with (and evaluation of) the experiment as a whole was consequently much lower.

3.2 Experiments Considered Collectively

Returning to Fig. 1, it can be seen that the response to item 14 (evaluating the experiment as a learning experience overall) was much more consistent with the response to the interest item than they were to the relevance item. This relationship between student interest and their overall evaluation of the experiment has been further investigated. By assigning each response a value (see Table 1) and calculating means, it is possible to plot the evaluation of the interest elicited by each experiment examined against its overall rating. This has been undertaken for the 19 experiments presently in the data-set, and the results are illustrated in Fig. 3. The correlation between these items is statistically significant ($t_{17} = 7.14$, $p < 0.00001$) and indeed remarkably strong ($R^2 = 0.79$), and indicates that there is a near-linear increase in the overall evaluation of the experiment as a learning experience with increasing levels of student interest.

To ensure that the observed result was not an analysis artifact relating to the grouping of responses by experiment (due to the fact that the number of respondents per experiment ranges from 17 to 143 with an average of 51 responses per experiment), the analysis was repeated for a near-complete set of 19 experiments involving 972 student responses. These results are shown in Fig. 4. Here, the size of the circle plotted represents the number of respondents corresponding to a given matrix element; the (A,A) element corresponds to a "strong interest - outstanding overall," or (+2, + 4) rating and a (E,E) element corresponds to a "no interest - worthless," or (−2,0) rating. It is clear that the qualitative conclusions supported by Figs. 3 and 4 are identical, and we can be confident that the result is not an analysis artifact.

Similar analyses have been undertaken for the first 12 items from the ASLE survey, and it has been found that the strongest correlations occur for items 1, 3, 6, and 12. This is an extremely encouraging result, as these items relate not to maintenance issues (such as the quality of notes or demonstrators or the clarity of assessment procedures) but rather to factors reflecting affective and cognitive

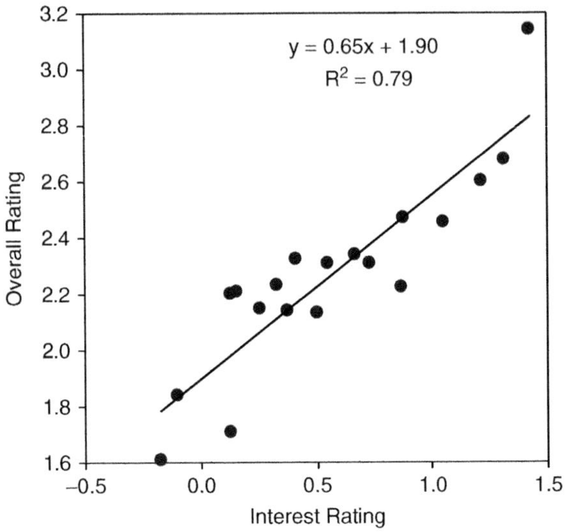

Fig. 3 Relationship between student interest score and overall score for all experiments presently in the ACELL ASLE data set, with each data point representing a single experiment

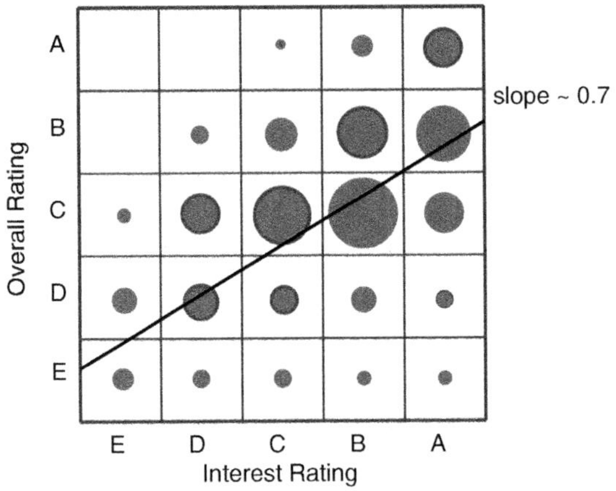

Fig. 4 Relationship between student interest score and overall score for all experiments presently in the ACELL data set. The data comes from all responses, with the area of each data point being directly proportional to the number of respondents represented

engagement and self-directed learning. The fact that students' evaluations of the learning aspects of a laboratory activity appear to derive from the high-level engagement and deep learning for which we strive strongly suggests that the laboratory is a valuable learning environment and that appropriately designed experiments can meet the challenges outlined in the introduction.

4 Conclusions

The ACELL project's ASLE survey data is providing a rich source of information that is allowing for the identification of the procedural and cognitive factors that influence student learning in the laboratory context. Strong correlations between students' overall rating of an experiment as a learning experience and the development of their (i) data interpretation skills, (ii) interest in the required activities, (iii) understanding of chemistry, and (iv) ability to take responsibility for their own learning have been identified. The data show that these correlations hold for traditionally less popular areas of chemistry such as thermodynamics, as well as for more popular sub fields of activity, and provide a valuable insight into the factors that significantly influence students' learning experiences.

Acknowledgements The ACELL project would not be possible without the financial support of the Royal Australian Chemical Institute and the Australian Government, through the Higher Education Innovation Program. The School of Chemistry and Physics at the University of Adelaide and the School of Chemistry at the University of Sydney continue to provide funding, staff and resource support to the project. Collection of data relating to the ACELL processes was authorized by the Human Research Ethics Committee at the University of Sydney, project number 12-2005/2/8807. The authors extend their gratitude to all ACELL and APCELL participants, without whose committed and enthusiastic involvement, this project would never have succeeded.

References

1. Barrie SC, Buntine MA, Jamie IM, Kable SH (2001a) APCELL: Developing better ways of teaching in the laboratory. Paper presented at the Proceedings of Research and Development into University Science Teaching and Learning Workshop, Sydney, NSW
2. Bucat RB (2004) Pedagogical content knowledge as a way forward: Applied research in chemistry education. *Chem Educ Res Prac*, 5:215–228
3. Marais P, Jordaan F (2000) Are we taking symbolic language for granted? *J Chem Educ*, 77:1355–1357
4. Royal Australian Chemical Institute. (2005). The Future of Chemistry Study: Supply and Demand of Chemists. http://www.raci.org.au/future/futureofchemistry.html. Accessed 13 May 2008
5. Hawkes SJ (2004) Chemistry is not a laboratory science. *J Chem Educ*, 81(9):1257
6. Baker AT (2005) Chemistry: Laboratory science or not? *Chem Aust*, 72(3):12–13
7. Morton SD (2005) Response to 'Chemistry is Not a Laboratory Science'. *J Chem Educ*, 82(7): 997
8. Sacks LJ (2005) Reaction to 'Chemistry is Not a Laboratory Science'. *J Chem Educ*, 82(7):997–998
9. Stephens CE (2005) Taking Issue with 'Chemistry is Not a Laboratory Science'. *J Chem Educ*, 82(7):998
10. Hegarty-Hazel E (Ed.) (1990) The Student Laboratory and the Science Curriculum. London: Routledge
11. Moore JW (2006) Let's go for an A in lab. *J Chem Educ*, 83(4):519
12. Bennett SW, O'Neale K (1998) Skills development and practical work in chemistry. *U Chem Educ*, 2(2):58–62
13. Boud D, Dunn J, Hegarty-Hazel E (1986) Teaching in Laboratories. Guildford, Surrey: SRHE and NFER-Nelson

14. Domin DS (1999) A review of laboratory instructional styles. *J Chem Educ*, 76:543–547
15. Shibley Jr I A, Zimmaro DM (2002) The influence of collaborative learning on student attitudes and performance in an introductory chemistry laboratory. *J Chem Educ*, 79(6):745–748
16. Psillos D, Niedderer H (Eds.) (2002) Teaching and Learning in the Science Laboratory. Dordrecht: Kluwer
17. Green WJ, Elliott C, Cummins RH (2004) 'Prompted' inquiry-based learning in the introductory chemistry laboratory. *J Chem Educ*, 81(2):239–241
18. Kirschner PA, Sweller J, Clark RE (2006) Why minimal guidance during instruction does not work: An analysis of the failure of constructivist, discovery, problem-based, experiential and inquiry-based teaching. *Educ Psychol*, 41:75–86
19. Mayer RE (2004) Should there be a three-strikes rule against pure discovery learning? The case for guided methods of instruction. *Am Psych*, 59:14–19
20. Skinner EA, Belmont MJ (1993) Motivation in the classroom - Reciprocal effects of teacher-behaviour and student engagement across the school year. *J Educ Psychol*, 85:571–581
21. Byers W (2002) Promoting active learning through small group laboratory classes. *U Chem Educ*, 6:28–34
22. Boulton-Lewis G. (1994) *Tertiary students' knowledge of their own learning and a SOLO taxonomy High Educ*, 28:387–402
23. Boud D, Cohen R, Walker D (1993) Understanding learning from experience. In Using Experience for Learning; Boud, D. Cohen, R. and Walker D. (Eds.); Society for Research into Higher Education and The Open University Press: Buckingham, pp 1–17
24. Johnstone AH, Al-Shuaili A (2001) Learning in the laboratory; some thoughts from the literature. *U Chem Educ*, 5(2):42–51
25. Vianna JF, Sleet RJ, Johnstone AH (1999) Designing an undergraduate laboratory course in general chemistry. *Quimica Nova*, 22(2):280–288
26. George B, Wystrach VP, Perkins R (1985) Why do students choose chemistry as a major? *J Chem Educ*, 62(6):501–503
27. Paris SG, Turner JC (1994) Situated motivation. In P. R. Pintrich, D. R. Brown and C. E. Weinstein (Eds.), Student Motivation, Cognition and learning (pp. 213–237). Hillsdale, NJ: Erlbaum
28. Coe EM, McDougall AO, McKeown NB (1999) Is peer-assisted learning of benefit to undergraduate chemists? *U Chem Educ*, 3:72–75
29. Deters KM (2005) Student opinions regarding inquiry-based labs. *J Chem Educ*, 82, 1178–1180
30. Teixeira-Dias JJ, de Jesus HP, de Souza FN, Watts M (2005) Teaching for quality learning in chemistry. *Int J Sci Educ*, 27:1123–1137
31. Gibbs G, Gregory R, Moore I (1997) Teaching More Students 7: Labs and Practicals with More Students and Fewer Resources. Oxford, UK: Oxford Centre for Staff Development
32. Bowen CW (1999) Development and score validation of a Chemistry Laboratory Anxiety Instrument (CLAI) for college chemistry students. *Educ Psychol Meas*, 59:171–185
33. Johnstone AH (1984) New stars for the teacher to steer by? *J Chem Educ*, 61:847–849
34. Johnstone AH (1997a) And some fell on good ground. *U Chem Educ*, 1(1):8–13
35. Johnstone AH (1997b) Chemistry teaching - science or alchemy? 1996 Brasted lecture. *J Chem Educ*, 74:262–268
36. Johnstone AH, Sleet RJ, Vianna JF (1994) An information-processing model of learning - Its application to an undergraduate laboratory course in chemistry. *Stud High Educ*, 19(1):77–87
37. Johnstone AH, Wham AJB (1982) Demands of practical work. *Educ Chem*, 19(3):71–73
38. Chandler P, Sweller J (1991) Cognitive load theory and the format of instruction. *Cogn Instr*, 8: 293–332
39. Kirschner PA (2002) Cognitive load theory: Implications of cognitive load theory on the design of learning. *Learn Instr*, 12:1–10
40. Paas F, Van Merrienboer J (1994) Instructional control of cognitive load in the training of complex cognitive tasks. *Edu Psych Rev*, 6:351–371
41. Sweller J (1994) Cognitive load theory, learning difficulty, and instructional design. *Learn Inst*, 4:295–312

42. Wickman PO (2004) The practical epistemologies of the classroom: A study of laboratory work. *Sci Educ*, 88:325–344
43. Lock R (1988) A history of practical work in school science and its assessment, 1860–1986. *School Sci Rev*, 70(250):115–119
44. Hodson D (1993) Re-thinking old ways: Towards a more critical approach to practical work in school science. *Stud Sci Educ*, 22:85–142
45. Marthie AM, Meester M, Maskill R (1993) The practical side of chemistry. *Educ Chem*, 30(6):156–158
46. Bennett SW (2000) University practical work: Why do we do it? *Educ Chem*, 37(2):49–50
47. Rigano D, Ritchie S (1994) Students' thinking in a chemistry laboratory. *Res Sci Educ*, 24:270–279
48. Markow PG, Lonning RA (1998) Usefulness of concept maps in college chemistry laboratories: Students' perceptions and effects on achievement. *J Res Sci Teach*, 35:1015–1029
49. Hofstein A, Navon O, Kipnis M, Mamlok-Naaman R (2005) Developing students' ability to ask more and better questions resulting from inquiry-type chemistry laboratories. *J Res Sci Teach*, 42:791–806
50. Hofstein A, Lunetta VN (1982) The role of the laboratory in science teaching: Neglected aspects of research. *Rev Educ Res*, 52:201–217
51. Hart C, Mulhall P, Berry A, Loughran J, Gunstone R (2000) What is the purpose of this experiment? Or can students learn something from doing experiments? *J Res Sci Teach*, 37:655–675
52. Hofstein A, Lunetta VN (2004) The laboratory in science education: Foundations for the twenty-first century. *Sci Educ*, 88:28–54
53. Nakhleh MB, Polles J, Malina E (2002) Learning chemistry in a laboratory environment. In J. K. Gilbert, O. De Jong, R. Justi, D. F. Treagust and J. H. Van Driel (Eds.), Chemical Education: Towards Research-based Practice (pp. 69–94). Dordrecht: Kluwer
54. Tobin KG (1990) Research on science laboratory activities. In pursuit of better questions and answers to improve learning. *School Sci Math*, 90:403–418
55. Buntine MA, Read JR, Barrie SC, Bucat RB, Crisp GC, George AV, Jamie IM, Kable SK (2007) Advancing Chemistry by Enhancing Learning in the laboratory (ACELL): a model for providing professional and personal development and facilitating improved student laboratory learning outcomes. *Chem Educ Res Prac*, 8(2):232–254
56. Jamie IM, Read JR, Barrie SC, Bucat RB, Buntine MA, Crisp GT, George AV, Kable SK (2006) From APCELL to ACELL - Expanding a multi-institution project for laboratory-based teaching and learning. *Aust J Educ Chem* 67:7–13
57. Read JR (2006) The Australian Chemistry Enhanced laboratory Learning project. *Chem Aust*, 73(1):3–5
58. Read JR, Barrie SC, Bucat RB, Buntine MA, Crisp GT, George AV, Jamie IM, Kable SH, (2006a) Achievements of an ACELL workshop. *Chem Aust*, 73(9):17–20
59. Read JR, Buntine MA, Crisp GT, Barrie SC, George AV, Kable SH, Bucat RB, Jamie IM (2006b). The ACELL project: Student participation, professional development, and improving laboratory learning. Paper presented at the Symposium Proceedings: Science Assessment, Sydney, NSW
60. Barrie SC, Buntine MA, Jamie IM, Kable SH (2001b) APCELL: The Australian Physical Chemistry Enhanced laboratory Learning project. *Aust J Educ Chem*, 57:6–12
61. Barrie SC, Buntine MA, Jamie IM, Kable SH (2001c) Physical chemistry in the lab. *Chem Aust*, 68(2):37–38
62. ACELL (2008) http://acell.chem.usyd.edu.au. Accessed 13 May 2008
63. Buntine MA, Read JR (2007) Guide to Content Analysis http://acell.chem.usyd.edu.au/Guide-to-Content-Analysis.cfm. Accessed 13 May 2008
64. Miles MB, Huberman AM (1994) Qualitative Data Analysis: An Expanded Sourcebook (2nd ed.). London: Sage
65. Ainley M, Hidi S, Berndorff D (2002) Interest, learning and the psychological processes that mediate their relationship. *J Educ Psychol*, 94:545–561
66. Hidi S, Renninger KA (2006) The four-phase model of interest development. *Educ Psychol*, 41:111–127

67. Hidi S, Renninger KA, Krapp A (2004) Interest, a motivational variable that combines affective and cognitive functioning. In D. Dai and R. Sternberg (Eds.), Motivation, Emotion and Cognition: Integrative Perspectives on Intellectual Functioning and Development (pp. 89–115). Hillsdale, NJ: Erlbaum
68. Schiefele U, Krapp A (1996) Topic interest and free recall of expository text. *Learn Individ Differ*, 8:141–160
69. Mitchell M (1993) Situational interest: its multifaceted structure in the secondary school mathematics classroom. *J Educ Psychol*, 85:424–436
70. Bergin DA (1999) Influences on classroom interest. *Educ Psychol*, 34:87–98
71. White J, O'Connor J, Mousley J, Cole M, MacGillivray H (2003) Rebuilding the Enabling Sciences: Reclaiming the Key to Unlock the Nation's Potential. http://www.raci.org.au/national/news/mediareleases.html. Accessed 13 May 2008

Chemistry Education for Socially Responsible and Sustainable Development: What are the Challenges for a Developing Country?

V. Hunma

Abstract Historically, chemistry education in Mauritius had a clear utilitarian and vocational focus: it was meant to support local agricultural activities. Its inferior position compared to 'higher classical attainments' was tacitly acknowledged. However, the introduction of formal chemistry courses associated with external examinations and scholarships raised its academic status and inadvertently steered the shift from a vocational to a purely academic focus. Moreover, the logistical constraints to effectively communicate the subject's inherent and interlinked macro, sub-micro and symbolic components rendered it bookish. In the process, the appreciation of the nature, the methods of science, the skills that should have been developed through scientific enquiry and other higher-order cognitive skills were neglected. This resulted in chemistry education being divorced from the developmental priorities of the country.

This paper explores the challenges facing Mauritius because of the dichotomy between the educational focus and the country's priorities and proposes corrective measures.

1 Introduction

1.1 Have We Become the Victims of Our Own Progress?

This author grew up in Bhopal and did her primary, secondary and most of her tertiary schooling in Bhopal. Unfortunately, we do not remember Bhopal for its very strong Begum Rulers or the beautiful lakes but for the chemical catastrophe that killed between 16,000 and 30,000 people and affected over 500,000 people [1].

In *Five Past Midnight in Bhopal* [1], Dominique Lapierre and Javier Morro recount this story. The government gave a cow and 0.5 acre (1 acre = 4046.86 m^2)

V. Hunma
Retired Principal Research & Development Officer, Mauritius Examinations Syndicate
Freelance Lecturer, Mauritius Institute of Education
e-mail: vandanahunma@yahoo.com

M. Gupta-Bhowon et al. (eds.), *Chemistry Education in the ICT Age*,
DOI 10.1007/978-1-4020-9732-4_35, © Springer Science+Business Media B.V. 2009

of land to grow fodder for the cow to each household of a village in Orissa. This cow was to produce new breeds that promised to give eight times more milk. The crossbred calves, unfortunately did not survive and it soon became difficult to feed the cows as the fodder crop became infested with black aphids.

Around this time, Union Carbide came to India to rescue the four million Indian farmers from the ravaging insects with methyl isocyanate (MIC) to be sold as 'Sevin' [1]. The rest we all know.

The intention here is not to paint a negative picture of science because of the most unfortunate, irresponsible and unpardonable neglect of safety procedures at the Union Carbide factory or to confirm Malthusian predictions, but to raise some fundamental and serious questions regarding development and its cost. Is chemistry all that evil? How have we become victims of our own success?

More recently, biofuels promised a renewable solution to the depleting fossil fuels but in the process brought about severe food shortages. On hindsight it is easy to say that the solution would have been viable if we had unlimited land …, if we had unlimited resources to grow…, if we did not have to replace food crops … and so on and so forth.

Clearly, there cannot be development if one irresponsibly disregards social, environmental and ethical implications of the application of otherwise sound scientific ideas.

1.2 Why Sustainable Development?

From many such incidents emerges the need for sustainable development. Brundtland Commission (1987) defines sustainable development as development that '*meets the needs of the present without compromising the ability of future generations to meet their own needs*' [2]. In most cases, the consequences of our actions manifest themselves within the same generation.

The situation can become critical particularly for small island developing states like Mauritius because of their inherent vulnerability; the Prime Minister of Mauritius, Hon. Dr. Navinchandra Ramgoolam pointed out during his address to the UN Assembly's annual high-level debate on global warming held on 28 September 2007. He further informed the Assembly that the small island developing states

> have limited capacity to withstand the negative effects of natural disasters and external economic shocks. …They face in their adaptation choices fundamental constraints of inadequate data and technical capacity and limited financial resources. [3]

Mauritius is already a victim of the adverse impact of climate change (we cannot forget the floods of 26 March 2008) and of fuel and food crises, to which it may not have contributed in any significant manner. The difficulties get further exacerbated because of our heavy dependence on imports of basic commodities and overseas markets for our goods, among others. The social responsibility angle of development has become crucial at least for the small island states.

Interestingly, Mauritius is not new to such challenges. In its short history, with human resources as its key asset, it has shown remarkable growth and development in various sectors, underpinned by education, hard work and research.[1] This success supported by quota and preferential trade agreements has gradually resulted in complacency. Thus, there is a need to take stock of the situation and refocus.

1.3 Education for Sustainable Development

The challenges that we face place an enormous responsibility on the education of our children. There is a need for education for sustainable development (ESD). In the context of the UNESCO United Nation's Decade of Education for Sustainable Development (2005–2014), UNESCO defines ESD as follows:

> Education for sustainable development, especially at the secondary level, … [involves] a process of learning how to make decisions that consider the long-term stability of the economy, ecology and equity of all communities. Building the capacity for such future-oriented thinking is a key task of education and training. [8]

Obviously, the emphasis is on developing the higher-order abilities crucial for sound decision making. Some of the core competencies for ESD[2] include critical thinking, futures thinking and systemic thinking. These are expected to provide the much needed 'helicopter view' that is vital for sound decision making.

Chemistry, as part of our everyday life as well as a central science, offers many possibilities to develop the ESD core competencies.

In this presentation, an attempt is made to examine the existing practices of secondary school chemistry to identify some directions to meet the challenges for sustainable development and raise some critical issues in this direction.

2 Existing Practices of Secondary School Chemistry

Given below are some research findings that focus on the current practices and provisions for school chemistry as well as some factors that affect these practices.

2.1 What Do Students Think of the Secondary School Chemistry Education?

A research study was carried out in 2006 and 2007 to find out the views of students regarding current practices and provisions. The sample of the study comprised 101 University of Mauritius (UOM) students and 108 MBBS Year 1 students of the

---

[1] Mauritius has a rich history of locally conducted scientific/sugar research that can be traced back to the eighteenth century. See [4–7] for details.

[2] Tilbury and Wortman [9] have delineated the core competencies for ESD as follows: envisioning or futures thinking; critical thinking; systemic thinking; building partnerships to work together and partnership in decision making.

Sir Seewoosagur Ramgoolam (SSR) Medical College who had successfully completed the Cambridge 'A' level chemistry.[3] The data was collected through questionnaires and interviews.

It is important to state here that the sample was not a representative sample of the school population that studied chemistry and comprised mostly those at the upper end of the spectrum. It was felt that the University students would be in a better position to reflect on their practices now that they were out of the school system and respond objectively, freely and also take into account the requirements of their current courses.

For the purpose of this paper, students' responses on three questions were considered. These questions required them to list topics they found difficult, plausible reasons for this difficulty and their suggestions for school chemistry.

Two interrelated and important aspects of secondary school chemistry emerged from the interviews and questionnaires.

2.1.1 Insufficient Links with Everyday Life

The first finding is about the weak links between the classroom chemistry and everyday chemistry (although all students recognised the importance of chemistry in everyday life). Classroom teaching involved examples from everyday life in the form of information. Most of these, however, were not dealt with at the depth that is required to understand the underpinning principles. After a mere mention of the subject matter's link to its application, the classroom teaching bogged down to theory mostly in its molecular and symbolic forms.[4] This rendered chemistry distant and 'irrelevant' to the students' needs and also affected their cognitive engagement in the tasks.

2.1.2 Understanding of Chemistry

The second finding relates to the exclusive focus on lower-order cognitive skills (LOCS) and knowledge inside the classroom. Limited illustrations and practical work hampered the students' understanding of chemistry and its applications in everyday life.

Practical work, on the other hand, required them to follow set procedures and drilling in order to correctly answer examination questions. Few practical sessions provided opportunities for them to design their own procedures to solve a problem. This narrow implementation of practical work was also underlined by an earlier study that involved observation of 'A' level practical classes [11]. Although practical work was geared towards examination requirements, surprisingly not all

---

[3] See Appendix for the sample of study.

[4] See Jonhstone [10] for the three forms of chemistry knowledge.

examinable objectives of practical work were provided for inside the science laboratory [12].

Most of the topics that students found difficult tackled chemistry either at the sub-micro and structural level where it was not easy to recognise how the structure and bonding influenced the properties or at the symbolic and mathematical level. They admitted that their difficulty with the physical chemistry topics was due to their insufficient mastery of mathematics and physics.

Sometimes the difficulty arose because of their failure to discern the links either to other related subject matter or to its applications. There were too many discrete chunks of information they had to grapple with.

They thus studied these topics (with or without understanding) to do well in examinations. The university lecturers also underlined, during informal conversations, students' inadequate knowledge and ability to handle the demands of the tertiary courses.

2.2 *What Do the Cambridge 'O' and 'A' Level Syllabuses Prescribe?*

The Cambridge 'O' and 'A' level chemistry syllabuses, on the other hand, prescribe higher-order cognitive skills (HOCS) and application of chemistry in everyday life. The aims include helping students

> become confident citizens in a technological world, able to take or develop an informed interest in matters of scientific import; recognise the usefulness, and limitations, of scientific method and to appreciate its applicability in other disciplines and in everyday life. [13]

The syllabus content further indicates the topics that are of everyday relevance and that can easily be linked to everyday applications.

However, examination reports confirm the narrow implementation of chemistry inside a classroom. Significant numbers of students reportedly face difficulty not only in applying their chemistry knowledge to solve problems but surprisingly also in handling basic skills [12]. The 'A' level November 2007 report for paper 5 Planning, Analysis and Evaluation [14][5] underlines students' difficulties with some of the basic skills and confirms the above findings:

> As in June 2007, many candidates had been well prepared for the examination but many others lacked the basic skills of graph plotting, drawing tangents, calculating gradients and evaluating the graphical evidence. The skills required in planning an experimental method remain a weakness for many candidates. Examiners would emphasise that such skills can only be adequately developed by practice in planning and implementing experiments in the laboratory. [14]

---

[5] The new format of paper 9701/05 was first examined in June 2007 and formed part of the HSC/A level examination for the first time in November 2007.

2.3 *What Does the Lower Secondary Chemistry Syllabus Prescribe?*

The Forms I, II and III chemistry textbooks [15–17] were examined to find out what they prescribe. This is because only a quarter of students opt for the science stream in Form IV. Some observations on the 'to the point' textbooks are as follows.

2.3.1 Subject Matter

The textbooks present chemistry subject matter with plenty of useful and colourful illustrations and exemplars in simple English appropriate to the level. Each chapter lists what a student would 'learn about' in it. Most of these 'learn abouts' and learning outcomes are addressed at knowledge level. The content is summarised under the heading 'points to remember' at the end of each chapter. Once again the emphasis is on the objectives of knowledge domain.

2.3.2 Exercises

Most of the exercises test knowledge and simple understanding of the subject matter. The answers are available in the preceding page(s).

2.3.3 Practical Activities

These aim at illustrating/confirming the content covered in the chapter. In the process, students will certainly develop manipulative skills and learn about safety procedures, practical equipment and procedures. However, there is no systematic and deliberate attempt at helping them develop skills and abilities needed to undertake scientific enquiry to critically analyse a situation, solve a problem and make decisions.

2.3.4 Everyday Relevance

There are many illustrations and examples from everyday life. These relate to students' everyday experiences. However, the description of these applications is at knowledge level with little scope and directions for critical analysis and scientific enquiry.

2.3.5 Historical Perspective

There is hardly any discussion of the historical perspective to indicate how this knowledge has come about and why has the science community accepted it. This can help them appreciate the nature and methods of science.

2.4 Secondary School Chemistry

Some inferences that may be drawn from the research carried out to develop an understanding of the existing practices of secondary school chemistry are as follows:

- Some LOCS and abilities are not sufficiently developed at the classroom level.
- HOCS and abilities are not sufficiently addressed.
- The opportunities to understand the nature and methods of science are inadequate.
- Everyday applications of chemistry are presented at knowledge level.
- The history of science is rarely discussed.

2.5 How Has Secondary School Chemistry Come to Assume Such a Narrow Role?

Historically, utilitarian considerations prompted the introduction of chemistry in the secondary school curriculum [11]. Common sense would indicate that this vocational bias would strengthen its links to the local needs and research. Unfortunately, this was not the case because of the supposedly less demanding nature of science subjects:

> The college is divided into a classical and a modern section, with a view of enabling the students to follow one or the other accordingly as he may be destined either for a learned Profession or for a career where the higher classical attainments are not considered indispensable. [18, p. 5]

This 'low status' of science partly contributed to the neglect of its provisions. There were limited facilities for 'hands on' experimentation. The content could not be illustrated and the teaching and learning became bookish [7].

The scheme of English scholarships introduced in 1813 to provide opportunities to study in a British university also contributed to making it theoretical and bookish [11]. The students of the Modern section, established in 1861, were eligible for these scholarships. However, it was only from 1894 when by the Articles I and II of Ordinance 1892 a scholarship was reserved for the best student of the Modern section [19]. This regulation, as evident from the reports of examinations, had an unfortunate and inadvertent repercussion on learning. The report of the 1925 scholarship examination points out:

> The answers indicated that the candidates had read widely but had not digested their reading. They half knew many things, but no single question in any part of the paper was really well done, and no point was thoroughly understood. Many of them seemed to suspect subtle and hidden difficulties in the questions themselves, clear and simple answers to simple and specific questions were the exception. Even the better candidates were lacking in a sense of proportion, as when they required an electric furnace in order to show the presence of phosphorous in bones and a blast furnace for demonstrating that there is carbon in steel. In general, the more elementary the question the poorer the answer it evoked. Not one candidate for instance knew a practical way of showing the presence of tin in a penny coin, and the majority proposed to dissolve the coin in hydrochloric acid.

> Similarly in the organic section, candidates who were able to write a structural formula of glucosazone were baffled when asked for experimental proof that there is chlorine in chloroform. [20, p. 67]

In this way, the secondary school science gradually distanced itself from the initial promise of vocational bias, became academic and remained unaffected by the scientific research of local import and in 1941, the Director of Public Instructions, Ward (1941), suggested:

> The emphasis in the science course should be shifted so as to provide better for the needs of a country in which the agriculturist is the chief user of a scientific knowledge and technique. I do not imply that commercialism should govern the curriculum. I do not suggest that the college should abandon its classical tradition. ... But I think the existing college curriculum goes almost to the extreme of remoteness from everyday life and could be recalled with advantage. [21, p. 40]

It would not be incorrect to conclude that the limited opportunities for tertiary education laid the foundations of an examination-driven and academically focused education system with the main aim of preparing for the 'high stakes' examinations. The deep-rooted strong academic traditions with examinations and scholarships still dominate the system.

3 The Challenges for Chemistry Education for Socially Responsible Sustainable Development

In principle, the responsibilities of secondary school chemistry are twofold. On the one hand, there is the need to help students master both the LOCS and HOCS inherent to the subject to operate at higher conceptual levels. On the other hand are the interlinked human, social, environmental and ethical considerations. The two positions are not mutually exclusive, and any artificial fragmentation results in not only presenting a distorted and narrow view of chemistry but also comes in the way of developing the core competencies for ESD.

The challenges for a developing country, however, are not limited to the formulation of a theoretical framework to address the two sides but extend to its effective implementation. How do we make it work? The difficulties arise mainly due to the following reasons.

3.1 Our Priority Is to Prepare for the Examination

It is the examination that drives what happens inside a classroom. Obviously, not all the intended objectives are implemented inside a classroom. However, what is surprising is that not all feasible objectives amenable to examinations are addressed inside the classroom.

3.2 Physical Facilities

Inadequate physical facilities to illustrate the underpinning theory make learning bookish. The use of information and communication technology (ICT) in learning is negligible.

3.3 Logistics Override Pedagogy

One example is the replacement of the 'O' level science practical examination due to malpractice in 1982 by a written alternative to practical paper. While we all recognise the weaknesses of the latter and the resulting unwanted repercussions at the classroom level, it is the logistical constraints that come in the way of introducing coursework assessment to test practical skills.

3.4 Child Is at the Centre of Education

Within this paradigm, we also hold learners responsible for their performance and not the practices that contribute to their performance. Thus, we do not question our practices and accept the results as irreversible. As a result, many students do not develop even the basic skills.

3.5 Management of Learning

With the main concern to maintain attendance, punctuality and discipline, most school managers overlook what happens inside a classroom when the teacher and students meet. What are the objectives? How are these objectives being addressed? What support, training, physical infrastructure do teachers require? What feedback is given to students and to what extent is this useful in improving their learning? In other words, the pedagogical issues are seldom addressed [22].

3.6 Inadequate Language Skills

We learn science in English which is a foreign language. For some mysterious reasons, many science students (including the very bright ones) believe that they do not require language skills, neglect languages, boast about it, '*I am not good at English*' and spend more time focusing on science lessons. Parents readily buy this view and further encourage it. This is unfortunate because we cannot deny the importance of language skills in understanding science and communicating it.

4 Recommendations

Education for sustainable development requires a different mindset characterised by a 'helicopter view' of the subject matter and its interactions with nature and society.

4.1 *Recognise the Differences Between Chemistry and School Chemistry*

Science operates in an open system and involves exploring and working with multiple unrelated divergent ideas to understand the unknown. It proceeds, as argued by Paul Feyerabend [23] in his *Against Method*, through 'epistemological anarchism'.

> Science is an essentially anarchistic enterprise: theoretical anarchism is more humanitarian and more likely to encourage progress than its law-and-order alternatives. This is shown both by an examination of historical episodes and by an abstract analysis of the relation between idea and action. The only principle that does not inhibit progress is: *anything goes.* ... The idea that science can, and should, be run according to fixed and universal rules, is both unrealistic and pernicious.

School chemistry, on the other hand is mostly about communicating the established body of knowledge or, in other words, a convergent view of science. Kuhn [24] describes the practices of '*normal science education*' as follows:

> The objective of a textbook is to provide the reader, in the most economical and easily assimilable form, with a statement of what the contemporary community believes it knows and of the principal uses to which that knowledge is put. Information about *how that knowledge was acquired (discovery)* and about *why it was accepted by the profession (confirmation)* would at best be *excess baggage.* (p. 186)

His observations on practices of science education are still valid at least in Mauritius where the access to secondary and tertiary education is controlled by a system of external examinations.

In such a system, it is much safer and 'fairer' to communicate the convergent view and not waste much time on the '*excess baggage*', irrespective of its educational significance. Let the examinees learn the expected and reproduce it. For obvious reasons, teachers would consider twice before encouraging students to think for themselves. One cannot take risks. The stakes are too high.

There is also a fear that left to explore by themselves, students may construct cognitive structures on their experience, folklore, fiction and knowledge base that may differ from the established knowledge. With large classes and little time and expertise for formative assessment[6], teachers may remain unaware of these 'alternative frameworks'.

Thus to start with, it is important to recognise the differences between the practices of chemistry and chemistry education. The commonalities that exist between the two blur the differences with the result that we believe that the two are one and the same thing and subsequently expect school chemistry to produce scientifically literate citizenry and specialist chemists who would be able to respond to societal needs, sustain development and leave a legacy that future generations would be proud of. This is not automatic, at least not for all students.

There is a need for deliberate efforts in this direction especially because science education does not remain a bastion of the select bright ones and the attempts are being made to increase access and encourage others.

4.2 Address the 'Excess Baggage'

It is crucial to address this '*excess baggage*' [24] to help students see how science knowledge has come about and why it was accepted. Research has established the importance of history and philosophy of science in understanding science and appreciating its nature and methods [25–27]. Students would come to appreciate that there is no one fixed scientific method that is applicable in all situations. It depends on the situation and varies with the available techniques and knowledge.

It is important to stress here that there is no harm in exploring and 'wasting time' if it can introduce them to the nature of science.

Moreover, going beyond the limits set to present the subject matter would introduce an authentic context and enhance cognition, interest and motivation. Teachers would be able to clarify issues and information and identify inconsistencies in students' learning that they had not previously considered. Instead we often leave unanswered questions for later when there would be more time and students more mature – '*you will learn it later*' or worse still, '*you will not understand at this stage.*'

4.3 Examinations

While it is neither possible nor desirable to do away with the examinations, it is important to underplay the competition aspect. This would help address one important core competency of ESD which is about building healthy partnership to work together and in decision making.

---

[6]During the research work carried out in August/September 2006, students pointed out that the assessment feedback was in terms of marks/grades/simple comments such as good, fair, can do better. … The directions to identify their weaknesses and to improve were minimal. Often students gave up and demonstrated poor confidence in their abilities to handle the demands with the very common, "I am not good at …."

It is also important to introduce coursework assessment, which offers a sound option to test a wide range of practical skills to replace the 'O' level written alternative to practical (ATP) paper.

4.4 *Best Practices*

The examples of best practices and how these can enhance cognition, interest and engagement are plenty and considerable work is being done. However, what is required is an effective management structure to ensure sound implementation of these practices at the classroom level. Some of these are presented below.

4.4.1 Learner-Centered Approach

Recently, some 40 PGCE students[7] allowed their chemistry classes to be directed by the needs and interests of their students and found the approach more effective in enhancing students' understanding and motivation in the task. Unfortunately, they also found the approach to be more time-consuming. The question is what is more important, the quality of learning outcomes or the volume of course content covered.

4.4.2 Context-Based Approach

The need to incorporate a context-based approach cannot be over emphasised, especially when we move towards the goals of mass education or 'science for all'. The Royal Society of Chemistry (RSC) [28] has produced some interesting material appropriate to the interests of students that contextualises chemistry without losing sight of the underpinning subject matter.

4.4.3 De-emphasising Memorisation and Encouraging Critical Thinking

It is crucial to discourage mere memorisation under any circumstance. It is important to move away from questions that ask students to give full name say of the inventor of Bunsen burner - this was one of the questions asked at a recent science quiz organised to encourage science learning. It sends very misleading signals for the directions that learning should take. A sound alternative would be to ask them to imagine how he arrived at the current model.

Research studies [12, 29] have shown that most tasks set at the classroom level are close-ended type with fixed answers. While one would readily concede that there is a need to confirm the established body of knowledge at the end of the lesson, and while one would be against the idea of a chaotic class either with students

[7] PGCE 2007–2008 chemistry cohort.

exploring and discovering 'science' on their own, one cannot but regret the absence of open-ended tasks. Alternative options and 'wrong' answers along with the reasons why these are not appropriate can be discussed and explored further. Some evaluation exercises have been included in the 'A' level paper 5 [13] from last year. These should be emphasised at other levels also.

4.4.4 Practical Work

There should be more opportunities for practical work at all levels to illustrate the abstract science as well as help students develop practical skills needed to undertake scientific enquiry, solve problems and make decisions.

4.4.5 Introduction of Formative Assessment

Research work by Black and Wiliam [30] and Black et al. [31] has not only established that formative assessment can improve standards but has also suggested the strategies for formative assessment. Some exploratory work [32] with primary school pupils also confirmed the pedagogical significance of self-assessment and introspection. It was only when the 10 year olds were asked to think about their learning after the lesson that they realised their difficulties and thereby provided useful directions to the teacher.

4.4.6 Classroom Discussions and Projects

These provide excellent opportunities for systemic and future thinking.

4.4.7 ICT

ICT can play an important role in ESD. First, it can enhance the understanding of the subject matter by presenting models of micro chemistry invisible to the naked eye and simulations of reactions that are too complex, fast, slow, dangerous, expensive, minute, … to demonstrate in the classroom.

Second, it offers an excellent tool to promote futures and systemic thinking by simulating complex interactions among multiple variables which are not easy to discern at one and the same time.

5 To Conclude

Given the central role chemistry has in everyday life, school chemistry offers immense possibilities to develop the core competencies for sustainable development. We have to decide whether we want to present it as a '*rhetoric of conclusions*' [33] for examination success as is the case today or whether we wish to go beyond to produce future generations suitably equipped to support the sustainable development agenda. The choice is ours.

Acknowledgement The author wishes to thank:

- UoM and IUPAC for this opportunity
- UoM – Prof. I. Fagoonee, Prof. A. Gurib-Fakim, Associate Prof. H. Li Kam Wah, Associate Prof. P. Ramasami, Associate Prof. Laulloo for their support in data collection
- SSR Medical College – Prof. S. Shukla and the staff of Anatomy department for their support in data collection
- Students of UoM and SSRMC for painstakingly filling in the questionnaire and answering my questions

Appendix

Sample of the Research Study

Table 1 gives the sample of the study and Table 2 their 'A' level chemistry results. Around 20% students did not give their chemistry results.

Table 1 Sample of the study

	University of Mauritius			SSR Medical College			
Year		Boys	Girls	Country	Boys	Girls	Total
2006	BSc 1	8	15		–	–	23
2006	BSc 2	2	–		–	–	2
2006	BSc 3	11	26		–	–	37
2007	BSc 1	11	14	India	16	16	57
2007	BSc 2	–	–	Mauritius	36	32	68
2007	BSc 3	2	12	South Africa	2	6	22
Sub total		34	67		54	54	
Total		101			108		209

Table 2 'A' level chemistry results of the sample

	University of Mauritius		SSR Medical College		
Grades	Boys	Girls	Boys	Girls	Total
A	4	4	30	27	65
B	8	20	16	18	62
C	4	18	4	6	32
D	1	6	–	–	7
Subtotal	17	48	50	51	166
Total	65		101		166

Two UoM chemistry practical classes were observed and 20 students selected randomly were interviewed. Thirty-five SSR Medical College students were selected by their lecturer for the interview.

References

1. Lapierre, D. and Morro, J. (2002). *Five Past Midnight in Bhopal*. UK: Scribner.
2. Brundtland Commission Report (1987). http://www.un.org/documents/ga/res/42/ares42-187.htm. Accessed 10 May 2008.
3. Government of Mauritius (2007). Prime Minister's Address at the UN on 28.09.2007. http://www.gov.mu/portal/goc/pm/files/onu07.pdf. Accessed 12 May 2008.
4. Public Relations Office of the Sugar Industry (PROSI) (1987). *Sugar in Mauritius*, Fourth Edition, (Revised). Mauritius: The Mauritius Chamber Of Agriculture.
5. Manrakhan, J. (1991). *The Mauritius School of Scientific Agriculture*. Reduit: University of Mauritius & EOI.
6. Storey, W.K. (1997). *Science and Power in Colonial Mauritius*. Rochester, New York: University of Rochester Press.
7. Hunma, V. (2002). Secondary School Science and Technology in Mauritius. *Science & Education*, 11(5), 497–511.
8. UNESCO United Nations Decade for Education for Sustainable Development (2005–2014). http://portal.unesco.org/education/en/ev.phpRL_ID = 27234&URL_DO = DO_TOPIC& URL_SECTION = 201.html. Accessed 12 May 2008.
9. Tilbury, D. and Wortman, D. (2004). *Engaging People in Sustainability*. Cambridge: IUCN.
10. Johnstone, A.H. (2000). Teaching Chemistry – Logical or Psychological? *Chemistry Education: Research and Practice in Europe*, 1(1), 9–15.
11. Hunma, V. (2001). Examinations, Scholarships and Higher School Certificate Science Practical Work. *Journal of Education*, 1(1), 11–20.
12. Hunma, V. (2003). Do 'A' Level Science Laboratory Practices in Mauritius Cater for the Requirements of Cambridge International Examinations? *Journal of Education*, 2 (1) 9–28.
13. University of Cambridge International Examinations (CIE) (2008). Chemistry 'O' and 'A' Level Syllabuses, http://www.cie.org.uk/qualifications/academic/uppersec/alevel/subject?assdef_id = 736 and http://www.cie.org.uk/qualifications/academic/middlesec/olevel/subject?assdef_id = 901. Accessed 10 May 2008.
14. University of Cambridge International Examinations (CIE) (2008). Examination Report of the November A Level Chemistry Examination, CIE Syllabus and Support CD for Centres in Mauritius, 9701_w07_er.pdf, Cambridge: CIE.
15. Ministry of Education and Human Resources (2004). *Science Form III*. Stanley, Rose Hill: EOI.
16. Ministry of Education and Human Resources (2005). *Science Form II*. Stanley, Rose Hill: EOI.
17. Ministry of Education and Human Resources (2006). *Science Form I*? Stanley, Rose Hill: EOI.
18. Royal College of Mauritius (1861). *Rules and Regulations of the Royal College of Mauritius*. Port Louis: General Steam Printing Company.
19. Royal College of Mauritius (1894). *The Calendar 1894*. Port Louis: Government Printing Establishment.
20. Royal College of Mauritius (1927). *Calendar of the Royal College of Mauritius 1925–27*. Port Louis: Government Printer.
21. Ward, W.E.F. (1941). *Report on Education in Mauritius*. Port Louis: Colony of Mauritius.
22. Hunma, V. (2005). *Assessment for Assessment's Sake or for Learning: What Can Trigger the Change?* Report of the Symposium on Assessment and Future Directions, Mauritius Examinations Syndicate, Reduit, 48–75.

23. Feyerabend, P. (1975). *Against Method.* http://www.marxists.org/reference/subject/philosophy/works/ge/feyerabe.htm. Accessed 5 May 2008.
24. Kuhn, T.S. (1977). *The Essential Tension.* Chicago: University of Chicago Press.
25. Matthews, M.R. (1994). *Science Teaching: The Role of History and Philosophy of Science.* New York: Routledge.
26. Niaz, M. (2001). How Important Are the Laws of Definite and Multiple Proportions in Chemistry and Teaching Chemistry. *Science & Education,* 10(3), 243–266.
27. Hodson, D. (1988). Towards a Philosophically More Valid Science Curriculum. *Science Education,* 72(1), 19–40.
28. Kind, V. (2004). *Contemporary Chemistry for Schools and Colleges.* London: RSC.
29. Hunma, V. (2007). Secondary School Chemistry, Unpublished research report.
30. Black, P.J. and Wiliam, D. (1998). *Inside the Black Box, Raising Standards Through Classroom Assessment.* London: School of Education, King's College.
31. Black, P.J., Harrison, C., Lee, C., Marshall, B. and Wiliam, D. (2002). *Working Inside the Black Box, Assessment for Learning in the Classroom.* London: School of Education, King's College.
32. Hunma, V. (2002). Self-Assessment for Enhancing Learning. *Mauritius Examinations Bulletin,* 9, 37–50.
33. Schwab, J.J. (1962). The Teaching of Science as Enquiry. In J.J. Schwab and P.F. Brandwein (Eds.) *The Teaching of Science,* (pp 3–103). Cambridge, MA: Harvard University Press.

Associative Learning Through Art Activities

Y. Dahlman and M. Boman

Abstract Specific art activities were introduced at the chemistry program at Uppsala University (UU) during the academic year 2006/07 at three different university levels: (a) the preparatory year for natural science students, (b) the seventh semester in the chemical engineering program, and (c) the Ph.D. program. During 2007/08, the seventh semester activities were repeated (with some alterations). The background is that one of the chemistry teachers realized that most students are capable of learning what the textbook says, but a considerable number still lack the necessary insight to comprehend the subject. Hence, he wanted to try a different pedagogical approach – one which not only permits students to articulate what they have learned, but also allows them to demonstrate their *understanding*.

Many of the students who have participated in our project have now been able to formulate what they do not understand and, by so doing, increase their knowledge. This paper will discuss why working with pictures has had this effect.

1 Introduction: The Chemist Who Could Draw

By the early nineteenth century, chemists had successfully managed to map out the structure of many organic compounds. One type of compound, however, resisted every such attempt. These were the so-called aromatic hydrocarbons. One of them was benzene, discovered in 1825. Up to that point all known molecules were structured like chains. The benzene molecule, however, seemed impossible to diagram.

Y. Dahlman (✉)
Swedish University of Agricultural Sciences Department of Urban and Rural Development, P.O. Box 7012, SE-75002, Uppsala, Sweden
e-mail: ylva.dahlman@sol.slu.se

M. Boman
Uppsala University, Department of Materials Chemistry, P.O. Box 538
SE-75121 Uppsala, Sweden

M. Gupta-Bhowon et al. (eds.), *Chemistry Education in the ICT Age*,
DOI 10.1007/978-1-4020-9732-4_36, © Springer Science+Business Media B.V. 2009

Forty years later the solution to the benzene molecule's structure allegedly came in a dream. According to the story, the chemist Friedrich August Kekulé dozed off and, as he slept, saw atoms dancing in front of his eyes. He watched the atoms combining into chains, twisting and turning like snakes. Suddenly one of the snakes bit its own tail, closing the chain – and that is how the ring shaped molecule was discovered. This is considered one of chemistry's great breakthroughs, but what enabled it to happen at that moment in time was Kekulé's ability to recognize it when something crucial appeared. He managed to transpose the vision of his dream into the realm of the laboratory, where his knowledge of facts alone had not sufficed.

Growing up, Kekulé had shown an extraordinary talent for drawing and was therefore encouraged to pursue a career as an architect. Although he began studying architecture, he changed a few years later to chemistry. It is probable that his artistic sensibility played a large part for his later discovery.

2 The Problem of a Contemporary Chemist: How Do You Talk About What You Do Not Know?

This project is based on concrete pedagogical problems that have arisen at the chemistry program at UU. Mats Boman, professor of materials chemistry has often taught students who were having difficulties understanding certain essential chemical phenomena. According to Boman, most students are capable of learning what the textbook says, but a considerable number still lack the necessary insight to comprehend the subject. Hence, he wanted to try a different pedagogical approach – one which not only permits students to articulate what they have learned, but also allows them to demonstrate their *understanding*. He chose to employ art activities, a method that produced good results at SLU, where it was represented by Ylva Dahlman's research [1–4].

Even though there are lots of examples where teachers claim the benefit of combining artistic work with academic knowledge, there is no complete theory that explains the connection. This lack of scientific theory has been discussed by Eliot Eisner among others [5].

3 In Practice

Specific art activities were introduced during the academic year 2006/07 at three different university levels: (a) the preparatory year for natural science students, (b) the seventh semester in the chemical engineering program, and (c) the Ph.D. program. During 2007/08, the seventh semester activities were repeated (with some alterations).

This article provides a comprehensive overview of the method employed with natural science students during their preparatory year, and with Ph.D. candidates. It includes a survey of some of the assignments carried out by seventh semester students.

Finally, a general model is presented of how and why working with pictures is important to our understanding of the world, as well as a description of some of the obstacles this kind of pedagogy may encounter.

The assignments given to students can be divided into three categories: exercises in (a) diffusion, (b) observation and (c) associative learning. Our focus here will be on the third category "associative learning."

The actual assignments given and the discussions that followed are the basis of the material presented here.

Natural science preparatory year	Seventh semester	Ph.D. candidates
Defusing exercises	Defusing exercises	Defusing exercises
Observational exercises	Observational exercises	Observational exercises
Associative learning	Associative learning	Associative learning

Natural science preparatory year	**Seventh semester**	**PhD candidates**
defusing exercises 	defusing exercises 	defusing exercises 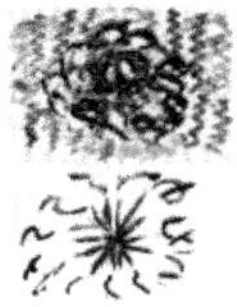
observational exercises 	observational exercises 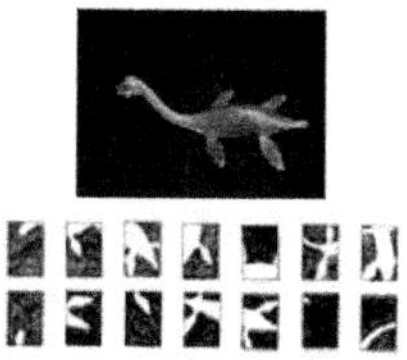	observational exercises
associative learning 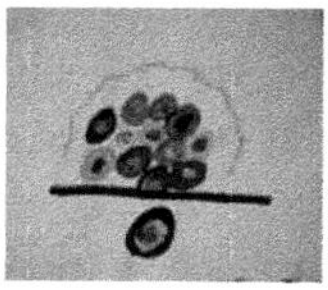	associative learning 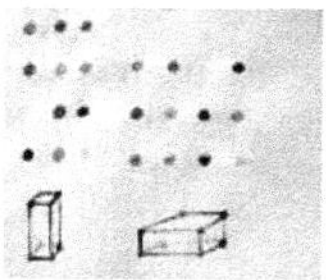	associative learning 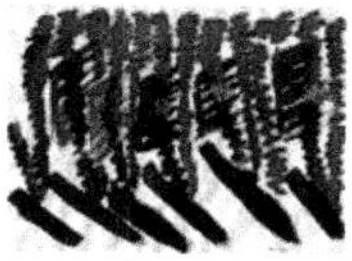

3.1 *Natural Science Preparatory Year*

Thirty-three of the natural science preparatory year students took part to some degree in the assignments, and 18 participated in the evaluation.

The evaluations themselves were based on observations made throughout the course, in addition to surveys and examination results.

The art activities that were introduced took place three times for a total of 6 h. They consisted of:

1. An introducing lecture and three defusing assignments
2. Two assignments dealing with *mole*, an important concept that many students had problem with and
3. An assignment to determine what the students found especially difficult about the course

3.1.1 Conclusions

When students are obliged to participate in an educational experiment, one can expect some to be skeptical from the outset. In this group, a number of students expressed concern that their time was not being used in the most efficient way. They felt that, in the end, traditional teaching methods were the best. In designing the project, the classroom teacher made the schedule and judged how students would best assimilate the required information. Despite such planning, there was outspoken grumbling that valuable time was being wasted drawing pictures.

The chemistry teacher claimed that the pictures gave him considerable insight into what the students had or had not understood, allowing him to adjust his teaching, which otherwise, he said, tended to keep pace with the best and most vocal students. By means of the pictures and classroom discussions, he felt he could get a much better idea of those who were doing poorly and see what was lacking in their process of understanding.

The widely diverted viewpoints held by the students and by the teacher may be a result of the teacher grasping the entire situation and knowing the objective. For their part, students could not comprehend the relevance of making pictures not directly connected to chemistry and had no confidence it would lead to something meaningful. We concluded that this was partly a result of the teacher's introduction being unclear. When assignments were more visibly linked to the subject, understanding increased and the reception was more positive. Nevertheless, it was surprising that two students thought it was completely, and four somewhat, pointless for them to air their problems with the course to their fellow students or their chemistry teacher. According to the instructor, however, such discussions were very rewarding and helped improve the effectiveness of his/her teaching.

We assumed that the reason most students were disinterested in any new pedagogical methods was because the course was only a basic requirement for university studies. We chose to drop natural science preparatory year students from the project.

3.2 *The Ph.D. Program*

Art activities were introduced for Ph.D. candidates at the midpoint of a course entitled "Electron Microscopy." It consisted of a painting class in the morning, followed by a lecture on high-resolution electron microscopy and an assignment related to the lecture. Students were asked to illustrate the relationship between lattice fringes and atomic planes. Their pictures then were reviewed and discussed the next morning.

Ten students participated in the exercises and evaluations, as well as in in-depth discussions with their teacher.

3.2.1 Conclusions

The pictures produced by the Ph.D. candidates indicated that they had understood the basic principles underlying the relationship between lattice fringes and atomic planes, but lacked knowledge of many significant details. On the Ph.D. level the concluding discussion was considered most rewarding. The indication was that although the participants initially had difficulty expressing themselves on the topic in everyday language, the exercises highlighted some very fundamental issues.

The instructor saw the use of pictures as a good way of uncovering gaps in the students' knowledge. Good classroom discussions were stimulated, something the students were not very accustomed to. This visual project helped both the students and the teacher to visualize where in the sequence of thought the understanding of the problem met a block. Here the visualizations allowed access to the pattern of each student's thoughts. Following these patterns led to solution of conceptual difficulties. Since it was now possible to see which parts of a pattern were missing or what the student had misunderstood, the instructor could focus on that specific element and quickly correct the difficulty.

We found this method to enjoy success and would have liked to develop it. Unfortunately, the format of the course was altered, making it impossible to repeat the project for a second year.

3.3 *Semester Seven in the Chemical Engineering Program*

3.3.1 Background

Material analysis 7.5 ECTS is a course in which so many difficult concepts are introduced that usually less than half of the students are successful in getting a passing grade. One such concept, "reciprocal space," is defined as a set of imaginary dots organized in such a way that a vector from one point to another coincides with perpendicular levels in normal space, and the separation of the dots is the same as the inversion of the distance between the levels. Typical distances in this instance are distances between atoms. Since the reciprocal space is an imaginary one, it cannot be depicted in textbooks, but is usually described through equations. The concept is complex and students are burdened with mathematical assignments in order to learn how to employ it. This creates a somewhat mechanical understanding, and students come away thinking they have understood the concept (in much the same way that most of us believe we understand temperature, although to really do so would require both knowledge of partial derivatives and entropy).

Usually the course on material analysis begins with Bragg's law defining diffraction in crystalline substances:

$$sin\Theta = \frac{K}{d}$$

Θ is the diffraction angle, K is a constant, and d is the distance between different θ atomic planes. The equation shows that if Θ is large, then d is small, and vice versa, i.e., they are the reciprocal of each other. To go deeper into this concept of reciprocal space the students have to define vectors and one way to do this is to use the following expression:

$$e^{iKR} = 1$$

K represents vectors in space, while R identifies vectors in space that are related to lattice point vectors.

For many years the Department of Material Chemistry at UU has tried to develop other ways to clarify these concepts so that more students could comprehend and employ this theory.

During the academic year 2006/07, art assignments were given five times for a total of 12 h. In 2007/08 it was reduced to three times amounting to 10 h in all. For the first year, the course was taught in English, and then in Swedish for the second year. A total of 22 students participated in the project.

The evaluation is based on observations made throughout the process. On the last day of the course, a written and oral evaluation was assigned. The chemistry teacher submitted a longer written reflection. Based on our experience during the first year, we made a few small changes. The following two exercises are ones that we chose to keep.

3.3.2 The Relationship Between Real Space and the Reciprocal Space

Description

The assignment was to make two pictures, one showing real space and direct lattice, and one showing reciprocal space and reciprocal lattice. The two pictures were to be made on the same sheet of paper to emphasize their relationship. They were painted with water-based tempera colors for a period of 30 min.

The pictures were then posted on a wall and each student was asked to describe what he or she had painted. The other participants were allowed to comment or ask questions. The chemistry teacher led the discussion and explained unclear points so that everyone could be able to understand the two concepts and the relationship between them by the end of the class.

Results

Among the pictures that were produced, a recurring concept was "opposite," but it was expressed in different ways. One example is seen in Fig. 1, where inside and outside are each other's opposites. Fig. 2 depicts a swirl of motion in which everything is turned inside out. Another approach is in Fig. 3, where what changes is size.

3.3.3 Kikuchi Lines

Descriptions

This exercise was divided into three parts: (1) reading an article about kikuchi lines, (2) painting a picture of kikuchi lines, and (3) attending a lecture/seminar. Thus, the assignment consisted of depicting the kikuchi lines, a specific phenomenon within transmission electron microscopy (TEM).

Kikuchi lines appear after multiple electron diffraction in TEM and are very useful to the operator in finding the orientation of a specimen. In order to understand Kikuchi lines it is necessary to understand diffraction and reciprocal space.

Fig. 1 Real space - reciprocal space. "Space hopper with many spikes." Real and reciprocal are conceived of as opposites, as in these space hoppers where the left one is on the outside and the right one is on the inside

Fig. 2 Real space - reciprocal space. The student has represented the process as a spider web where the viewer is drawn into the formula and everything is turned inside out. The picture illustrates the mathematics behind singularity when approaching zero

Since not everyone might be expected to be familiar with the phenomenon, the exercise began by having students read an article on the topic and begin painting when they felt ready. In the first year, 30 min was devoted to reading and painting. As this turned out to be too brief, in the second year 1 h was scheduled, which proved adequate.

Again, the pictures were posted on the wall and each participant described their painting (Figs. 4–7). The university's expert on kikuchi lines, Professor Klaus Leifer, was an invited guest. He then gave a lecture clarifying misunderstandings that the students had shown through their pictures.

The evaluation is based on running observations throughout the project, as well as on written and oral feedback.

Results

On the feedback forms, one of the participants wrote of being unfamiliar with the phenomenon of Kikuchi lines before the class, but stated that he/she knew much more about them after the painting and discussion sessions. Another participant claimed to have understood the concept no better after the exercise than before. It is not clear whether this student meant that nothing was added

Fig. 3 Real space - reciprocal space. The small cube on the left shows something real that changes throughout the process, so that everything long becomes short, everything blue becomes red, and everything red becomes blue (in the reciprocal picture on the right). The sun is like the real: we can see it without having to do any processing. The moon is like the reciprocal: it takes something real for us to see it. It is there but we cannot see it without the sun. In small box, dots are red and sides are blue

to what he/she already knew, or because the outcome was a total lack of understanding. Being given the introductory article to read was appreciated by the students.

The exercise using kikuchi lines proved to be a very successful way of arranging the seminar and it made much clearer to students how kikuchi lines originate. We were encouraged to continue improving the exercise: The chemistry instructor expressed interest in making it a permanent part of the course.

Fig. 4 Real space - reciprocal space. "The picture shows the little I've understood," said the student. He first made the green (+) lines to show the symmetric kikuchi map he remembered from the lab work. The rest is a confusion of chaotic lines representing all that he does not understand

3.3.4 Conclusions

In the course evaluation forms handed out to students, three questions dealt specifically with the acquisition of knowledge through working with images. The following are examples of answers students gave to these questions:

What have you understood as the purpose of the work with images?

- "Understanding the difficult theory of this course by asking new questions in connection with the painting. To see things in a different way."
- "Sort of giving an increased understanding by using a different sense. Learning to put important (chemical) things on paper and by doing so getting a picture that one can relate to. Getting an increased understanding of complex things, getting it on paper, and being able to explain it to others."
- "To observe without having prejudices about how something looks. Building bridges between parts of the brain that are normally not connected."

Do you think the purpose has been fulfilled? Justify your answer.

- "The discussions with the teachers were usually good. We got involved in a different way and dared to ask more questions than we generally do in the lectures."
- "Yes, I've learned a lot from the chemistry image exercises (reciprocal, kikuchi), but not as much, however, from when we draw portraits of each other (observational

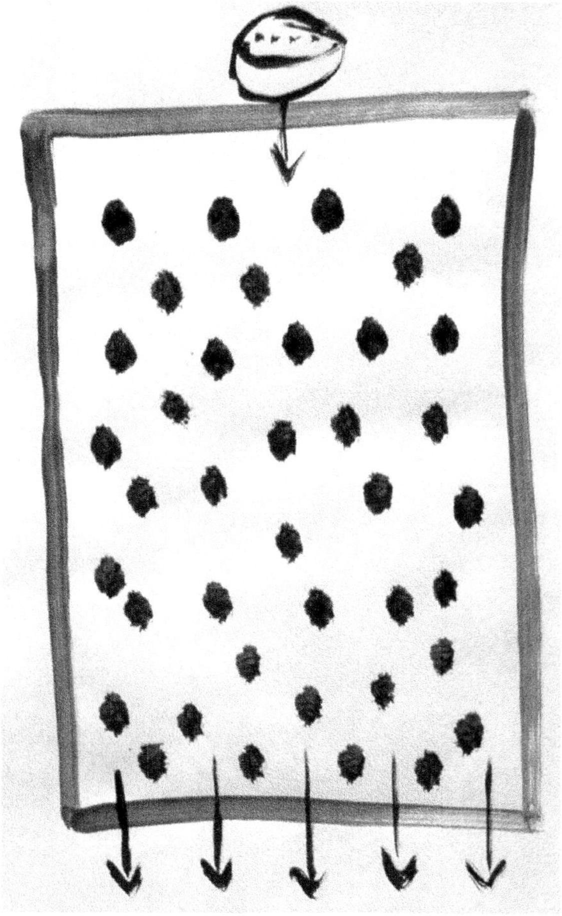

Fig. 5 Kikuchi lines. The process is visualized as a labyrinthine game. The electron is the ball, bouncing in a variety of ways and emerging unpredictably after it has undergone diffraction. The spots form a pattern without any particular order, similar to the kikuchi lines. The student claimed not to have understood what the kikuchi lines are

exercises). This may well have given me increased understanding in some way, but not as clearly and obviously as the chemistry image exercises."

- "It has helped me to understand TEM, but I would have learned the same thing by spending the same amount of time just reading about TEM."

Do you think that working with images can affect your understanding of chemistry and other subjects within science?

- "Absolutely! By looking at an image, it is easier to relate a concept to something. It is also easier to explain something if you have an image to show. This may be because it's easier for me to copy images to my brain than text. … I do not know."

Fig. 6 Kikuchi lines. The striped yellow field at the top shows electrons; the spotted part is the sample itself; the green shows the difficult theory that the student could not comprehend, and the red/black part represents the kikuchi lines

- "No, or well, it may be helpful but it is an inefficient way of learning."
- "The images can be a way to discuss difficult concepts and abstract ideas more easily. However, I wonder if the discussions aren't more important than the drawing itself."

4 Concluding Thoughts

Most students appreciated working with images, but resented the time taken away from traditional classroom teaching. Students expressed blind faith in the familiar pedagogical methods of lectures and reading textbooks. The same was true for

Fig. 7 Kikuchi lines. The student described the green kikuchi lines as a road leading to something more interesting, an exciting way of putting the kikuchi lines into use. "A kikuchi road map," the professor named it

teachers, who would have to drop something from their present course outline to accommodate the new approach.

Kekulé was skilled at drawing. He had the ability to utilize his imagination and the images from his dreams seriously in his work as a chemist. Many of the students who have participated in our project have now been able to formulate what they do not understand and, by so doing, increase their knowledge. Why did working with pictures have this effect?

Creating a picture means translating one's imagination into a concrete object. This does not mean that one can simply depict an idea. Rather, the imagination is channeled into articulating in a mode other than the verbal expression of thoughts or ideas. A picture captures a moment in the ongoing process of imagination. The act of drawing transforms such non-verbal forms of experience into artifacts that are possible to reflect upon. This is similar to what Michael Polanyi named the act of tacit knowing [6].

Art activities often require that seemingly incompatible categories of experience are being connected, one world thus being transposed into another. This is done by overcoming the resistance to leaving old and familiar categories. This theory is parallel to theories of metaphors presented by George Lakoff and Mark Johnson [7]. The range of imagination increases, more alternatives come to hand, and with them enhanced ability to formulate and solve problems. In the end, the world appears richer and more complex. When such a process succeeds and the world is accepted in a new representation, our knowledge has expanded and our relation to the world has deepened. It is in this way that knowledge is generated through engaging with pictures.

Note: Kindly contact the corresponding author if you need the colour figures.

References

1. Dahlman, Y. 1998. Bildtänkandets pedagogik (The Pedagogy of Pictorial Thinking), Pedagogiskt utvecklingsarbete, SLU
2. Dahlman, Y. 2000. Bildtänkandets pedagogik 2: Intevju med Betty Edwards (The Pedagogy of Pictorial Thinking 2: Inteview with Betty Edwards). Pedagogiskt utvecklingsarbte SLU
3. Dahlman, Y. 2004. Kunskap genom bilder: En studie i hur studenter inom natur- och samhäll svetenskapliga utbildningar fördjupar sin ämnesförståelse genom arbete med bilder. (Knowledge through pictures: A study of how pictorial practice affects understanding in the field of study for students of natural and social science). Acta Universitatis agriculturae Sueciae. Agraria, 448, SLU Sveriges lantbruksuniversitet
4. Dahlman, Y. 2007. Towards a theory that links experience in the arts with the acquisition of knowledge. *International Journal of Art Design Education*. 26:3, 274.
5. Eisner, E. 1998. Does experience in the arts boost academic achievement? *Journal of Art and Design Education*. 17:1.
6. Polanyi, M. 1983. *The Tacit Dimension*. 2nd ed. Glouchester, MA: Peter Smith.
7. Lakoff, G. and Johnson, M. 1980. *Metaphors We Live By*. Chicago, IL: The University of Chicago Press.

Self-Reflection on Professional Competences in the Master Program for Chemical Engineers

J. Reijenga and E. Vinken

Abstract This paper presents an investigation into the quality of self-reflection during industrial internships by chemical engineering students at the Eindhoven University of Technology in The Netherlands. The quality of the self-reflection reports written at the end of a compulsory industrial internship is directly related to the amount of guidelines provided to the students. Furthermore the self-reflection improves if the students are stimulated to seriously consider their own performance and learning goals at various stages during the internship. Based on these findings a new procedure with extended guidelines employing specific, measurable, achievable, relevant and time-bound (SMART) goals is introduced for the writing of self-reflection reports.

1 Introduction

Self-reflection is generally defined [1] as the capacity of an individual to exercise introspection and the willingness to learn more about one's fundamental nature, purpose, and essence. Human reflection or introspection entails the examination, consideration, and serious thought concerning one's own thoughts and feelings [2]. In a didactical context self-reflection involves a student's ability to be critical towards his own strengths and weaknesses on an academic, social, and professional level.

Self-reflection on professional competences has received wide attention in a number of fields such as vocational training [3], continuing education [4], and other professional working environments. Vocational training concerns the learning aspects of the teaching profession, whereas in continuing education professionals switch between an educational setting and a working environment. In a professional working environment, the organization aims to learn from past experiences, realizing

J. Reijenga (✉) and E. Vinken
Department of Chemical Engineering and Chemistry, Eindhoven University of Technology
P.O. Box 513, 5600MB Eindhoven, The Netherlands
e-mail: j.c.reijenga@tue.nl

M. Gupta-Bhowon et al. (eds.), *Chemistry Education in the ICT Age*,
DOI 10.1007/978-1-4020-9732-4_37,

that this learning process encompasses all individuals making up the organization. However, bringing theory into practice is not a trivial matter; something we also experienced in the internship in our master program.

The master program at the Department of Chemical Engineering and Chemistry at the Eindhoven University of Technology in The Netherlands includes a compulsory 3-month full-time industrial internship, either in the Netherlands or abroad. The educational goals are manifold [5] and comprise of the student obtaining experience in a professional, multidisciplinary environment as a chemical engineer. The student needs to learn how to solve problems by integrating (previously obtained) knowledge, skills, and critical inventiveness. At the end of the internship, the student is expected to be able to report, both written and orally, in a professional working environment.

The company at which the internship is performed actively grades the student on several criteria, together with making an overall grade recommendation. Of late, self-reflection on academic and professional competences also comprises an essential aspect in this internship. Reflection on competences has become increasingly important in the educational system on our campus, but departments are autonomous in its implementation. For example, the newly installed Department of Industrial Design has so widely implemented the system of self-reflection on competence-based learning, that the traditional system of numerical grading has been abolished [6]. Most other departments use competence-based learning in certain educational settings only, such as group work.

Self-reflection is a relatively new concept in our regular educational system for chemical engineers where it was introduced together with competence-based learning and the use of portfolios in our bachelor program in 2004. Since 2001, we have been investigating the added value of asking the students to write reflection reports after completion of their (compulsory) industrial internship in the master program. The present contribution presents the results of these self-reflection reports and how the quality of the reports is influenced by the guidelines given to the students.

2 Methodology

2.1 Group Divisions and Self-reflection Guidelines

We monitor 99 industrial internships over a 5-year period where three groups of students are (approximately chronologically) distinguished: those with very concise guidelines, those with extended guidelines, and those with extended guidelines plus additional self-grade forms, referred to as groups 1, 2, and 3, respectively. The concise guidelines merely requested the students to write a reflection on their (educational) goals in no more than 500 words. Group 2 was asked to write a self-reflection on their internships in which they were asked to discuss the following aspects/questions in 300 to 500 words.

1. What did you learn about yourself in a working environment?
2. Did you discover unsuspected talents?
3. Did you meet unexpected challenges?
4. How did you deal with them?
5. Which personal improvement points remain?
6. How did the internship broaden your view on future career opportunities?
7. The self-reflection must not be about the specific company, but about you.

These questions are referred to as questions R1 to R7 throughout this paper. The third group of students, in addition to the extended guidelines discussed above, was given a self-grading form (Appendix A) at three points in time: before commencing with the internship, midway through the internship, and after completion of the internship. By doing so, the students in group 3 were encouraged to estimate their performance based on their self-knowledge and expected challenges and to compare their own idea about their competences with that of their coaches and mentors.

3 Grading of the Student's (Academic) Performance

In our master program, the company internship projects are authorized on the basis of a document describing the subject, problem analysis, deliverables, etc. Additionally, certain details concerning the industrial supervisor are needed to ensure proper supervision and guidance on a sufficiently academic level. The hosting companies are asked to grade the student's performance on 12 different professional competences on a scale of A through E. This form is shown in Appendix A. The university department internship coordinator grades the student on the basis of the company recommendation, the technical report, the reflection report, an interview discussing these reports, and an oral presentation for a general student audience. In determining the final grade for the student, the overall company grade is taken as a guideline, from which is deviated only in a minority of cases.

4 Grading Self-reflection Reports

The self-reflection reports are graded between 0 and 5 for each of the first six questions posed above where 0 implies that little/nothing is written in the reflection report concerning a particular question, and 5 implies that excellent remarks are made concerning this question. Question R7 is also scored between 0 and 5 where 0 means nothing is written on the company itself and 5 means a great deal is written on the company. A good reflection report should therefore have high marks for questions 1–6 and a low mark for question 7. What we understand/look for in the reflection reports concerning each of the seven questions is discussed in Appendix B.

We consider a reflection report sufficient if the numerical average of questions 1–6 is at least 3.

The grading of reflection reports from the three student populations is analyzed and correlated with a number of other statistics and performance indicators. All data were collected in Microsoft Excel and imported into Statgraphics Centurion XV for further analysis [7]. We performed ANOVA tests, multiple range test or Kruskal-Wallis tests, and *t*-tests on the gathered information.

5 Results and Discussion

The most important hypothesis to test is if the quality of the reflection reports correlates with the group number, and therefore also with the amount of guidelines given to the students when writing their reflection reports.

The self-reflection reports for the three groups are analyzed semiquantitatively for each of the seven guideline questions discussed above. The quality of each report is best summarized by taking the average of R1 through to R6. One clearly observes an overall improvement in the quality of the reflection reports with increasing group number. According to ANOVA analysis (Table 1 bottom row), the *F*-ratio equals 13.07. (this is the ratio of the between-group estimate to the within-group estimate). Since the *P*-value of the *F*-test (0.000) is far less than 0.05, there is a statistically significant difference between the means of R1_6 of the three populations at the 95.0% confidence level. Multiple range tests indicate that all three pairs of groups (1–2, 2–3 and 1–3) show statistically significant differences at the 95.0% confidence level. This is illustrated in Fig 1. Thus, three homogeneous groups of means are identified, within each of which there are no statistically significant differences. For each group standardized skewness and standardized kurtosis are within range expected of a normal distribution, a requirement for using the multiple range test. The method used to discriminate amongst the means is Fisher's least significant difference (LSD) procedure.

The above described finding is the most important outcome of the present investigation: group 2 writes better self-reflections than group 1, and group 3 writes better self-reflections than group 2. There is, however, ample room for further future improvement, because, as shown by the individual scores in group 3 (see Table 1 last column), only about 50% of the students score 3 or higher (on a scale between 0 and 5) for R1_6.

A comparison between the three groups on specific points R1 through R7 was also carried out. Numerical results are presented in Table 1. The first and second column designates the grading criteria for the self-reflection report. ANOVA results are summarized as *F*-ratio and *P*-value of the null hypothesis of no significant differences between the means of the three groups. The multiple range test indicates whether the differences of the means of pairs of groups are significant at the 95.0% confidence level. The "≈" sign in all cases except R7 actually means "smaller than", but not significantly at the 95.0% confidence level. In all cases

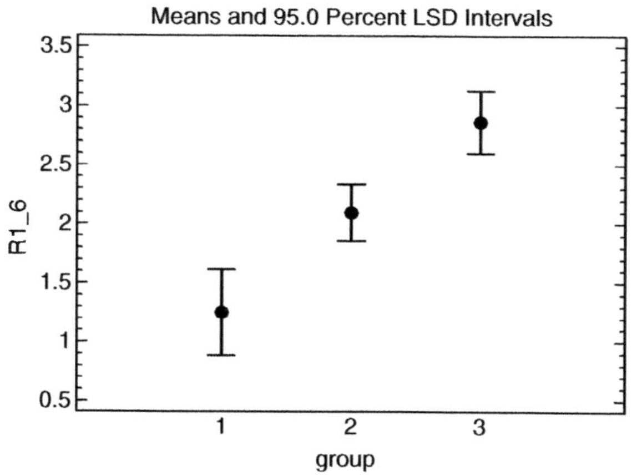

Fig. 1 Average and 95% LSD confidence interval of the quality of self-reflection reports of groups 1, 2 and 3. See text for further explanation

Table 1 Results of ANOVA and multiple range test or Kruskal-Wallis test of significant differences (SD) between three groups at the 95.0% confidence level. See text for further details

Code	Description	*F*-ratio	*P*-value	1 *n* = 19	SD	2 *n* = 44	SD	3 *n* = 36	Percentage of >=3 group 3
R1	Learn about yourself	5.69	0.0046	2.105	<	2.614	<	3.528	86%
R2	Unsuspected talents	12.17	0.0000	0.842	<	2.045	<	3.389	72%
R3	Unexpected challenges	2.18	0.1191	1.789	≈	2.136	≈	2.778	64%
R4	How to deal with them	0.85	0.4296	1.263	≈	1.659	≈	1.972	36%
R5	Improvement points	9.50	0.0002	0.368	<	1.454	<	2.639	58%
R6	Career view	6.22	0.0029	1.105	<	2.659	<	2.889	69%
R7	Company info	15.04	0.0000	3.474	>	2.25	>	0.861	8%
R1_6	Average of 1.6	13.07	0.0000	1.246	<	2.095	<	2.866	50%

except R3 and R1_6, one or more of the groups show non-Gaussian distribution, as observed from a standardized skewness and/or kurtosis outside the range −2 to + 2. In these cases, we used the Kruskal-Wallis test to compare the medians instead of the means.

More extended guidelines (1,2), and in addition forcing the students to grade themselves during different stages of the internship (2,3), clearly have a positive effect on points R1 (learn about themselves), R2 (unsuspected talents), R5 (remaining improvement points) and R6 (view on future career) in the self-reflection report. Unfortunately, the students have not significantly improved their ability to describe unexpected challenges, and how they have dealt with them (R3 and R4).

When grading the reflection reports, we suspected a gender effect. Earlier studies have indicated gender differences in self-efficacy [8] in adolescents, and the influence of performance feedback [9]. More recently, gender effects in perceived self-efficacy in the context of self-regulated learning have been reviewed [10]. Academic self-efficacy beliefs are considered significant during all phases of self-regulation: forethought, performance and self-reflection. In group 3, students fill in self-grade forms where self-efficacy is of primary relevance, some gender effect on the overall quality of the reflection report might be detectable. Mean values for most scores seemed in favor of females. All aspects listed in 1 for group 3 were investigated for significance. From t-tests there was no statistically significant gender difference between the means at the 95.0% confidence level. F-tests also yielded no statistically significant gender difference between the standard deviations of the two samples at the 95.0% confidence level. We have therefore been unable to determine any gender effects in self-reflection. In relation to this, we have to point out that in our case the numbers are relatively small, as only six out of 36 students in group 3 are female.

In spite of the fact that a considerable improvement in average quality of reflection reports has been achieved, thanks to detailed guidelines and the help of self-grade forms, we still observe that a significant fraction in group 3 scores below standard. In Table 1, last column, we list the percentage of students in that group, scoring 3 or higher (that would mean 6 or higher on our normal grading scale from 0 to 10). As mentioned, several of the distributions were very broad and non-Gaussian and three of these show an extreme skewness. From graphical representations we suspect a tendency for bimodal distribution, for at least some of the points concerned.

In educational settings, bimodal distributions of scores generally require special attention. Provided that a test does indeed assess the appropriate parameters, a bimodal distribution of results in our view can only be acceptable in a test that serves the purpose of preselection for subsequent courses in a curriculum. In this case, however, the self-reflection report is made at the end of the master program, right before or even after the graduation project. If the ability to carry out a good self-reflection of professional competences would be one of the primary learning objectives of the curriculum (which is not yet the case in our department), then a bimodal distribution would not be acceptable, because we would then graduate engineers that do not fulfill this particular learning objective.

6 Conclusions

The work presented here illustrates how the quality of self-reflection by chemical engineering students is related to the guidelines provided to the students. The most important conclusion is that the quality of the self-reflection improves significantly if the students receive clearly formulated guidelines from which it is evident what is expected from a self-reflection report. Further improvement is still needed because a large minority of students still do not fulfill the minimum requirement for writing a good self-reflection report. Apparently not all of the students are able to relate most of the competences in Appendix A, with the appropriate self-reflection report questions in Appendix B. The results obtained from the self-reflection reports, together with an assessment of the reflection techniques employed by other departments at our university, such as Industrial Design, have led us to further accentuate the guidelines given to the students when writing their reflection reports.

7 Recommendations

Improvements on the quality of the self-reflection reports are a direct consequence of the guidelines given to the students. From the results presented here it is evident that the students in group 3 performed much better in their self-reflection. To ensure that the quality of the self-reflection reports remains high, and to stimulate the further improvement of the self-reflection reports, it is decided to change the procedures and guidelines to the students slightly. We consider it likely that the quality of the self-reflection will improve if the students set personal goals before starting their internship. These so-called SMART [11] goals are a way of evaluating if the set objectives are appropriate for the individual student and the internship. SMART objectives are specific, measurable, achievable, relevant and time-bound. Before commencing with the internship the student needs to answer questions regarding himself and his views on an internship. Half way through the internship the student should obtain interim feedback from the company supervisor, and relate this to his SMART objectives, and possible readjustment of these. At the end of the internship the student should reflect on his own performance during the internship.

Appendix A - Criteria for Grading of the Student by the Company

The industrial supervisor is asked to grade the student's performance on each of the following aspects on a scale of A to E, where A is excellent and E is insufficient.

1. Attitude and dedication to work
2. Quality of the work done
3. Amount of work done
4. Relation with other people
5. Ability to work with a minimum of instruction and supervision
6. Ability to criticize own work and results
7. Scientific knowledge
8. Creativity

9. Language fluency
10. Communication, verbal (+ presentation)
11. Communication, written (+ final report)
12. Overall performance

Appendix B – Criteria for Grading the Self-reflection Reports

Under the guideline questions presented in this paper we look for the following when grading the self-reflection reports.

1. *What did you learn about yourself in a working environment?* Here we look for a general discussion on how the student functioned in the company. Did he or she participate in meetings, did he or she learn specific industrial problems such as time versus money issues, safety issues, etc?
2. *Did you discover unsuspected talents?* Here we look for any mention of what the student is good at. This can be communication skills, knowledge on the subject, etc. Also any other skill that may have been developed during the internship.
3. *Did you meet unexpected challenges?* Here we look for problems during the internship such as poor planning, language and communication problems, delivery time of equipment, etc.
4. *How did you deal with them?* Here we specifically look for actions the student undertook to solve the problems mentioned in question 3.
5. *Which personal improvement points remain?* Here we look for an honest look at what the student still needs to improve. This can be presentation skills, technical writing skills, people skills, assertiveness, etc.
6. *How did the internship broaden your view on future career opportunities?* Here we look for comments on how the internship helped the student to form an idea of what he or she would like to do after completing his master degree. Examples can be: do a Ph.D., go to work for a multinational, go abroad, or that the internship did not have any significant influence on the choice for a future career.
7. *The self-reflection must not be about the specific company, but about you.* Here we specifically judge if the student has written a reflection on himself or herself, and not a description of how the company operates, what the company manufactures, what the student did after working hours, etc.

Acknowledgements Invaluable suggestions on statistics by Dr. Koo Rijpkema (Department of Mathematics and Computer Science) are gratefully acknowledged. The authors thank Yi Zhang for data-entry work and Dr. Lucas Asselbergs (Department of Industrial Design) for inspiring discussions on reflection.

References

1. http://en.wikipedia.org/wiki/Human_self_reflection
2. Compact Oxford Dictionary (2005), Oxford University Press, Oxford

3. Lowe, M.; Rappolt, S.; Jaglal, S.; Macdonald, G.J. (2007) The Role of Reflection in Implementing Learning from Continuing Education in Practice. Continuing Education in the Health Professions 27:143–148.
4. Korthagen, F.A.J. (2004) In search of the essence of a good teacher: Towards a more holistic approach in teacher education. Teaching and Teacher Education 20:77–97.
5. http://edu.chem.tue.nl/goals
6. http://w3.id.tue.nl/en/education/
7. http://www.statgraphics.com/documents.htm
8. http://www.des.emory.edu/mfp/BanEncy.html
9. Schunk, D.H.; Lilly, M.W. (1984) Sex Differences in Self-Efficacy and Attributions: Influence of Performance Feedback. Journal of Early Adolescence 2:203–213.
10. Pajares, F. (2002) Gender and Perceived Self-Efficacy in Self-Regulated Learning. Theory Into Practice 2:116–125.
11. Drucker, P.F. (1954) The Practice of Management. Harper & Row, New York

The NAWIlino-Box – A Science Exploration Kit and Its Use in Practical Teacher Training

L. Bröll, J. Friedrich, and M. Oetken

Abstract Based on the fact that many teachers in primary school do not feel competent to teach science, the NAWIlino-Box was designed at the University of Education in Freiburg as part of a current thesis project. Especially pupils in primary school are interested in scientific topics. The NAWIlino-Box picks up this interest and illustrates scientific coherence with conceptually clear experiments. Pupils can use the NAWIlino-Box during their whole primary school time. Key content like sustainability is just as important as the ability to inspire children for scientific experiments. With our portable learning laboratory pupils get the possibility to expose themselves to science for longer periods, so that they can explore and encourage their own skills. Although the choice of experiments implemented had originally been geared towards the curriculum of the state of Baden-Württemberg, the NAWIlino-Box has no features barring its use on a national level, as shown by a comparison of federal curricula in Germany. To complement each experimental unit, we developed individual solution sheets and didactical and methodical advice for integration of the material in class.

1 Introduction

In the wake of nonsatisfying results, German pupils scored in studies like Third International Mathematics and Science Study (TIMSS), Programme for International Student Assessment (PISA) 2000 and also Internationale Grundschul-Lese-Untersuchung (IGLU) 2003, many activities for conquering these academic shortcomings were discussed. Particularly research into the proficiency level in

---

L. Bröll (✉), J. Friedrich and M. Oetken
University of Education Freiburg, Institute for Biology, Chemistry, Geography and Physics
Department of Chemistry, Kunzenweg 21, D-79117, Freiburg, Germany
e-mail: broell@ph-freiburg.de

M. Gupta-Bhowon et al. (eds.), *Chemistry Education in the ICT Age*,
DOI 10.1007/978-1-4020-9732-4_38, © Springer Science+Business Media B.V. 2009

primary school indicates a need for action within the domain of math and science lessons to enable pupils to reach higher levels of competency than they do today [1]. If we have a closer look at the results of PISA 2006, then it appears that "Germany advanced across from 2003 not only in the international ranking, but there is also a substantial growth regarding scientific competence." [2].

These changes have already arrived in primary schools; the newest curricula of several German states specifically mandate scientific topics for teaching. In southern Germany's Baden-Württemberg, for example, several science topics are now an integral part of the curriculum, as are certain experiments to be carried out in class [3].

2 Motivation for Developing the NAWIlino-Box

The development of the NAWIlino-Box is based on a questionnaire-based empirical study [4], which was conducted in the summer of 2006 within the administrative school district of Freiburg. The point of origin for this empirical study was conversations with primary school teachers, which clearly showed a lack of confidence in their ability to teach science and also a deficit regarding the teaching material that primary schools have at their disposition for teaching science. On the other hand, there was a lot of positive response regarding professional training in teaching science.

The results of our empirical study affirm the following statement. Especially teachers who have not studied science at university level should have the opportunity for professional training, in order to enable them to provide quality teaching to children regarding scientific topics. Professional training should consist of two parts, with the first part teaching scientific know-how and a second part covering practical advice for the actual teaching of science in primary school. This professional training should also ease and take into account the anxiety of teachers in bringing up new scientific subjects with pupils. Last but not least, primary schools need to be adequately equipped to carry out experiments in the first place.

In Germany, there are two sorts of science kits at the moment: commercial boxes from publishing houses, which are not originally made for use in school, but for interested children to experiment at home. On the other hand, there are boxes from universities or other educational institutions, which cover several topics from the curriculum [5–7]. The NAWIlino-Box covers all topics, which are thematised in the curricula.

3 The Topics

To classify the box contents, the following topics were chosen: sound and hearing, light and seeing, separation methods, different forces like magnetism, but also classical topics like fire, water and air. Altogether there are about 100 experiments. The following pictures (Figs. 1–3) show some examples.

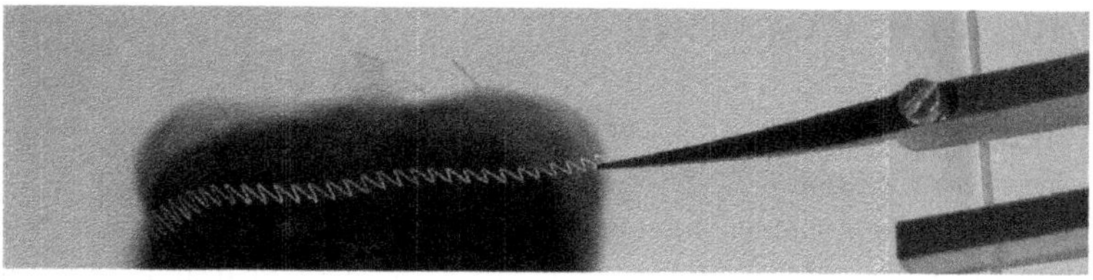

Fig. 1 Showing sound waves

Fig. 2 Dissolving salt in water, and getting it back

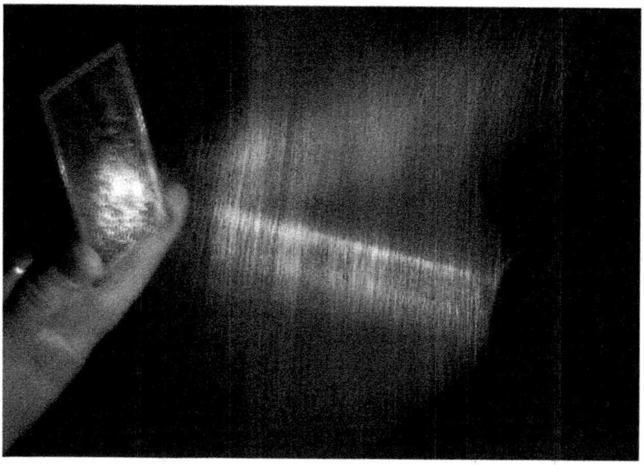

Fig. 3 Reflecting light in a mirror

4 Experiment Worksheets

The worksheet design was orientated at the cognitive and affective requirements of pupils in primary school.

First, all materials needed for the experiment are named and drawn by small handmade drawings (Fig. 4). Because of this iconic demonstration, children will grasp the experiment more easily. The American psychologist Bruner [8] distinguishes three forms of representation, with which humans open up their environment: the enactive one (through action), the iconic one (using pictures or visuals) and the symbolic one (representation in words or language). These forms of representation are also very important for learning in school. If pupils have to learn scientific topics on the abstract symbolic form, it is helpful to also give them the possibility to use the enactive and iconic forms.

For all experiments it was important to give short instructions and to draw them if necessary (Fig. 5).

All worksheets are also saved as Microsoft Word documents on CD-ROM. All experiments and worksheets were tested by teachers in a first step in assessing their suitability for primary school.

Each worksheet was complemented with a solution sheet (Fig. 6).

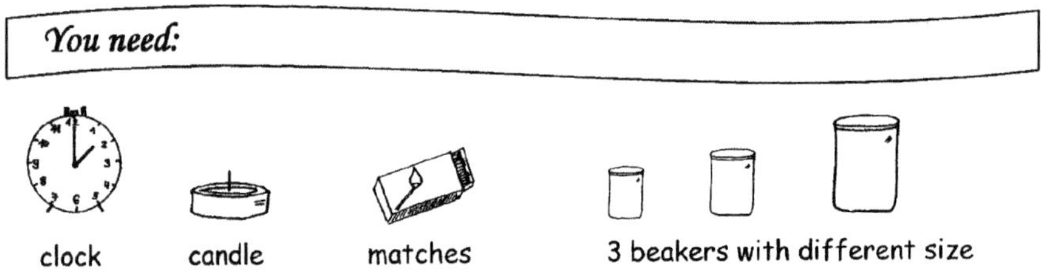

Fig. 4 That is what you need for the experiment

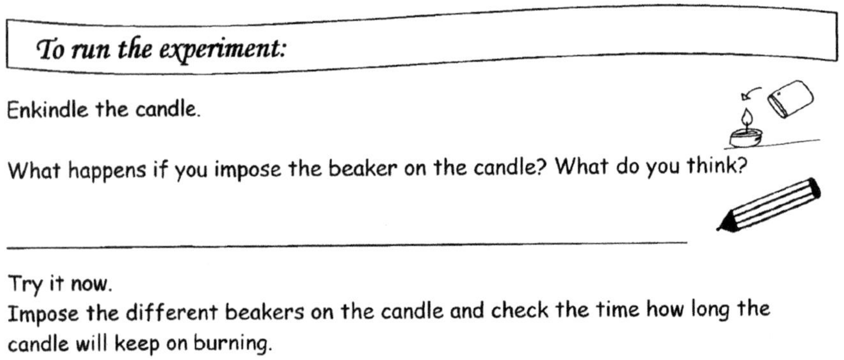

Fig. 5 That is what you have to do in this experiment

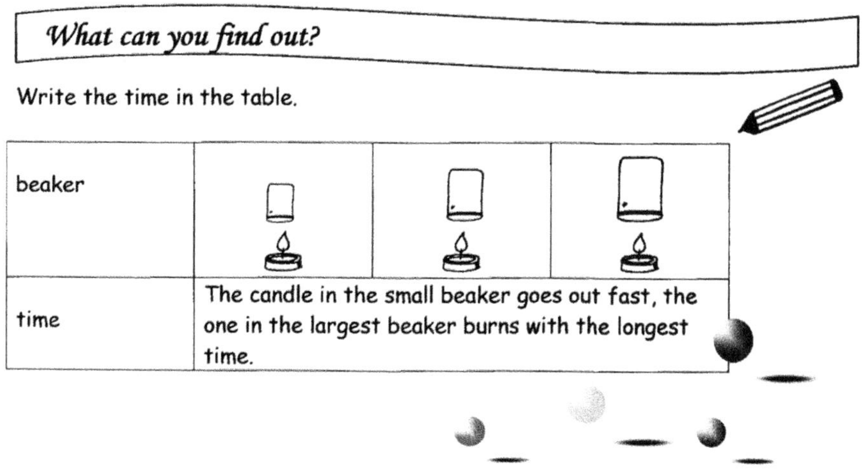

What can you find out?

Write the time in the table.

beaker			
time	The candle in the small beaker goes out fast, the one in the largest beaker burns with the longest time.		

Fig. 6 Example of instructions

The benefit of these documents is that teachers get scientific background information along with didactical and methodical advice for thematisation in class. The material also brings up possible pit-falls and false beliefs children may develop about the respective topics.

5 The NAWIlino-Box and Its Physical Setup

The required material is stored in colored boxes (Fig. 7).

A single box is used for each topic (Fig. 8). To see what should be in the box, every box is labeled with a description of its contents.

6 Handling in Class

Handling in class can take place in different ways. Teachers can either construct a working project (for example, a carousel activity) or can experiment together with the pupils. These two approaches are proven methods, considered by primary school teachers to be practicable and suitable. Because the experiments are up to the standard of an experiment for primary school, they can be repeated several times, so that questions and thoughts regarding the experiment can be addressed by repeating the experiment. Another option would be that every child in class does the same experiment at the same time. In this case, children need to bring the required materials like balloons or plastic cups in the class themselves.

Fig. 7 The NAWIlino-Box

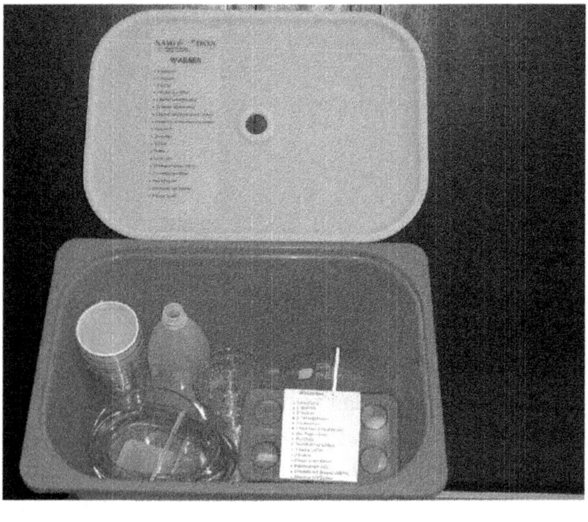

Fig. 8 Material for topic "Water"

7 Professional Training for Primary School Teachers

Because a lot of primary school teachers have not studied science at university level, we developed a professional training concept based on the NAWIlino-Box, covering all important aspects of effective professional training [9]. Primary school teachers get the possibility to make all the experiments themselves. Theoretical information and possibilities for implementation in school are also thematised. Teachers who took part in this particular professional training process in turn instruct their colleagues, so that in the end there are a lot of people being able to work with the NAWIlino-Box.

8 Evaluation

To prove the effectiveness of the NAWIlino-Box and related professional training, we started an empirical study in November 2007. This study will be finished in July 2008. A pilot survey took place in September 2007.

With this study we intend to answer these questions:

- Is the NAWIlino-Box a viable setting for teaching scientific subjects in primary school?
- Do teachers feel more competent to teach science in primary school after having taken part in professional training related to NAWIlino-Box?
- Are scientific topics also thematised by experiments?

The self-evaluation of teachers is a very important indication for answers here, because any teacher is self-conscious of his/her efficiency, since humans only act if they believe they are able to [10,11].

Therefore, we developed a questionnaire with 78 items by consulting experts. The questions of this survey addressed the following items:

- Expertise regarding non-scientific aspects of social studies
- Expertise for teaching science
- Expertise regarding scientific aspects of social studies
- Ability to establish the basic conditions for doing experiments
- Own experimental skills and reliabilty to do experiments
- Didactical and methodical competence to do experiments with pupils
- Personal commitment to carry out experiments

The survey also asks for self-evaluations regarding curriculum topics.

In autumn 2007 the pilot survey was carried out using pre-post design. Before the first and also after the last professional training session, teachers were asked to complete the above questionnaire. The teachers assigned a score of agreement to the answers on a four-point Likert scale (from 1 = full agreement to 4 = no agreement at all).

A statistical *t*-test was used to investigate the comparative changes.

9 Results

In a pilot survey with 15 primary school teachers attending professional training, we obtained the following results.

The results show, that attending professional training leads to an increase in competence. There was a significant change, $t(10) = 3.331$, $p = 0.008$, $d = 1.174$, with respect to confidence in teaching science (see Fig. 9).

Certain conclusions can be drawn by looking at the aspects covering didactical competence. All aspects except *personal commitment to do experiments* show significant change (see Fig. 10): *ability to establish the basic conditions for doing experiments*, $t(7) = 2.393$, $p = 0.048$, $d = 0.815$; *own experimental skills and confidence to do experiments*, $t(10) = 3.952$, $p = 0.003$, $d = 1.301$; *didactical and methodical competence to do experiments with pupils*, $t(9) = 3.508$, $p = 0.007$, $d = 1.404$.

An improvement was also observed regarding the curriculum assessment (see Fig. 11): *Confidence in teaching curriculum topics*, $t(9) = 2.590$, $p = 0.029$, $d = 0.692$); *Knowledge of curriculum experiments*, $t(10) = 5.918$, $p < 0.001$, $d = 2.226$.

	AM (SD)			*t*-test		
Dimensions	t_1	t_2	*d*	*t*	df	sig.
Confidence in teaching non-science aspects of social studies	2.26 (0.60)	2.16 (0.46)	0.187	1.000	9	0.343
Confidence in teaching science	2.59 (0.26)	2.18 (0.42)	1.174	3.331	10	0.008**
Confidence in teaching science aspects of social studies	2.79 (0.50)	2.60 (0.55)	0.361	1.747	10	0.111

Fig. 9 Results of the *t*-test for specialized competence

	AM (SD)			*t*-test		
Dimensions	t_1	t_2	*d*	*t*	df	sig.
Ability to establish the basic conditions for doing experiments	2.45 (0.39)	2.08 (0.51)	0.815	2.393	7	0.048*
Own experimental skills and confidence to do experiments	3.05 (0.40)	2.34 (0.66)	1.301	3.953	10	0.003**
Didactical and methodical competence to do experiments with pupils	2.76 (0.48)	2.02 (0.57)	1.404	3.508	9	0.007**
Personal commitment to do experiments	2.39 (0.34)	2.09 (0.49)	0.711	1.921	10	0.084

Fig. 10 Results of the *t*-test for didactical competence

	AM (SD)			t-test		
Dimensions	t_1	t_2	*D*	*t*	df	sig.
Confidence in teaching curriculum topics	2.42 (0.50)	2.05 (0.54)	0.692	2.590	9	0.029*
Knowledge of curriculum experiments	2.70 (0.42)	1.66 (0.51)	2.226	5.918	10	<0.001**

Fig. 11 Competence regarding the curriculum

10 Short Discussion of the Results

These results identify a trend - professional training seems to increase self-confidence in teaching science in primary school.

More results with larger samples are expected from our main study in July 2008, in which 120 teachers are set to take part. This upcoming study will also investigate the influence of additional facts on perceived self-confidence of teachers.

The development of the NAWIlino-Box and related professional training with this science exploration kit seems to be a good opportunity to enhance teachers' competence to teach science in school. First results show this in a very clear way. Results of the main study shall be published soon.

Acknowledgements We would like to thank "Fonds der chemischen Industrie," "Lernort Labor," "badenova," and "Landesstiftung Baden-Württemberg" for supporting this work.

References

1. Prenzel M, Geiser H, Langeheine R et al. (2003) Das naturwissenschaftliche Verständnis am Ende der Grundschule. In: Bos W, Lankes EM, Prenzel M (eds.) Erste Ergebnisse aus IGLU. Waxmann, Münster
2. Prenzel M, Artelt C, Baumert J et al. (2007) Die Ergebnisse der dritten internationalen Vergleichsstudie. Zusammenfassung. http://pisa.ipn.uni-kiel.de. Accessed 24.01.2008
3. Ministerium für Kultus, Jugend und Sport Baden-Württemberg (2004) Bildungsplan 2004. Grundschule. Philipp Reclam Jun., Ditzingen
4. Bröll L, Friedrich J, Oetken M (2007) Naturwissenschaftliche Bildung im Primarbereich?! Eine Untersuchung zur Bedeutung und Realisierung naturwissenschaftlicher Inhalte in der Grundschule. PdN-ChiS 6: 36–41
5. http://experimentierkistenzubehoer.de. Accessed 14.01.2008
6. http://www.nat-schuelerlabor.de. Accessed 14.01.2008
7. http://ddsu.uni-muenster.de/www/dooperationsprojekte/projektuebersicht.pdf. Accessed 14.01.2008
8. Bruner JS (1971) Toward a Theory of Instruction. Belknap Press of Harvard University Press, Cambridge, Massachusetts

9. Lipowsky F (2004) Was macht Fortbildungen für Lehrkräfte erfolgreich? Befunde der Forschung und mögliche Konsequenzen für die Praxis. Die Deutsche Schule 4: 462–479
10. Bandura A (1977) Self-efficacy: Toward a unifying theory of behavioral change. *Psychological Review* 84: 191–215
11. Bandura A (1992) Exercise of personal agency through the self-efficacy mechanism. In: Schwarzer R (ed.) Self-efficacy: Thought Control of Action. Hemisphere, Washington, DC

Assessment of Mercury Pollution at Mare Chicose Landfill in Mauritius

V. Dookhun and K. Mahadeo

Abstract Modern society has resorted to the extensive use of chemicals either natural or synthetic over the last several decades. These chemicals are utilized for controlling diseases, increasing food production, and to provide convenience in our daily lives. Ironically, many of these well-intentioned chemicals are now wreaking havoc around the world, threatening the health of wildlife and people. Mercury, one of those chemicals, has become a prime subject as far as its release to the environment is concerned. This project is about assessing the release pathway of mercury from the landfill and determining the status of mercury pollution there. Mercury is not produced in Mauritius and the sources were mainly from waste consumer products and industrial releases. The amount of mercury entering Mauritius through consumer products and raw materials for industries imported since the last 5 years was determined and eventually, the mercury reaching the landfill was estimated. The laboratory analysis for landfill leachate and ground water revealed no presence of mercury above 0.0002 mg/l and 0.0001 mg/l, respectively. Simulation results for workface releases showed that the workers' exposure were significant (Figs. 1 and 2).

1 Introduction

Mercury is ranked third by the CERCLA priority list of toxic substances and has been found to be present in the environment in concentration that could be harmful [1]. Prevention measures for mercury pollution are primordial and the first proactive step is to identify the status of mercury pollution in high-risk areas. Though mercury is quite useful, it has long been known as a persistent, bio-accumulative toxic pollutant that adversely affects the central nervous system. Persistent, bio-accumulative,

V. Dookhun (✉) and K. Mahadeo
Department of Chemical and Environmental Engineering, Faculty of Engineering
University of Mauritius, Reduit, Mauritius
e-mail: v.dookhun@uom.ac.mu, mahadeo_krish@yahoo.com

M. Gupta-Bhowon et al. (eds.), *Chemistry Education in the ICT Age*,
DOI 10.1007/978-1-4020-9732-4_39, © Springer Science+Business Media B.V. 2009

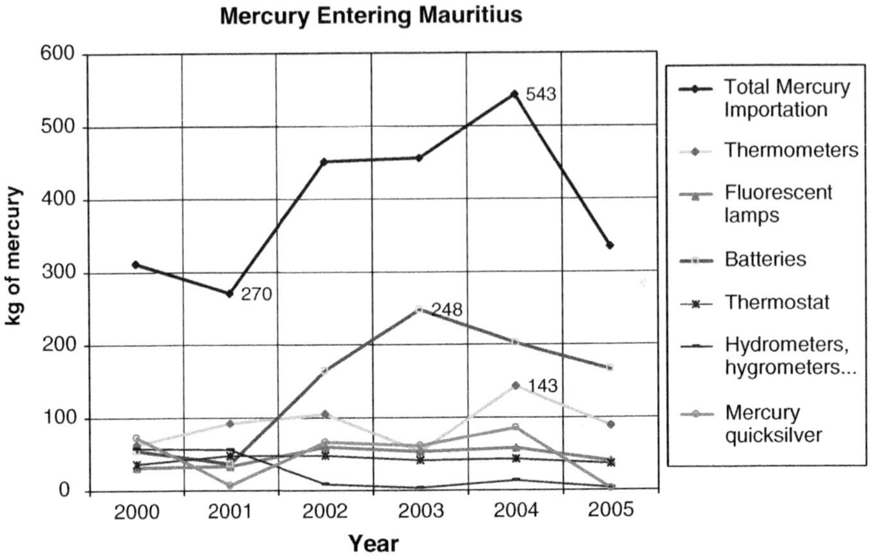

Fig. 1 Mercury Consumption for Mauritius 2000–2005

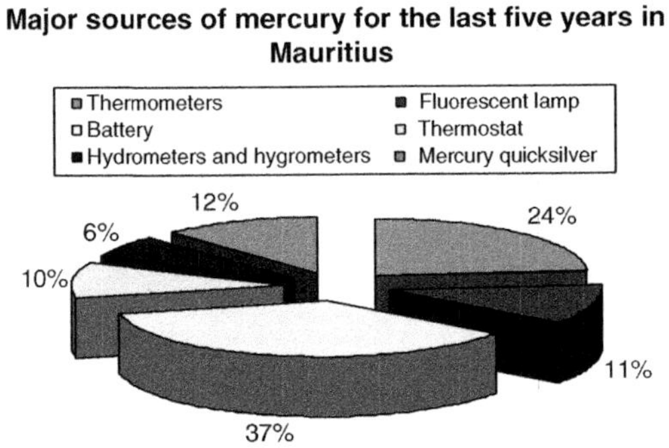

Fig. 2 Major sources of mercury in Mauritius

and toxic pollutants (PBTs) are long-lasting substances that can build up in the food chain to levels that are harmful to human and ecosystem health. These contaminants can be transported to long distances and move readily from land to air and water. Because of their persistence and bio-accumulative properties, they do not break down easily and are particularly difficult to clean up.

Thus this project aimed at assessing the release of the pollutant mercury to the atmosphere from the landfill and finding the most probable pathway through which mercury from the wastes would integrate the environment. Total mercury releases from waste at the landfill were also quantified on an annual basis.

2 Literature Review

2.1 *Mercury and its Compounds*

The presence of mercury in the atmosphere is attributed to two sources: Natural sources - mainly from volcanoes; and anthropogenic (man made) sources – from excessive usage of mercury containing products and burning of fuel. Mercury is also famous as being a high volatile metal that produces toxic vapour. The air in equilibrium with liquid mercury will contain 14 $mgHg/m^3$ of air at 20°C and the maximum allowable level of mercury vapour in the air known as the *threshold limit value (TLV)* is equal to 0.05 $mgHg/m^3$ of air [3].

2.2 *Fate of Mercury in the Environment*

The most common natural forms of mercury found in the environment are metallic mercury, mercuric sulphide, mercuric chloride and methyl mercury. Some microorganisms and natural processes can change the mercury in the environment from one form to another. Elemental mercury in the atmosphere can undergo transformation into inorganic mercury forms, providing a significant pathway for deposition of emitted elemental mercury. Natural transformations and environmental pathways of mercury are very complex and are greatly affected by local conditions. To assess the environmental fate and the impacts of anthropogenic mercury emissions, researchers must examine a range of biogeochemical interactions affecting mercury in its different physical states and chemical forms. Understanding the relationships between local conditions and mercury levels in various environmental media and living organisms is critical to predicting changes in concentration and form. There are two main types of reactions in the mercury cycle that convert mercury through its various forms: *oxidation/reduction* and *methylation/de-methylation*.

Mercury entering a landfill will be subjected to a number of reactions. Firstly, part of the mercury would evaporate due to the high volatility of the element coupled with the working face that is the compaction of the wastes. The remaining mercury would be oxidised to ionic mercury (monovalent or divalent), that would either precipitate reacting with sulphate ions, or percolate in the leachates, or undergo the methylation process and be liberated through the form of methyl mercury in the landfill gas.

2.3 *Legislations Relative to Mercury in Mauritius*

Standards for the maximum permissible concentration of mercury present in a substance to be discharged or disposed to the environment have been developed.

The maximum acceptable level of mercury concentration for different environmental media such as drinking water, surface water and foodstuffs is also regulated (Table 1).

Table 1 Standards for the permissible amount of mercury in environment for Mauritius (From Ministry of Environment of Mauritius)

Environmental media	Concentration
Drinking water	0.001 mg/l
Surface water	0.1 μg/l
Coastal water quality	0.0005 mg/l
Foodstuffs (Fish)	1 ppm

Source: Ministry of Environment of Mauritius

Table 2 The maximum permissible concentration of mercury allowed in waste discharges to the environment for Mauritius (From Ministry of Environment of Mauritius)

Environmental media	Effluent discharge onto	Concentration
Water point sources	Land	0.005 mg/l
	Underground	0.005 mg/l
	Surface water course	0.005 mg/l
	Ocean	10 μg/l
Waste disposal restrictions	Treated wastewater for irrigation use	0.02 mg/l

Source: Ministry of Environment of Mauritius

3 Methodology

Major anthropogenic sources of mercury are expected to finish their course at the landfill, hence Mare Chicose, the only sanitary landfill in Mauritius was chosen as the study case.

A list of imported products were screened and sorted into groups according to their different compositions of mercury. The contributions of each product to the introduction of mercury in the island were calculated, and the total annual mercury consumption for Mauritius was determined by summing them up together with the yearly amount of mercury quicksilver imported. This activity was carried out for the last 5 years so that a clear indication of the variation mercury consumption could be obtained. Performing an inventory at the Mare Chicose landfill to determine the amount of mercury entering is a very difficult task as waste comes to the landfill in commingle form and they are only visually inspected by a tipping officer at the tipping floor to see whether any unauthorized wastes are being brought in. In such circumstances, the inventory for mercury input at the landfill can be carried out as per the UNEP Toolkit for identification and quantification of mercury releases [6]. According to this tool, a low-end default factor is set to indicate a low-end estimate for the mercury input to the source category (but not the absolute minimum), whereas a high-end factor will result in a high-end estimate. The low-end input factor is relevant for situations where substantial parts of the waste products with high mercury concentration (thermometers, batteries, dental amalgam wastes, switches, etc.) have been sorted out of the waste for separate treatment, and will therefore be present in lower numbers in the municipal waste (Table 3). The high-end input factor is expected to be relevant for situations where no such sorting

Table 3 Preliminary input factors of mercury in municipal landfill [6]

Material	Default input factors; g Hg/metric ton waste; (low end - high end)*1
Municipal solid waste (general "household' waste)*1	1-10

takes place and most of the product waste with high mercury concentrations is therefore present in the municipal waste.

Even though no sorting of the 'mercury' containing products takes place in the Mauritian system, the lower end of the input factor was chosen due to the low imports value of mercury consumer products and low annual mercury consumption compared to the annual waste generation. Mercury analysis of the leachate was performed according to Method 7470 from the US EPA using cold vapour atomic absorption. Sampling protocol was as specified by the USEPA Paper - Sample Submission Procedures for the 'Analytical Services & Quality Assurance Branch Laboratory' Revision 9 August 2005. Grab sampling was considered convenient means of sampling, as it is desired that the mercury level in the leachates should at any time and for any volume be compliant with the standards (<0.005 mg/l).

The concentrations of mercury in the landfill gas were estimated with the aid of a simulation program 'LandGEM' that was developed by the office of Research and Development United States Environmental Protection Agency (US EPA). Land GEM is based on a first-order decomposition rate equation for quantifying emissions from the decomposition of landfilled waste in municipal solid waste (MSW) landfills. The software provides a relatively simple approach to estimating landfill gas emissions. Model defaults are based on empirical data from US landfills. Field test data can also be used in place of model defaults when available. This software was also useful to forecast the future levels of mercury generation from LFG.

4 Results and Discussion

From a preliminary screening of importation data from the governmental central statistics office, it was observed that mercury was imported as mercury quicksilver and mercury colloidal suspension. Imported goods that contained mercury were: mercury vapour lamps, TV camera tubes, clinical thermometers, batteries (button cell), fluorescent lamps and electrical switches. Mercury could also be present in coal that is imported to be used for power generation and other consumer products (e.g. bleaching creams).

The total mercury consumption varied from 270 to 543 kg of mercury annually with a mean value of 393 kg. Moreover, 'mercury' battery consumption of the island underwent a considerable increase from the year 2001 and was at its peak in 2003 with a total contribution of 248 kg of mercury. The general decline in the trend can be associated partly with the government decisions to prohibit use of button cell batteries and mercury usage in paints (except road marking paints) and

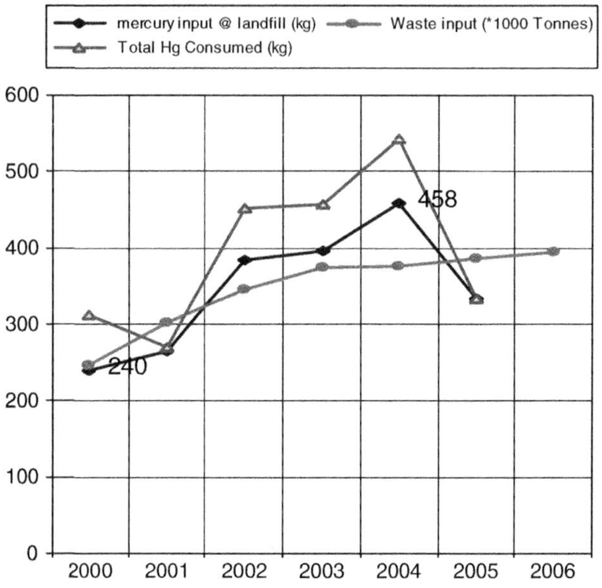

Fig. 3 Mercury input to the landfill 2000–2005

pesticides. Further, for the 2000–2005 period, it was observed that batteries had been the largest contributor for mercury pollution (37%), followed by thermometers (24%). The other contributors were: Thermostat (10%), mercury quicksilver (12%), lamps (12%) and hydrometers, hygrometers and pyrometers (6%).

The annual mercury introduced to the sole landfill estimated as a function of the total wastes going to the landfill and the overall consumption of mercury was found to vary from 240 to 458 kg with a mean value of 346 kg of mercury (refer to Fig. 3). From this result it can be deduced that most of the mercury in imported goods finish their course at the landfill. The difference of 47 kg is due to the mercury quicksilver that is dedicated for industrial use and laboratory analyticals.

5 Results for Mercury Concentrations in Leachates

From the mercury analysis carried out on the landfill leachate samples, none of them detected for mercury concentrations above 0.0002 mg/l.

6 Results for Mercury Concentrations in Landfill Gas

Graph Number 1 from Fig. 4 below shows the simulated results for landfill gas generation at Mare Chicose. Table 4 indicates the input parameters for the different possible models. Input parameters from model 4 were considered appropriate for Mare Chicose landfill.

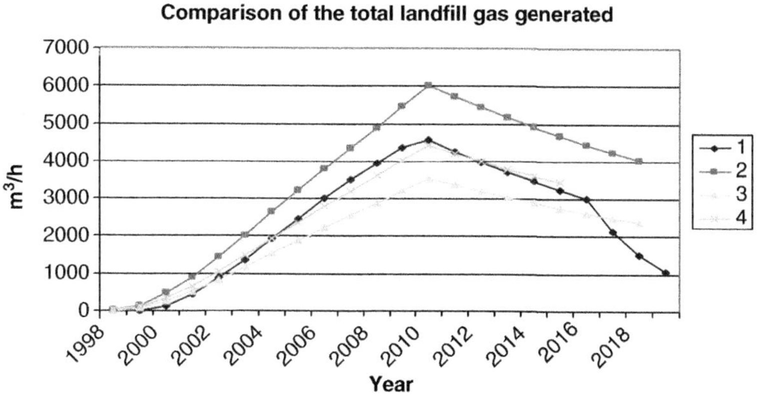

Fig. 4 Comparison of the model landfill gas generation

Table 4 Variation of the input parameters of the models

Input parameters defining model	Model 2	Model 3	Model 4
Methane generation rate **k** (year $^{-1}$)	0.05	0.05	0.05
Methane generation capacity $\mathbf{l_0}$(m^3 /Mg)	170	100	100
NMOC concentration (ppmv as hexane)	2,400	600	600
Methane content (% by volume)	50	50	40

Changes to landfill operation, such as operating under wet conditions through leachate recirculation or other liquid additions, will result in generating more gas at a faster rate but with this software such operations were not considered [7].

The amount of mercury that would be liberated according to that model from the landfill gas is as shown below in Fig. 5. Presently, the annual mercury emission is expected to be about 0.068 kg through the landfill gas.

Mercury releases from the working face of the landfills were more than tenfold higher than the mercury releases with flared land fill gas [6]. Thus, it is assumed that the release from the working face is ten times greater than the mercury released from LFG. This would imply that a worker would be at significant risk of exposure to mercury.

7 Conclusion

Mercury from a landfill can be released through three main pathways to the environment:

- To the air – through landfill gas and the working face at the landfill
- To water – through leachates from accumulated waste
- To the land – directly from mercury containing wastes

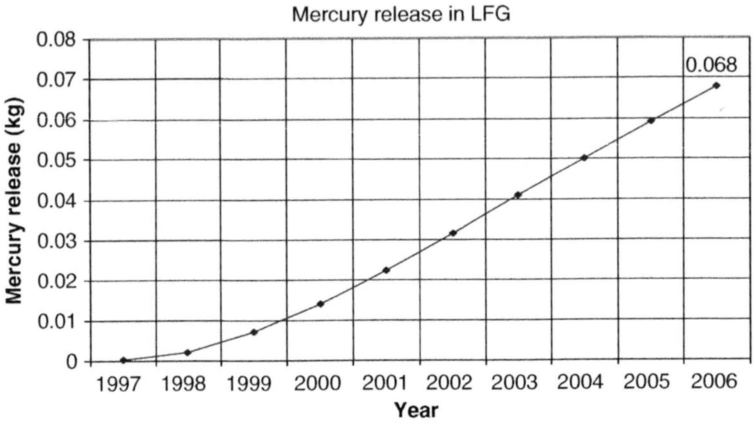

Fig. 5 Total mercury emitted from landfill since its inception

Analysis of the leachates and the groundwater has shown that mercury release to water is very low and would never reach a concentration greater than 0.0001 mg/l in the forthcoming years if the waste input trend stays the same at the Mare Chicose landfill.

The key outcomes were that fortunately, the condition at the landfill was mostly alkaline whereby the formation of toxic organo-mercury compounds are inhibited and most of the mercury is trapped in the buffer zone that contains the waste piles but however, mercury does enter the environment from the landfill through the working face, LFG and leachates. Even though mercury disposal through a controlled landfill releases very small amount of mercury to the air and water, the mercury accumulation in land can be sufficient to pose serious health hazards. Mauritius is diversifying its Solid Waste Management strategies by moving from the traditional landfilling system to Waste to Energy, Recycling and Composting alternatives. The situation of mercury pollution in the landfill as forecasted does not include changes in waste patterns due to introduction of these technologies and hence pollution might be reduced from this source in the future years. Most mercury going to the landfill is being trapped to the land but if ever an incineration facility is set up with no sorting of waste and improper end of pipe treatment, mercury in the waste will be liberated to the environment.

8 Recommendations and Future Works

Performing a material balance for mercury at the landfill would definitely help to establish the amount of mercury that is being accumulated in the landfill. Analysis of landfill sediment within the landfill and nearby areas can also be undertaken to refine the material balances. Future projects can also assess workface liberations of mercury.

Acknowledgements We are very thankful to Mr. Jokhun and Mr. Radha, the technical staff of University of Mauritius; Mr. Ghoorun, Technical Manager at Mauritius Standards Bureau; Mr. P. Kowlessur, Director of Ministry of Local Government; Mr. D. Dookhee, Mr. B. Beerachee and Mrs. S. Ramracheya, the technical team of the Solid Waste Management Division for their support, guidance and advice.

References

1. Boening D W (2000) Ecological effects, transport, and fate of mercury; a general review. Chemosphere Vol. 40: 1335–1351.In UNEP, 2002. Global Mercury Assessment. Geneva: UNEP: 80.
2. COWI (2002): ACAP and Danish EPA Reduction of Atmospheric mercury emissions from Arctic countries – Questionnaire on emissions and related topics. In UNEP(2005) Toolkit for identification and quantification of mercury releases: 234.
3. O'Neil P (1985) Environmental Chemistry. London.
4. UNEP (2002) Global Mercury Assessment report, Geneva. UNEP.
5. UNEP June 2004. Regional Awareness-Raising Workshop on Mercury Pollution - A global problem that needs to be addressed; Pretoria, South Africa, 1–4 June 2004. Geneva: UNEP.
6. UNEP (2005) Toolkit for identification and quantification of mercury releases, Geneva: UNEP: 237.
7. US EPA (2005) Landfill Gas Emission Model Version 3.02 User's Guide. US EPA Research Office. Washington DC: 3.
8. World Health Organization (1989). Mercury, Environmental Aspects. Environmental Health Criteria. Vol. 86. WHO.
9. (2006) http://atsdr.cdc.gov/cercla/index.html. ATSDR, 2005. CERCLA Priority list of hazardous substances. [Online]. Accessed 20 December 2006.

Index

D

GPSR Compliance

The European Union's (EU) General Product Safety Regulation (GPSR) is a set of rules that requires consumer products to be safe and our obligations to ensure this.

If you have any concerns about our products, you can contact us on ProductSafety@springernature.com

In case Publisher is established outside the EU, the EU authorized representative is:

Springer Nature Customer Service Center GmbH
Europaplatz 3
69115 Heidelberg, Germany

Batch number: 10755284

Printed by Printforce, the Netherlands